T0132905

Natures of Colonial Change

NEW AFRICAN HISTORIES SERIES

Series editors: Jean Allman and Allen Isaacman

David William Cohen and E. S. Atieno Odhiambo, *The Risks of Knowledge: Investigations into the Death of the Hon. Minister John Robert Ouko in Kenya, 1990*

Belinda Bozzoli, *Theatres of Struggle and the End of Apartheid*

Gary Kynoch, *We Are Fighting the World: A History of Marashea Gangs in South Africa, 1947–1999*

Stephanie Newell, *The Forger's Tale: The Search for Odeziaku*

Jacob A. Tropp, *Natures of Colonial Change: Environmental Relations in the Making of the Transkei*

Natures of Colonial Change

*Environmental Relations in the
Making of the Transkei*

⌒

Jacob A. Tropp

OHIO UNIVERSITY PRESS
ATHENS

Ohio University Press
The Ridges, Building 19
Athens, Ohio 45701
www.ohio.edu/oupress

© 2006 by Ohio University Press

Printed in the United States of America
All rights reserved

Cover: Photograph courtesy of Cape Town Archives Repository,
Jeffreys Collection

Ohio University Press books are printed on acid-free paper ⊛ ™

13 12 11 10 09 08 07 06   5 4 3 2 1

Library of Congress Cataloging-in-Publication Data

Tropp, Jacob Abram.
  Natures of colonial change : environmental relations in the making of the
Transkei / Jacob A. Tropp.
    p. cm. — (New African histories series)
  Includes bibliographical references and index.
  ISBN-13: 978-0-8214-1698-3 (cloth : alk. paper)
  ISBN-10: 0-8214-1698-7 (cloth : alk. paper)
  ISBN-13: 978-0-8214-1699-0 (pbk. : alk. paper)
  ISBN-10: 0-8214-1699-5 (pbk. : alk. paper)
  1. Indigenous peoples—Ecology—South Africa—Transkei—History. 2. Indige-
nous peoples—South Africa—Transkei—Social conditions. 3. Forest ecology—
South Africa—Transkei. 4. Landscape changes—South Africa—Transkei. 5. Eu-
rope—Colonies—Africa. 6. Transkei (South Africa)—Colonization. 7. Transkei
(South Africa)—Environmental conditions. I. Title.
  GF758.T76 2006
  333.75'130968758—dc22
                                                          2006021402

Contents

List of Illustrations   vii

Acknowledgments   ix

Abbreviations   xiii

Introduction   1

PART I

Chapter 1. Tensions in the Colonial Restructuring of Local
Environmental Authority, 1880–c. 1915   31

Chapter 2. Environmental Entitlements in the New Colonial Order,
1888–c. 1905   63

Chapter 3. Shifting Terrains of Wood Access in the Early Twentieth
Century, 1903–1930s   89

PART II

Chapter 4. Remapping Historical Landscapes
Forest Species and the Contours of Social and Cultural Life   125

Chapter 5. The Python and the Crying Tree
*Commentaries on the Nature of Colonial and
Environmental Power*   145

Conclusion   160

Notes   167

Bibliography   235

Index   263

Illustrations

FIGURES

0.1. Hut-building in the Engcobo District, Transkei, c. 1954   4

0.2. Woman weaving hut door made of saplings, c. 1930s   5

0.3. Pulling a loaded sledge, c. 1930s   6

0.4 View of some KwaMatiwane mountains, late 1990s   20

2.1. Woman carrying a headload of fuelwood, with
Gulandoda Mountain in the background, c. 1959   82

2.2. Bringing the harvest home in a sledge, c. 1954   83

3.1. Mpondo hut-building, c. 1907   94

3.2. Man building a sledge with saplings, c. 1930s   105

3.3. Women carrying heavy poles in the Engcobo District, c. 1950   113

4.1. Boys demonstrating stick-fighting in the Engcobo District, c. 1971   129

4.2. An apparently posed example of *ukufutha*, c. 1930s   139

MAPS

0.1. Colonial annexation of the Transkeian Territories in the
late nineteenth century   17

0.2. KwaMatiwane   20

TABLES

3.1. Hut Wattle Purchases from Government Plantations, 1899–1930   99

3.2. Annual Fuelwood Sales, 1910–30   111

Acknowledgments

THERE ARE MANY WAYS to measure how much has gone into this book's development, but I'll start with the personal. When I first visited South Africa and the Transkei in 1992—a trip that really sparked my interest in pursuing historical research on the Eastern Cape—I was also intensely curious about the long-term future of my relationship with my girlfriend at the time. Some thirteen years later, I find myself not only completing a book on the Transkei but very happily married to that former girlfriend, Elizabeth Herrmann, and together raising our four-year-old daughter, Rosa, and our infant son, Samuel. In the intervening years, there have also been significant losses—of family members, mentors, and friends. Among other things, these different experiences and the complex emotions they evoke have taught me important lessons which in turn have informed the writing of this present work: that memories and the unanticipated directions of life continually make the past novel, meaningful, and anything but simple.

I am deeply indebted to the many individuals and institutions that have contributed to this book's development over the years. The University of Minnesota Graduate School, the MacArthur Program at the University of Minnesota, the Joint Committee on African Studies of the Social Science Research Council and the American Council of Learned Societies (with funds provided by the Rockefeller Foundation), and Middlebury College together provided vital financial support for this project. While pursuing my doctorate at the University of Minnesota, many fellow students pushed my thinking in new directions both inside and outside of the classroom. I am particularly grateful for the opportunities to learn from and with Heidi Gengenbach, Amy Kaler, Premesh Lalu, Matt Martin, Marissa Moorman, Maanda Mulaudzi, Wapumuluka Mulwafu, Agnes Odinga, Derek Peterson, Daviken Studnicki-Gizbert, and Guy Thompson. Various faculty members inspired and helped refine the early evolution of this project into a dissertation, and I would especially like to thank Jean Allman, the late Susan Geiger, Allen Isaacman, and

Ben Pike for immersing me in the complexities and challenges of African Studies, and Jean O'Brien-Kehoe for guiding me through the exploration of comparative themes in Native American history.

My research experiences in South Africa would not have been possible without the generosity, support, and warmth of many people. Lwandlekazi de Klerk, Veliswa Tshabalala, and Tandi Somana provided invaluable skills and insights as research assistants at different stages of my fieldwork. Staff at the South African government archives in Cape Town and the University of Transkei's Bureau of African Research and Documentation were particularly accommodating. Various individuals and families also graciously opened their homes and their social worlds to me in the Eastern Cape and in Cape Town, including Gerard Back, Tessa Dowling, Sue and Jim Gibson, Premesh and Vivienne Lalu, Hugh and Monica Macmillan, Mandla Matyumza, Maanda Mulaudzi and Amy Bell-Mulaudzi, Sophie Oldfield and David Maralack, Pule Phoofolo, and Lance van Sittert. And I am forever thankful to the various men and women in the Eastern Cape who so generously shared their time, their lives, and their thoughts with me in interviews and whose words have challenged and expanded my understanding of the Transkei's history.

I am also very grateful for the critical comments and constructive suggestions many colleagues have offered as I have presented parts of this evolving research at various annual meetings of the African Studies Association and the American Society for Environmental History, the 2003 International Conference on Forest and Environmental History of the British Empire and Commonwealth at the Centre for World Environmental History (University of Sussex), the Northeastern Workshop on Southern Africa (Burlington, Vermont), the History Department Seminar Series and the Programme for Land and Agrarian Studies at the University of the Western Cape, the Institute of Social and Economic Research at Rhodes University, and the History Department's "Empires, Colonies, Nations" Seminar Series at McGill University. Special thanks go to William Beinart, Ben Cousins, Derick Fay, Nancy Jacobs, Thembela Kepe, Premesh Lalu, Gregory Maddox, Jim McCann, Lungisile Ntsebeza, Lance van Sittert, Richard Tucker, and Andrew Wardell for their feedback, insights, and encouragement. I would also like to thank the editors of the *International Journal of African Historical Studies* at Boston University's African Studies Center for permitting me to reuse material from my previously published article "The Python and the Crying Tree: Interpreting the Tales of Environmental and Colonial Power in the Transkei," *IJAHS* 36, no. 3 (2003): 511–32.

Gill Berchowitz, Allen Isaacman, Jean Allman, and two anonymous reviewers for Ohio University Press offered crucial suggestions for revising the manuscript that have truly helped me transform my reading and writing of this material, several times over. Gill has been stellar throughout, patiently walking me through the production process and fielding my never-ending questions. I also want to thank the University of Fort Hare Library and Ronald Ingle for their assistance and permission to use photographs from the Piper Collection, the Cape Town Archives for permission to use photographs from the Jeffreys Collection, Conor Stinson at Middlebury College for producing the maps, and the production staff at Ohio University Press for all of their efforts. My gratitude and debts to Jean and Allen are many years deep and go well beyond the realization of this book. Thank you for your continuous confidence in my abilities (despite my own occasional doubts) and your enormous generosity and care as mentors, colleagues, and friends. Special thanks to Jean for helping me learn by example how teaching and scholarship can truly inform each other. And to Allen, who has most closely influenced this project from my fledgling ideas in a graduate seminar to its final throes as a manuscript, I send out my heartfelt appreciation of your frank, perceptive, and always constructive comments on my work; your caring guidance in my professional development; and your thoughtful understanding of and support during the highs and lows of my personal life.

Finally, I'll end these acknowledgments where I began, with my wife Liz and our children. This book owes much of its existence to Liz's patience, endurance, grace, and emotional support throughout, despite the many challenges along the way. Rosa's dancing and art projects and Samuel's constant exploring of his new world have further buoyed my work and provided much needed perspective on the important things in life. My love to you all.

Abbreviations

BBNA	Cape of Good Hope, Department of Native Affairs, Blue Book on Native Affairs
CCF	Chief Conservator of Forests
CLPW	Department of Crown Lands and Public Works
CMK	Chief Magistrate of East Griqualand
CMT	Chief Magistrate of the Transkeian Territories
CTA	Cape Town Archives Repository, South Africa
CTA AGR	Archives of the Secretary for Agriculture
CTA CMK	Archives of the Chief Magistrate of East Griqualand
CTA CMT	Archives of the Chief Magistrate of the Transkeian Territories
CTA FCE	Archives of the Conservator of Forests, Eastern Conservancy
CTA FCT	Archives of the Conservator of Forests, Transkeian Conservancy
CTA FDU	Archives of the District Forest Officer Umtata
CTA FKS	Archives of the District Forest Officer Kokstad
CTA NA	Archives of the Secretary for Native Affairs
CTA PMO	Archives of the Prime Minister's Office
CTA 1/ECO	Papers of the Resident Magistrate of the Engcobo District
CTA 1/KNT	Papers of the Resident Magistrate of the Kentani District
CTA 1/MQL	Papers of the Resident Magistrate of the Mqanduli District
CTA 1/TSO	Papers of the Resident Magistrate of the Tsolo District
CTA 1/UTA	Papers of the Resident Magistrate of the Umtata District
DFO	District Forest Officer

DWAF	Republic of South Africa, Department of Water Affairs and Forestry
FCT	Conservator of Forests, Transkeian Conservancy
KWT	Kingwilliamstown
LMA	Department of Lands, Mines, and Agriculture
NAR	National Archives Repository, South Africa
NAR NTS	Archives of the Native Affairs Department
NAR FOR	Archives of the Forest Department
PAR	Pietermaritzburg Archives Repository, South Africa
PGC	Pondoland General Council
PGC	Proceedings and Reports of Select Committees at the Session of the Pondoland General Council
PM	Prime Minister
PRO	Public Record Office, Kew Gardens, England
RM	Resident Magistrate
SAL	South African Library, Cape Town, Manuscripts Collection
SLMA	Secretary for Lands, Mines, and Agriculture
SNA	Secretary for Native Affairs
TTGC	Transkeian Territories General Council
TTGC	Proceedings and Reports of Select Committees at the Session of the Transkeian Territories General Council
UAR	Umtata Archives Repository, South Africa
UAR CMT	Archives of the Chief Magistrate of the Transkeian Territories
USA	Under-secretary for Agriculture
USNA	Under-secretary for Native Affairs
UCT	University of Cape Town, Manuscripts and Archives Library
UTTGC	United Transkeian Territories General Council
UTTGC	Proceedings and Reports of Select Committees at the Session of the United Transkeian Territories General Council

Introduction

In the decade or so since the formal unraveling of apartheid in South Africa, some of the nation's most persistent and intractable tensions have revolved around natural resources. In the territories formerly managed by successive white governments as African labor "reserves," "Bantustans," or "homelands," resource problems have been particularly acute. Decades of state-sponsored segregation, resource deprivation, economic impoverishment, and political dis-enfranchisement for the African majority have resulted in urgent demands by rural populations for more equitable, democratic, and sustained access to and control over vital livelihood resources. As the ANC-led regimes under Mandela and Mbeki have responded to such colonial and apartheid legacies, the state's prioritization of particular resource policies and development strategies has in turn helped generate new sources of frustration and tension among rural African communities, so-called traditional authorities (chiefs and head-men), and various branches and levels of government.

Within the boundaries of the former Transkei, once the largest homeland within the apartheid system but formally incorporated into the Eastern Cape Province since 1994, struggles and negotiations over resources have intensified in numerous spheres and in often dramatic ways. The redrawing of administrative boundaries and the installation of new local governmental structures have rankled many traditional leaders, who see such moves as threats to their

1

personal influence in local resource allocation and management.[1] In some cases, the resulting fractures in the political landscape have created new opportunities for environmental access for rural residents. Communities along the southern Wild Coast, for instance, made headlines in the early 1990s as they staged aggressive "invasions" of state-protected park areas and boldly exploited their reserved marine resources.[2] As these and other communities have asserted their resource rights through various formal and informal channels, they have participated in wider negotiations and debates with diverse government agencies, NGOs, and academics over intersecting and often competing interests in local and regional development, environmental conservation, and popular resource entitlements.[3]

Such developments in this period of untangling colonialism and apartheid have their parallels in what might at first glance seem an unlikely point of comparison: the original expansion of colonial rule in the late nineteenth century. In this earlier era, a tectonic shift in a political regime—the Cape Colony's conquest of independent African societies between the Kei River and the southern limit of the Natal Colony, the "Transkei"—and the introduction of new state policies and institutions similarly resulted in critical transformations in how differently positioned actors used and negotiated natural resources in the region. The colonial process of selectively removing or incorporating chiefly power in different locales and installing new political hierarchies generated, as have the changes of recent years, new jockeying over how "traditional" authorities and government representatives should assume responsibilities for environmental control and management. Different arms of the colonial state likewise debated with each other and local authorities over how and for whom the Transkei's resources should be "developed" or "conserved." At the same time, rural residents similarly negotiated shifts in local political, economic, and biophysical landscapes to find new avenues of resource access and to assert their own perspectives on their changing livelihood needs and environmental entitlements.

More than just paralleling contemporary situations, however, the formative period of colonialism in the Transkei put in place particular structural relations of power and resource access that have continued to haunt the region's environmental and developmental predicaments. Amid the many efforts and struggles in South African society to dismantle colonial legacies and "reincorporate" the populations of the former homelands into the body politic, it thus seems particularly crucial to understand such foundational moments in the making of the region's deeply rooted problems. Looking at the early evolution of the Transkei can illuminate some of the colonial contradictions and constraints that have had such long-term and ongoing impacts on local people's lives, particularly by revealing the ways in which issues of environmental use and control originally became entangled in deeper colonial transformations and experiences.

This book takes on this challenge by exploring how changes in environmental access played a key role in colonial dynamics and everyday social interactions in the Transkei in the late nineteenth and early twentieth centuries. My primary focus here is on forests, for the colonial government's most intrusive environmental interventions and some of the most intense environmental negotiations of this period revolved around the Transkei's forest resources. The colonial reshaping of forest access from the 1880s onward was deeply interwoven into the larger development of a new colonial political and economic order. Shifts in environmental control and rights under the successive Cape Colony and Union governments were implemented, experienced, and negotiated amid the major forces transforming local political economies—the restructuring of African political authority, the establishment of colonial governance and law, and the gendered socioeconomic and ecological pressures transforming the Transkei into a labor "reserve" and driving the expanding out-migration of African men. At the same time, changes in forest access touched the lives and livelihoods of rural men and women across the region in more direct ways. The overwhelming majority of Africans in the late nineteenth and early twentieth centuries relied on wood sources for fuel, for building livestock kraals and huts on their homesteads, and for manufacturing implements used in producing and transporting agricultural goods, such as hoes and sledges. Many others utilized forest lands for crop cultivation and livestock grazing, hunted wildlife in local forests and woodlands, and exploited other forest resources for food, medicine, healing, and a host of other social and cultural purposes. Colonial restrictions on forest use represented varying constraints on these and other practices.

Such changes in the early colonial period also continue to resonate in the memories and personal narratives of many people in the former Transkei today. As I interviewed elders in the Umtata, Tsolo, and Engcobo districts in the late 1990s, the topic of colonial changes to forest access often elicited quite emotional responses, as many individuals linked their more recent experiences of environmental dispossession and subordination under apartheid and their ongoing daily struggles after 1994 with longer-term processes of colonial disempowerment rooted in the late nineteenth and early twentieth centuries. Reflections on resource access in the past and present often critically invoked a history of colonialism setting in motion local people's loss of direct environmental control and constraints on their ability to employ a variety of meaningful forest resources in their everyday livelihoods and cultural practices,

Figure 0.1. Hut-building in the Engcobo district, Transkei, c. 1954. *Photograph courtesy of University of Fort Hare Library, Piper Collection*

from selecting desirable tree species for fuel or building materials to accessing certain ritually important landscapes.

Yet despite the significance of colonial-era shifts in resource access, in the lives and thoughts of Africans both then and today, scholars have given this topic relatively little attention. To be sure, various writers have generally abandoned the simplified and self-congratulatory versions of state resource management in the region spun by officials themselves in the colonial and apartheid eras, narratives which environmental specialists all too often perpetuated, sometimes verbatim. In such accounts, Africans' ecological "destruction" in the Transkei in the late nineteenth century necessitated the colonial "protection" of forests and trees from "extinction." Colonial foresters, driven solely by their commitment to environmental conservation and equipped with their unique expertise and farsightedness, fortunately rescued the situation and put resource management on a proper course in the succeeding decades.[4] The dramatic changes in South Africa since the early 1990s have forced a major rethinking of such statist narratives and a retooling of the exclusionary and nondemocratic direction of past forestry policies. More participatory and development-linked forestry programs for the populations of the former homelands, and ongoing resource conflicts in the former Transkei, including land restitution claims involving state-controlled forests, have directed more researchers to consider the long-term historical roots of many contemporary

Figure o.2. Woman weaving hut door made of saplings, c. 1930s. *Photograph courtesy of Cape Town Archives Repository, Jeffreys Collection*

Figure 0.3. Pulling a loaded sledge, c. 1930s. *Photograph courtesy of Cape Town Archives Repository, Jeffreys Collection*

problems of forest access and control. Primarily focusing on contemporary land reform, resource management, and development issues, such accounts generally treat the early Transkeian period as historical background.[5]

By contrast, historians have been relatively quiet when it comes to exploring environmental themes in the colonial Transkei. The overwhelming majority of writing on the environment has instead focused on the contending discourses, practices, and local experiences of socioeconomic "development" schemes in subsequent decades. As a host of writings has examined, from the late 1930s onward South African state authorities responded to severe soil erosion and other environmental problems in the Transkei and other African "reserves" by implementing "betterment" and "rehabilitation" measures — schemes ostensibly designed as ecological conservation strategies but integrally tied to state concerns over racial segregation, the continued supply of cheap African labor, and political stability. Particularly after the Nationalist Party came to power and the Bantustan system was erected, the state instituted a reengineering of the African countryside, uprooting and forcibly removing populations, reorganizing local livelihoods, and often spawning the violent response of rural communities against representatives and symbols of the apartheid state.[6] While the drama and trauma of these tumultuous events of more recent decades has understandably drawn scholars' attention, this historiographical emphasis on the political struggles of the apartheid era has also helped obscure earlier experiences of state environmental intervention in the countryside.[7] The following discussion reorients this vantage point, exploring the colonial antecedents of the Transkei's interwoven social and environmental problems and revealing that significant negotiations surrounding natural resource access have a much longer history in the region than previously recognized.[8]

Historical writing on the colonial Transkei itself, on the other hand, has only occasionally, and rather unevenly, investigated environmental themes directly. Like other South African radical scholars of the 1970s and 1980s, Colin Bundy and William Beinart, in their pathbreaking work on the social history of the Eastern Cape and Transkei, tended to treat environmental concerns as relatively minor players on the larger stage of political-economic dynamics.[9] Certain ecological factors might make cameo appearances here and there, but the driving engines of their narratives were the structural forces of colonial expansion and settler capitalist accumulation, the capacities of African societies to weather the transformation of their economies into underdeveloped labor reservoirs for white farming and mining, socioeconomic differentiation within African societies, and the rural politics surrounding certain colonial policies.[10] By contrast, more recent work by William Beinart, Lance van Sittert, Richard Grove, and others has more seriously investigated environmental concerns in important new ways—covering such diverse topics as forest conservation, invasive plant species, fauna preservation, "vermin" control, pasture degradation, and veterinary practice—but often the Transkei is relegated to the sidelines of the history of the Cape Colony.[11] Moreover, in a striking departure from the radical scholarship they followed, these studies have primarily focused on official, scientific, and settler communities' thoughts and practices, providing little information on the historical transformations affecting the livelihoods and responses of the majority populations in the Eastern Cape and Transkei.[12]

This study injects environmental issues far deeper into the colonial past and the social-historical fabric of the Transkei. Variously situated actors in rural communities negotiated important changes in resource access from the earliest days of colonial administration of the region, and these responses were tied to wider transformations and experiences in the making of the colonial Transkei. Exploring such themes can thus advance a much more thorough and critical process of interrogating and unpacking the reigning logics of state- and settler-centered histories of resource management in the Eastern Cape and South Africa.

The following chapters pursue a central question: how does looking at the various social interactions surrounding environmental access reshape our historical understanding of the Transkei? Put differently, what new insights does an exploration of environmental relations bring to the region's colonial history? In probing this question, I have intentionally shifted away from a conventional narrative of "European imposition and indigenous response," since changes in resource access in the colonial Transkei involved a much broader, more complex constellation of political, economic, social, cultural, and environmental dynamics than such a "symmetrical" story allows.[13] Instead, the discussion here centers on the deeper social relations and processes implicated in forest access from the 1880s to the 1930s, as different groups of historical actors, African and non-African alike, brought the particular and myriad changes affecting their lives to bear on their negotiation of forest access. Forest access in the Transkei thus serves as a window onto differently situated people's historical perspectives on, and responses to, both their shifting biophysical environments and the changing nature of power relations and everyday life in the colonial period.

In framing the discussion around the notion of access, I draw from the theoretical insights of Jesse Ribot, Nancy Peluso, and Sara Berry. Ribot and Peluso have together recently charted a synthetic conceptual framework for what they term natural resource "access analysis," moving beyond past discussions of tenure rights and claims to resources to a broader perspective on "the multiplicity of ways people derive benefits from resources, including, but not limited to, property relations."[14] Studying access thereby becomes a wider investigation into the many social "means, processes, and relations" by which actors are able to enjoy what Ribot and Peluso term the "benefit flow" from particular resources, all situated within the multiple webs of power and political-economic dynamics of particular historical contexts. Similarly, Berry's discussion of agrarian change in multiple African settings focuses on how different groups exploit and invest in various social relations in order to gain and control access to resources, extending the analysis beyond formal economic and legalistic interpretations of environmental regulation into deeper, more intricate social dynamics of power and culture.[15]

Understanding the many forces that shaped forest access in the colonial Transkei requires attention to such broader and deeper contours of change. The following chapters examine how people negotiated forest access from their various positions in the evolving colonial order and the diverse stakes they brought to these negotiations over time. Changes in forest access posed new and shifting opportunities, threats, and often ambiguous futures for differently positioned magistrates, foresters, African chiefs, headmen, and male and female commoners. Besides pursuing particular material benefits from gaining, controlling, and/or maintaining forest access,[16] actors in the colonial Transkei also strove to enhance their wider power, authority, influence, and legitimacy through such negotiations. When different men and women in particular locales responded to shifts in local forest management or constraints on their environmental practices, they thus drew from the multiple dimensions of their life circumstances.

I also employ the term "negotiation" throughout this book, borrowing and building from Nancy Rose Hunt's usage in her recent study of birth medicalization in the Belgian Congo. Hunt employs a dual sense of the term: the more "classic sense of adversarial parties bargaining over contested ground, of mediation, arbitration, and sometimes even compromise," and the more "processual, performative connotation for negotiating—as in traveling, making a turn, veering—as is appropriate for rendering a situation of everyday 'making do.'"[17] Bringing such definitions of social interaction to my historical interpretation of the Transkei purposely shifts away from a sole focus on environmental contestation.[18] The colonial restructuring of forest access in the region not only involved conflicts between and within official circles and African communities—whether expressed in formal debates, public protests, or everyday interactions and acts of resistance—but also shifting alliances, compromises, and multidirectional decisions and practices which are difficult to fit into neat categories.[19]

Moreover, as Hunt's second definition connotes, resource access negotiations moved in ways not fully anticipated, controlled, or predetermined by any of the various actors involved, as they each coped with their situations and relationships in a fluid colonial context. Examining negotiation intents and strategies not only allows us to see how shifting power dynamics structured and constrained the actions of differently positioned people in unequal ways, but also how individuals creatively responded to these conditions—from high-level colonial authorities jockeying for bureaucratic power, to forest guards and headmen exploiting their forest gatekeeper status for personal dividends, to men and women in particular locales hiding illegally harvested resources. Expanding on the everyday performative sense of negotiation, I also look at how individuals contended with the diverse and complex dimensions of their daily lives and how they then brought such particular yet multiple interests and meanings to their responses to changes in resource access. For example, when rural residents in some locales at the turn of the century violently attacked the bodies and property of African forest guards, they simultaneously performed dramatic acts of resistance against the colonial state's domination of local environments and exacted personal revenge on individuals with whom they were entangled in personal and financial disputes.

In approaching and exploring negotiations in such ways, this book offers new directions for scholars of South Africa and the wider continent to conceptualize both the interconnections between environmental and social histories and the dynamics of colonial interactions. My analysis intentionally engages with the vast amount of scholarship on colonialism in Africa and beyond concerned with how various "colonizing" and "colonized" groups "invented," "imagined," and "conversed" about "tradition" in the making of colonial legal and political institutions.[20] While prolific, this literature has tended to neglect the significant environmental dimensions of many such interactions in the early colonial era, with most Africanist writings focusing primarily on the contestation of "customary" land ownership and use and "traditional" authority during the "high" period of colonialism and anticolonial resistance.[21] This is an unfortunate lacuna, since understanding the complex making and negotiation of diverse environmental "traditions" in the transition from precolonial life to colonial rule can potentially offer critical perspectives on the discourses employed in contemporary development debates, a point which I revisit for the South African context in this book's conclusion.

The following discussion responds to this relative silence in two major directions. First, I explore how changes in African chiefs and headmen's involvement in environmental control and their ambiguous and intercalary position in the evolving colonial order were at the center of complex power struggles from the earliest days of political-ecological restructuring in the Transkei.[22] Within African communities, various actors brought a wide array of contending social, economic, and environmental stakes to their negotiations of local environmental authority. As I examine in detail for areas in the Tsolo, Umtata, and Engcobo districts at the turn of the century, such diverse factors as ethnic and religious politics, competition for local leadership positions, class differentiation, ecological change, and colonial administrative schemes all intersected and informed various individuals' allegiances and alliances concerning chiefs and headmen's influence over local resource access.

African authorities' environmental power was also negotiated between and among colonial officials and chiefs and headmen. In the late nineteenth century, forest officials and magistrates deployed competing images of chiefs' precolonial resource control as ways to legitimize their particular claims to the reins of colonial forest management.[23] Chiefs and headmen also asserted their own claims to "customary" legitimacy as environmental managers of community resource access, exploiting their indispensability to official schemes and pursuing their particular political, economic, and ecological interests—whether it be strictly enforcing government forest restrictions, shielding local residents from interventions and local representatives of the colonial state, or using their position to derive personal benefits beyond official control. Although relegated to an increasingly subordinate role in resource management at the turn of the century, chiefs and headmen found formal and informal avenues for protecting their ability to interpret their environmental authority and prerogatives in their own localized ways, often at the expense of residents in their wards.[24]

This book further examines another important yet underdeveloped field in African history writing—the negotiations around "customary" environmental practices and the meaning of environmental entitlements in the colonial era. In the last decade or so, scholars have more closely examined the societal institutions that have structured rights and access to natural resources in various African settings. Working from concepts originally employed in analyses of famine, these writings have opened up new ways of conceptualizing environmental entitlements and their location in complex social processes and relationships.[25] Anthropologists and geographers have made the most productive use of such insights, particularly by exploring the gendered dimensions of development and state environmental management schemes and their impact on local resource tenures, entitlements, and practices. Such writings have skillfully moved beyond simplistic and static renderings of gender and environmental relations—often employed by state, NGO, and local actors—to explore the political, economic, and ecological struggles surrounding men and women's changing access to their natural surroundings.[26]

Despite this flurry of scholarship and the increasing examination of gender in the history of human-environmental relations in Africa,[27] however, surprisingly little historical analysis has been devoted to questions of gender and environmental rights.[28] Southern Africanist scholars have scrutinized the gendered effects of state forestry policies in African communities, for instance, yet similar questioning has been neglected for the colonial antecedents of these dynamics.[29] And while some recent writing in South Africa's growing environmental historiography begins to incorporate gender issues, many historians are still shying away from seriously investigating issues of gender and resource access.[30] Through such neglect many opportunities are lost to explore what light such themes might shed on gendered histories more generally in the South African past.

This study helps to fill such gaps by exploring the colonial restructuring of Africans' popular forest rights in the Transkei and the role "custom" and gender played in negotiations over both state policies and the wider colonial transformations in which these policies were embedded. As in other colonial spheres, different arms of the Transkeian administration at first bickered over the scope and pace of governmental interventions, as foresters' and magistrates' contending policy priorities—conservation versus "native policy"—often collided at the turn of the century.[31] Yet in one respect, at least, officials increasingly reached consensus: the utility of environmental restructuring to driving African men into labor migrancy and tying African women to a "subsistence" economy in the "reserves." Beginning in the 1880s and 1890s, officials transformed existing institutions—such as those tied to gender, class, age, and marital status—which structured men and women's access and entitlements to crucial natural resources, particularly wood.[32] Through selective representations of Africans' "customary" gender roles and "subsistence" practices, officials attempted to legitimize their efforts to restrict and charge Africans for access to forest resources.

Yet differently situated men and women invoked their own understandings of "customary" practices and resource entitlements as they critically engaged with colonial restrictions.[33] In the earliest years of colonial forestry, amid the internal squabbling among officials, popular discontent with government restrictions and Africans' regular claims to and exercise of more expansive resource entitlements in fact helped shape the broader trajectory of government forest policies, belying officials' own retelling of this history.[34] People also contested the gendered orientation of forest regulations, particularly regarding fuelwood access, through formal protest and in daily practice, demonstrating their dissatisfaction with state interpretations of their "customary" gender roles and "subsistence" rights. Even as Africans in the Transkei faced an increasingly authoritarian and bureaucratic state in the early twentieth century—with greater influence over local environments, and growing interests in reducing popular forest rights and intervening in women's environmental practices in particular—men and women in many locales continued to negotiate and reshape the meaning of entitlements in ways they could most effectively control, whether selectively harvesting particular tree types or reorganizing their wood-collection practices. At the same time, in all of these negotiations, individuals contended with shifting forest rights in ways that best responded to their particular needs and experiences of deeper social, economic, and ecological changes. Negotiating colonial forest restrictions was often embedded in individualized experiences of wider transformations of the colonial era—impoverishment, increasing ecological scarcity, broadening state power, and the various pressures driving male out-migration and leaving women to carry the heaviest burdens of rural work.

While men and women negotiated the meaning of new forest rights policies in their livelihood practices, they also asserted alternative perspectives on the very meaning of forest resources themselves. In exploring this realm of resource negotiations, I am suggesting a more expansive definition of access than that theorized by Ribot, Peluso, and Berry. Although these writers point to the social processes by which people are able to derive benefits from resources, it is important to delve further into how different actors historically understood the specific meanings of their changing landscapes and how such perceptions then contributed to their particular ways of negotiating resource access. Ribot and Peluso's emphasis on "benefit flows," for instance, does not adequately grapple with the complex ways in which the nature of a resource "benefit" or value can be contested terrain itself. Building from expanding literatures on local ecological knowledge, "peasant science," and African landscapes, this book reflects critically on how our understandings of Transkeian environments and resources over time are fundamentally altered when we move beyond dominant state conceptualizations to recognize Africans' own historical perspectives on their social and natural surroundings.[35]

As the colonial state demarcated the Transkei's environments according to its own definitions of nature, differently situated rural residents interpreted such actions according to the various social, economic, and cultural meanings of particular forest resources in their lives and livelihoods. The significance of specific forest species in an individual's agricultural repertoire (chapter 3) or in a group's ritual practice (chapters 4 and 5) often profoundly influenced how particular actors viewed and negotiated both state efforts to restructure access to these resources and the expansion of colonial influence more generally. In my research, information from oral interviews and such written sources as Africans' petitions to colonial officials, court testimonies, and Xhosa-English dictionaries compiled in the early twentieth century revealed how very specifically and selectively people interacted with their changing environments in different historical locales. People's use of language particularly reflected this reality. The Xhosa term *amahlathi* (forests, sing. *umhlathi*) repeatedly showed up in written and oral sources when people were asserting broader claims to wooded areas and all of the diverse resources available within them, such as trees, forest plants, potential grazing areas, cultivable lands, etc. Yet these sources also regularly used much more specific labels to identify the key types of trees, plants, roots, and grasses they routinely exploited and preferred in daily practice in particular historical settings.

Such selectivity has helped me recognize that local historical understandings of "valuable" forest resources not only diverged widely from official perceptions but also varied quite markedly across space and time and from individual to individual. Exploring such details has thus led to ways of disaggregating and historicizing the "indigenous knowledge" differently situated Africans in the Transkei employed as they utilized their environments and responded to government interventions into them.[36] People in the colonial Transkei engaged their particular natural surroundings in highly differentiated ways, bringing their specific life circumstances and positions (in terms of gender, class, age, status, and ethnicity) to bear on their interests in resources and their availability.[37]

Pursuing these themes has been especially helpful in interpreting the deeper social and cultural dimensions of everyday resource use in the colonial Transkei—a type of analysis generally absent in much of the South African literature, particularly environmental histories of the Cape in the nineteenth and twentieth centuries.[38] As Tamara Giles-Vernick has recently explored in her research among Mpiemu groups in central Africa, human-environmental interactions can play crucial roles in the definition and development of persons, not just environments.[39] In the Transkei, particular forest resources were similarly pivotal in social and ritual interactions of young males as they attained manhood, in the ceremonies surrounding the development of newborns, and in communications between the living and the dead. But beyond incorporating specific trees and plants into the rituals of such major life transitions, differently situated individuals further relied upon particular species to contend with their more mundane, everyday needs of healing and protection and to mediate everyday social interactions and tensions— to magically inflict harm on others or defend oneself, to charm a potential lover, or even to sway the mind of a colonial official when necessary. Such themes are not conventional in African environmental histories. Yet investigating such specific historical stakes in forest resources seems essential to understanding how different individuals approached and responded to local changes in species access within the wider contexts of their daily lives.

CONTEXTUALIZING THE STUDY

In probing such varied forest negotiations of the late nineteenth and early twentieth centuries, important contours in the Transkei's social and environmental development in this period have shaped the particular spatial and temporal parameters of my analysis. The physical geography of the area eventually comprising the Transkeian Territories—some 40,000 square kilometers—was and continues to be extremely diverse.[40] The great escarpment, climbing up to some 2,700 meters in the Drakensberg Range, today lines the western boundary between the former Transkei and Lesotho. Just below this escarpment is a highland zone, ranging from 1,300 to 2,000 meters and including areas from rolling grass-covered country to treeless mountains in the Mount Fletcher and Matatiele districts. In the central inland districts a minor escarpment running roughly parallel with the Indian Ocean coast separates the higher plateau and the lowland plateau extending from its eastern base. This lesser escarpment, historically often referred to as the Zuurberg, reaches its highest altitude at about 1,400 meters and serves as the catchment area for some of the largest rivers in the former Transkei, which eventually run their course to the sea. In the early colonial period, much larger afromontane forests also stretched along the seaward side of this range.[41]

Below this escarpment lies an undulating lowland plateau stretching down to the beginning of the coastal belt at roughly 700 meters—generally flat, but occasionally dotted with hills and mesas in the southwest, becoming much more mountainous as one travels northward, and dissected by deep river valleys and streams in certain areas. This plateau receives much less precipitation than either the escarpments to its west or the coastal belt to its east, affecting the vegetation in the region at the onset of colonialism: deciduous woodland and scrub dotted the grasses of the flatlands, while denser clusters of various types of trees crowded river and stream banks. The coastal belt added even greater levels of complexity to the Transkei's vegetation in the late nineteenth century. From the lowland plateau to about 300 meters lay a strip of coastal scarp forest, which then transitioned into a series of lower coastal forest types, some quite extensive, becoming progressively more diverse and subtropical in the northern parts of the coast towards Natal.[42]

When colonialism first expanded into the various settings of this complex ecological mosaic, multiple African societies inhabited the region.[43] Certain polities—such as the Bomvana, Mpondo, Mpondomise, Thembu, Xesibe, and Xhosa chiefdoms—had long histories in the Transkei, while the political tumult of the early decades of the nineteenth century brought new communities—including the Bhaca, Ntlangwini, and Hlubi—migrating southward from Zulu territory. Sotho, Kwena, Tlokoa, and Griqua groups who had traversed the Drakensberg later joined these polities, settling in the northern highland territory that would become known as East Griqualand. In the southern parts of the Transkei, some refugee communities had moved into Gcaleka Xhosa territory, negotiated opportunities for land acquisition in the neighboring eastern districts of the expanding Cape Colony, and eventually became known as the Mfengu.[44]

The expansion of colonial influence in the broader region in the mid- to late nineteenth century had varied effects on these different polities and groups.

Mfengu and Xhosa communities living in closest proximity to the Cape experienced the forces of colonial expansion earlier and more immediately than other areas. A brutal series of Cape-Xhosa wars, their culmination in the late 1850s with the tragic Cattle-Killing episode, and the defeat of Gcaleka and Ngqika Xhosa chiefs and their followers led to the dispersal and then eventual resettlement of Xhosa communities in the southern coastal area of their former territory between the Kei and Mbashe rivers in the mid-1860s. In the meantime, thousands of Mfengu and Thembu people from the Eastern Cape were resettled in the now "vacant" lands of the northern and central parts of the former Gcaleka stronghold: a group of Thembu chiefs and their followers relocated in the Drakensberg foothills, in what became known as Emigrant Thembuland, and the Mfengu settled below them, in the territories renamed Fingoland (Mfenguland) and the Idutywa Reserve.[45] British government agents were appointed to reside in these territories, and diplomatic and trade relations were established with a few other chiefdoms in the Transkei. The majority of polities in the region, however, continued to live free of colonial control prior to the early 1870s.

In 1872, after the Cape Colony assumed responsibility for self-government, momentum increased for incorporating the various polities and populations between the Kei and the Natal border into the colonial fold.[46] Yet formal annexation was only really able to accelerate across the region following two tumultuous events: the Ninth Frontier War in 1877–78, pitting the Cape primarily against the Gcaleka, and the rebellion against the colonial government waged mostly by Mpondomise, Thembu, Qwathi, and Sotho groups in 1880–81. Colonial success in these wars not only enabled annexation to proceed but heightened official interests in securing political control over these populations. After defeating the Gcaleka in 1878, Cape authorities began consolidating the colonial administration of "native affairs," establishing three chief magistracies: one for the "Transkei" proper, comprising Gcalekaland, Fingoland, and the Idutywa Reserve; one for "Thembuland," comprising Thembuland, Emigrant Thembuland, and Bomvanaland; and one for East Griqualand.[47] In 1879, the formal annexation process began—incorporating Fingoland, the Idutywa Reserve, and East Griqualand—until the rebellion of 1880–81 slowed down the process, and Cape authorities responded with a more tentative and gradual approach to colonial expansion and its impact on political stability in the region. Only over the next several years were other territories successively annexed. By 1887 Pondoland was the only remaining region independent of colonial control; after a long period of negotiation, the Cape then finally annexed Pondoland in 1894 and completed the colonization of all African polities between its borders and those of Natal.[48] Adminis-

Map 0.1. Colonial annexation of the Transkeian Territories in the late nineteenth century. *Map by Conor Stinson*

trative consolidation proceeded in 1902, when the chief magistracies of the various separate regions were amalgamated into the position of chief magistrate of the Transkeian Territories, based at Umtata.

While colonial authorities began establishing formal political control in these newly annexed territories, they also gradually put in place mechanisms for regulating local environments in the Transkei. As Richard Grove has described, forestry in the Eastern Cape and Transkei grew out of the broader development of British imperial conservationism in the Cape and beyond. As official interests in "scientific forestry," the moral "salvation" of South African landscapes, and the sustainability of settler society all merged in the Cape, the formalization of forest administration accelerated, particularly in the 1880s.[49] Following the appointment of the first superintendent of woods and forests to oversee the entire colony and the expansion of conservation efforts in the Eastern Cape, colonial foresters then began to take a closer look at and deeper interest in the vast forest tracts beyond the Kei River, particularly along the coast and in the inland mountain ranges. As more territories in this region were brought under colonial influence, and particularly as this "opening" of the frontier enabled migrating sawyers and settlers from the Eastern Cape to exploit local forests, officials became increasingly concerned with regulating popular forest use, both African and non-African.[50] Some magistrates introduced and attempted to enforce some minor restrictions on forest use on their own, but it was only with the formal establishment of the Transkeian Conservancy in the late 1880s that the state seriously expanded its intervention into the region's forests.[51]

The interconnections between these various developments and timelines have influenced the periodization of this study. The book focuses most centrally on the years roughly between 1880 and 1915, the formative period of colonial environmental restructuring and negotiation in the Transkei. The 1880s were a crucial time when colonial political control and environmental interventions emerged and began merging in the region. It was only following the last major military confrontations between the Cape government and African chiefdoms during the late 1870s and early 1880s that magistrates were able to expand colonial rule and their interventions into African socioeconomic and environmental practices more seriously, negotiating such changes with local African authorities. Colonial conservation efforts also expanded in the 1880s, and, as forest officials began sharing the responsibilities of environmental management with magistrates, controlling African resource access became more heatedly contested both within and between official and popular circles. Through the next three decades of negotiation and conflict at various levels, the juridical, political, and economic foundations of state forestry were firmly established in local communities.

While the structural building blocks of state forestry were established in the period ending at about 1915, significant contestations and negotiations over their meaning extended well beyond this point. The book explores how different groups within the state and rural communities perceived, negotiated, and responded to the prior transformations affecting resource access as they continued to shape local experiences amid new changes in the 1920s and 1930s. I have ended the story roughly at 1940, for the late 1930s represented a turning point in the relationship between the state and rural Africans, particularly surrounding the willingness and capacity of the South African government to intervene in and transform rural livelihoods. In 1936 the Natives Trust and Land Act, as part of the culmination of the segregationist "Native bills" advanced by South African state leaders over the course of the previous decade, enabled the government to function as the "trustee" of African people more comprehensively than ever before. Three years later the first "betterment" proclamation was enacted, leading to the more systematic restructuring of local forest use in the Transkei from the 1940s onward, with officials utilizing key provisions in the new regulations to restrict popular access to forest tracts, expand fencing of these areas, and establish extensive afforestation operations.[52]

While the broad temporal framework described above captures changes in the Transkei in an overall sense, it is also necessary to understand how forests were negotiated in the late nineteenth and early twentieth centuries in more specific historical settings. In the early stages of this project, I became particularly interested in focusing on the history of the Tsolo District. With its sizeable mountain forests and its perpetual reputation in official forestry reports as a "hotspot" of forest conflicts at the turn of the century, the district seemed to make a useful case study, bringing into relief in particular ways the broader colonial contestations of the Transkei. Learning much more about how Africans living in the Tsolo and neighboring districts perceived and utilized their forested mountain landscapes in the late nineteenth and early twentieth centuries, however, has reshaped my approach. Examining issues of forest access within the confines of district boundaries helps explain some of the political machinations surrounding natural resource use, yet it does not adequately capture the ways in which most historical actors perceived and manipulated their local environments.[53]

Through the process of interviews and archival research it has become clearer that the afromontane forests straddling the Tsolo and the adjoining Umtata and Engcobo districts were the center of an interconnected ecological and socioeconomic zone of activity during this period, involving European, African, and some small "coloured" communities living both adjacent to the hills and in outlying lowlands.[54] For the purposes of this study, I have called this zone "KwaMatiwane," derived from the KwaMatiwane Range situated roughly at its center (see map 0.2). Following Kate Crehan's approach, I recognize that the notion of "KwaMatiwane" is in certain ways arbitrary and partial, that social and ecological dynamics of course continued beyond the imposed boundaries of this map.[55] However, like Crehan, I view this conceptual map as a means of illuminating "certain features and certain relationships but one that makes no claims to re-create that reality in its totality."[56] Within this historical-geographical zone, the negotiation of divergent forest practices and interests shaped local experiences in important and visible ways during the formative colonial period.

KwaMatiwane's geographical diversity had an important influence on this history.[57] The defining geologic formation of the region, the minor escarpment, is the catchment area for such significant rivers as the Umtata and the Mbashe. The region receives on average anywhere from 650 to 1,000 millimeters of rain, most of which falls during the summer months; in the winter, while the settlements and farms of the low-lying valleys experience the dry season, the highest peaks of the range are often covered in snow. Because of the moister environments of the mountains, during the winter months African

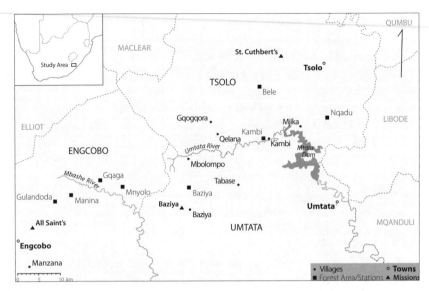

Map 0.2 KwaMatiwane. *Map by Conor Stinson*

Figure 0.4. View of some KwaMatiwane mountains, late 1990s. *Photograph by Eliza-beth Herrmann*

stockowners living below the escarpment have historically exploited the more palatable grasses of the range. As colonialism began in the area, vast hardwood forests also existed in the mountains up to roughly 1,200 meters in altitude, particularly in zones along the southern and eastern edges of the plateau, where they retained moisture and were protected from frost and damaging "bergwinds."

Because of the sizeable and dense forests in the region, colonial conservation efforts were particularly directed at local ecologies from the late nineteenth century onward. Soon after colonialism was expanded to local Mpondomise, Thembu, Qwathi, and Mfengu communities in the 1870s and 1880s, they began to feel some of the first colonial interventions into local environments in the Transkei, particularly following the 1880–81 rebellion that temporarily destabilized the region. In the rebellion's immediate aftermath, official interventions into local environmental practices were bound up in colonial stabilization plans. Official interests in "conserving" local forests and restricting African access then quickly expanded, particularly as European settlement and economic demands for timber expanded locally and in neighboring areas of the Cape Colony. By the 1890s and early 1900s, KwaMatiwane experienced the establishment of some of the first government forest reserves in the Transkei and the expansion of forest patrols in local areas. Colonial forest consolidation proceeded in the late 1900s and early 1910s, finalizing the demarcation between state-controlled areas and forests under local African authority. In the following discussion, I look in detail at how these developments in KwaMatiwane, combined with other social, economic, and ecological changes particular in time and space, helped shape differently situated residents' experience and negotiation of forest access over time.

SOURCES, METHODS, AND MEMORIES

I have utilized a variety of oral and written source materials as diverse entry points into the histories of such negotiations in the Transkei and KwaMatiwane. The South African government archive repositories in Cape Town, Pretoria, Umtata, and Pietermaritzburg hold a mine of detailed and previously neglected information on the workings of colonial forest administration. Archival correspondence and records of the Forest, Agriculture, Native Affairs and other departments in these centers as well as the Public Record Office in London offered valuable insights into the personal, political, economic, and cultural motivations for environmental restructuring in Transkeian communities and the internal battles waged by administrators over such interventions. Prior to this project, researchers working on the Transkei's history have left the Forest Department files largely untouched, a sign of the broader, artificial segregation of environmental and social fields of inquiry in much South African historical scholarship until relatively recently.

Besides their value as windows onto the dynamics of official discourse and action, these documents also contain important descriptions of environmental conditions, contestations, and protests in specific locales. The magisterial archives—containing testimonies by African men and women in criminal and civil court records, minutes of public and private meetings, petitions, and correspondence—as well as such official publications as commissions of inquiry, annual reports on "native affairs," and the minutes of the Transkeian Territories General Council and Pondoland General Council sessions, all provided unique insights into inter- and intracommunity struggles over political-ecological change and the local negotiation of colonial rule. Although the production and recording of such sources cannot be detached from their historical location in structures and relations of colonial power,[58] they offer details and perspectives on social and ecological change that are vital to recovering this period of the Transkei's past.

Several other types of written sources also provided unique historical insights. Private letters, diaries, and memoirs of prominent colonial officials—both published works and unpublished material at the government archives at Pietermaritzburg, the Manuscripts and Archives Library at the University of Cape Town, and the Cory Library at Rhodes University—offered important information on the personalities and relationships which affected the formulation and implementation of state policies, details often not visible in official correspondence now housed in archives. Contemporary travelers' accounts, English and Xhosa language newspapers, and missionary texts were also extremely helpful in documenting local historical changes and contestations and both European and African perspectives on them. Some of the missionary writings are particularly noteworthy. In my research into KwaMatiwane, I have explored two underutilized but extremely valuable sources for interpreting Transkeian history. The published accounts of Moravian missionaries, one of the earliest groups to be active in Thembuland in the late nineteenth century, include significant descriptions of Thembu and other KwaMatiwane communities from the 1860s onward. In addition, the manuscripts collection at the South African Library in Cape Town holds a wonderful ethnographic source—an unpublished Xhosa-English dictionary written by Robert Godfrey, a Free Church of Scotland missionary who was active in the Tsolo District and elsewhere in the Transkei in the early to mid-twentieth century. Godfrey was a keen observer and recorder of natural history in the region, and much of his own research into local environments and cultural and environmental practices, as well as the insights of African assistants in the Tsolo District, are included in the text.[59] Moreover, unlike other dictionaries which historians of the Eastern Cape and Transkei often reference for cultural data,[60] Godfrey's manuscript provides clues to locating the information it holds more specifically in time and place, regularly noting the sources of his definitions and differentiating between the terms, expressions, and practices of various ethnic groups in the Transkei.[61]

In addition to exploring this cluster of written sources, I have also pursued an oral research program. Although my research includes Europeans' views on the restructuring of forest management, interviewing elders in KwaMatiwane offered an opportunity to move beyond the colonial categories and perspectives embedded in the archival record, to understand how differently situated Africans experienced and remembered this history in ways not reflected in written sources.[62] In late 1997 and early 1998, with the aid of successive research assistants (Lwandlekazi de Klerk, Veliswa Tshabalala, and Tandi Somana), I interviewed men and women in the KwaMatiwane region and some outlying areas for their memories of and perspectives on the meaning of forests and colonial forest interventions in the early twentieth century. Most of the interviews focused on elderly individuals who had grown up in the KwaMatiwane region, who could personally remember their own experiences in the formative years of colonialism, or at least could recall the stories about these times told by their elders. Rather than focusing on one particular locale, I spoke with people who had lived in various and diverse settings within KwaMatiwane—surrounded by mountain forests, in the flat and relatively treeless lowlands below, at varied distances from government-run forests and tree plantations, etc.—to get some sense of how local differences in environmental conditions and proximity to forest resources affected men and women's experiences during the early colonial period. I also made a concerted effort to talk with elderly women as well as the prominent men to whom most people led me, since women's experiences and perspectives were often particularly opaque in the archival sources. For the most part interviews were conducted in isiXhosa, although some informants wandered freely back and forth between English and isiXhosa, with Lwandlekazi, Veliswa, and Tandi successively and indispensably providing assistance in interpreting and subsequent translating and transcribing.

These oral sources were invaluable in providing unanticipated insights into various subjects, perspectives, and daily experiences, forcing me to critically reflect on both official viewpoints embedded in the archival record and the limited categories I had brought to the interviews.[63] Much of my discussion of gender relations, social differentiation, and the deeper cultural dimensions of resource use—underreported or intentionally silenced in colonial accounts—would not have been possible without these personal narratives. At the same time, I have not viewed the combined pursuit of oral and written sources as merely an additive process of collecting complementary "facts" here and there. Historical patterns can become visible when different types of sources and the information they convey converge, but their collisions and divergences can also be revealing.[64] Particularly when exploring how colonial officials and African residents negotiated the meaning of the government's expansion into certain ritually significant landscapes in KwaMatiwane (chapter 5), investigating the interplay of oral and written accounts became crucial.

Oral sources also presented their own challenges and limitations.[65] Given the time period of my study and the age of most informants, there were automatically gaps of memory. Only the oldest informants had personal memories which reached as far back as the 1910s and early 1920s, and although some individuals did recall their elder relatives' reminiscences of the early colonial period, most men and women I interviewed spoke from their own experiences of the 1930s, 1940s, and onward. Having grown up when the state's domination of local forests was already firmly established, many informants described forest relations at the turn of the century in ways which reflected the quite different environmental and social realities a generation or two later.[66]

Interpreting these oral sources further required sensitivity to the contexts and processes through which informants represented the early colonial era. The identities, experiences, and situations of myself, research assistants, and informants, together with the relationships which developed through the interview process, shaped the production of historical narratives in the field.[67] Two brief examples can illustrate some of these dynamics. When Tandi Somana and I began interviewing one seventy-seven-year-old woman, she was hesitant and immediately called for her son to be present among these "strangers." Despite my attempts to convince her that I was in no way connected with the South African government, nor even South African, she was naturally concerned that a white person had driven up to her home in a peri-urban location outside of Umtata, when historically such a rare event would usually mean trouble with the law or the arrival of unfortunate news. During our conversation, when I inquired about how she evaded colonial forest restrictions as a young woman, she interrupted my questioning to express her concerns about divulging too much information. After her son allayed her fears, she explained that she worried she might have to go to jail for admitting to illegally collecting firewood from a government forest several decades ago![68] By contrast, in another part of the Umtata District, one prominent man was very eager to share his own stories and connect me with elders in outlying areas, to talk about colonial forest interventions and a more recent dispute over state forest lands. As eventually became apparent, he wanted to utilize this dispute, and my research into it, as a way of contesting local community authorities and asserting his own political leadership and legitimacy.[69] In these and other instances, my identity, my work, and the research context directly influenced how individuals felt threatened or saw opportunity in my presence, and offered or withheld information, often in ways I could not fully anticipate or see.[70]

For many of the men and women I interviewed, traumatic experiences of state environmental intervention in more recent decades—particularly intrusive "betterment" and "rehabilitation" schemes during the apartheid era— further influenced their historical narrations of the colonial period. This was especially the case for people who lived in the Gqogqora location of the Tsolo District in the early twentieth century. When I began my research, I pursued this and adjacent mountain forest areas in the district as one of the focal points of forest contestation in the early colonial period. Published reports and archival records had already revealed that forests in these locales were at the center of some of the most enduring and often violent conflicts over resource management between the colonial state and Transkeian communities at the turn of the century. Early on in my fieldwork, however, I soon learned how dramatically social and environmental landscapes in the Gqogqora location had been altered since then. During the late 1950s and early 1960s, the apartheid state embarked on an aggressive afforestation program in the area, culminating in the complete removal of the entire population of the location—nearly 1,900 men, women, and children—to several peri-urban locations on the outskirts of Umtata, and its replacement with large-scale exotic tree plantations.[71]

As my research turned to interviewing men and women in these resettlement areas—physically removed from the "mnemonic surroundings" and "the topography that helps to uphold memory"[72]—their accounts of the early colonial era presented unique challenges and opportunities for historical interpretation. In all cases, informants certainly perceived and represented their local landscapes "as saturated with power, meanings, and historical struggles for land rights."[73] Many men and women were particularly sentimental when describing life in their former "home,"[74] and narrating the histories of their youth became ways to critique the state's role in the subsequent decline of their landscapes, livelihoods, communities, and traditions. Such narratives idealized life and resource access before betterment or resettlement, which were represented as the major turning points in the well-being of their social, spiritual, and natural worlds.[75] Yet within some of these very same interviews, and in conversations with others removed from the area, divergent representations of the past also emerged: the displacements of the apartheid era were only a continuation of processes of disempowerment and rural decline dating back to the earliest days of colonial expansion.[76]

In the analysis that follows, I am less concerned with validating or denying the apparently contending historical truths embedded in these testimonies than with gleaning from them particular ways of reinterpreting dominant narratives of state environmental intervention in the Transkei's history.[77] While informants from Gqogqora and other areas remember and "deploy" the more remote past as a "resource" in varying ways, they also provide important correctives to state- and settler-centered narratives in which popular experiences of displacement and disempowerment have been silenced.[78] In certain cases, particularly in some informants' critiques of the colonial, apartheid, and even postapartheid governments' domination of forest control and their ability to extract payments from rural people for wood access, I have been able to find revealing continuities in oral sources and the written records of African protests in the early twentieth century (chapter 3). Moreover, in narrating their various histories, interviewees often offered important clues about the significance of particular forest sites and species in certain locales, insights I could then use to ask new questions of the written evidence for the earlier colonial period. Individuals' traumatic experiences and memories of state intervention in more recent decades might overshadow certain prior historical developments but could also bring into sharper focus the historical meanings of resources to which they no longer had access.

ORGANIZATION OF THE BOOK

The book is organized into two parts. Part I examines forest negotiations through the lens of particular political and socioeconomic changes—shifts in environmental authority (chapter 1) and environmental entitlements (chapters 2 and 3)—combining a look at broader patterns in the Transkei with a focus on the unique historical trajectories in KwaMatiwane. Part II (chapters 4 and 5) explores in deeper relief the social and cultural meanings differently situated actors in KwaMatiwane gave to specific resources and how such stakes informed their negotiation of resource access over time. These two parts are meant to illuminate one another, to suggest how access to forests involved diverse dimensions of changing power relations and everyday experiences in the colonial Transkei.

Chapter 1 explores the negotiated shifts in environmental authority accompanying the reshaping of political ecological landscapes in KwaMatiwane and the wider Transkei from the early 1880s to the mid-1910s. As officials stabilized colonial control by remapping the administration of local populations and environments over time, political negotiations over changing local power dynamics ensued at multiple levels—between and among demoted African chiefs, newly appointed headmen, and rural residents, and between these groups and representatives of the colonial state. With the expanded presence of colonial forest personnel and the shifting role of headmen in managing forest use, variously situated individuals played out contending political, economic, and environmental interests in the local shape of forest authority, often in very personalized ways.

Chapter 2 looks at the restructuring of Africans' environmental entitlements across the Transkei, from the launch of comprehensive restrictions for the region in the late 1880s to roughly the first few years of the twentieth century, when the consolidation of Forest Department control would begin to shift state impositions on popular forest rights. The chapter examines how popular protest and evasion, interdepartmental squabbling, and Africans' alternative invocations and pursuit of their entitlements complicated the course of colonial resource policies and their enforcement. The discussion particularly focuses on how differently positioned people critically responded to government restrictions on their "subsistence" and "customary" practices, their gendered orientation, and their implication in wider colonial transformations. In daily practice, African men claimed resource access entitlements through their tax participation in the broader colonial system, and women asserted their right to harvest and utilize fuelwood in the most self-beneficial economic ways possible.

Chapter 3 proceeds with the discussion of entitlements by looking at how the restructuring of forest access changed with the consolidation and expansion of colonial control over Transkeian forests beginning in 1903. I analyze how government interests in more systematically "weaning" Africans from indigenous forest use and inducing them to purchase their regular wood needs from the state contributed to and coincided with mounting socioeconomic stresses and problems of ecological scarcity in many areas in the early twentieth century. Men and women contended with constraints on wood access and availability as they also negotiated their local experiences of poverty, resource pressures, and the expanding out-migration of male workers. People responded to their changing personal circumstances in multiple ways. Many protested their growing dependence on exotic tree plantations and the state's profit motives amid their own impoverishment. And as the state increasingly moved to restrict African women's resource rights in the late 1920s and 1930s, women coped by adapting their everyday forest practices.

In Chapter 4, the book then moves into a discussion of the meanings of particular forest resources in different people's changing lives and their implication for understanding the meaning of resource access over time. The chapter surveys the ways in which men and women in KwaMatiwane historically incorporated forest species into various social and cultural practices in the late nineteenth and early twentieth centuries, very selectively pursuing certain local trees and plants as resources for healing, protection, rituals defining personhood and marking life cycle stages, and charms for dealing with everyday social interactions. In certain cases, the sources allow me to further explore how such particular interests shaped individuals' actual physical negotiation of the local forest restrictions they faced. Chapter 5 then looks in greater depth at the particular dynamics of power and culture surrounding access to one renowned tree at one Tsolo District site in the late nineteenth century. While colonial observers minimized the cultural nature of their interventions into this setting, local residents commented critically on the extension of government power into the resources and sites essential to rituals of public healing. These responses suggest the negotiations over the meaning of environments and environmental power surrounding European expansion in the Transkei more generally in the early colonial period. In the book's conclusion, I then point to the salience of many of the historical issues raised in the preceding chapters for grappling with the ongoing challenges of dismantling colonial legacies in South Africa today.

Part I

1 ⇝ Tensions in the Colonial Restructuring of Local Environmental Authority, 1880–c. 1915

As THE COLONIAL GOVERNMENT appropriated Transkeian resources through annexation and the establishment of new administrative institutions, intense disputes emerged over the nature and scope of environmental authority. Paramount chiefs were demoted, commoner headmen were elevated to new ranks, magistrates took over the helm of newly demarcated districts, and the Forest Department increasingly exerted its influence on the lives of African communities. From the 1880s to the mid-1910s, as colonial personnel, chiefs, headmen, and commoners negotiated control over natural resources across the Transkei, they simultaneously contested the meaning of these collective transformations in local political and environmental relations.

In KwaMatiwane negotiations over environmental authority were embedded within the region's unique histories of political and environmental restructuring. Following the final major Transkeian rebellion of the nineteenth century, officials expanded the colonial domination of both local populations and resources, viewing them as mutually supportive agendas. From the early 1880s onward, controlling forests became a key ingredient in official strategies to stabilize the social order, consolidate colonial rule, and resettle local populations. The settlement of diverse populations into newly created subdistrict locations and the appointment of government-salaried headmen to oversee them served the twin causes of political stability and expanded resource regulation. As colonial conservation was formally established and expanded from the late 1880s onward, officials more systematically empowered and obliged headmen to help enforce governmental forest schemes in their locations. Over the course of the 1890s, the government also expanded the activities of European forest officers and African forest guards in local areas, who worked alongside headmen in controlling location forest access until the late 1900s, as a new management policy emerged for location forests. From this point forward, foresters' patrols were confined to only demarcated government reserves, and headmen took over the daily supervision of newly defined "headmen's forests" in their wards.

Differently situated chiefs, headmen, and commoners brought their wider personal experiences of social, economic, and ecological transformation to bear on their responses to these changes in environmental authority. The complex relocation policies of the postwar 1880s, including the resettlement of economically and culturally differentiated groups into common locations, exacerbated ongoing tensions and magnified such lines of difference. As such groups competed over land and forest resources, intralocation factions settled their scores in heated disputes over headmen's political and environmental authority. African authorities' ambiguous participation in colonial forest policies particularly generated uneasy and unpredictable relations with both location residents and officials, as newly subordinated chiefs and newly appointed headmen often exploited their hand in forest control to derive various political and economic benefits. With the expansion of forest patrols and personnel in the region in the 1890s and early 1900s the politics of resource negotiations became even more complex and intense, as headmen jockeyed to maintain their influence over popular resource access and location residents felt and expressed their deeper frustrations with the colonial state, particularly in a period of acute rural stress, through struggles with forest officers and guards. At the same time, headmen, residents, and forest guards each brought deeper personal histories and vendettas to their negotiation of these shifting power relations.

Over the course of the late 1900s and early 1910s, as location forests were placed under headmen's exclusive control, negotiations over forest access continued to involve a combination of various social, economic, and environmental stakes for differently situated actors. Officials still faced ambiguously positioned headmen who often exploited their limited sphere of authority for the benefit and protection of themselves and the people of their wards. Simultaneously, many men and women contended with headmen who extracted demands and constrained local forest access according to personal interpretations of their environmental prerogatives.

REBELLION AND ITS AFTERMATH:
STABILIZING COLONIAL RESOURCE AND POPULATION CONTROL

By the late 1870s colonialism had already begun reshaping the political landscape of KwaMatiwane and adjacent areas. From the early 1860s onward the

Thembu paramount Ngangelizwé developed increasingly closer ties with Cape authorities, strategically weathering multiple frontier wars in the wider region, and by 1876 Thembuland was divided into four new magistracies.[1] Across the Umtata river, the Mpondomise chiefdom under Mditshwa began negotiating for colonial "protection" in the early 1870s, particularly as a means to defend against raids from neighboring Mpondo groups. By 1877, the chiefdom came under complete colonial control when the Cape formally annexed the entire area of East Griqualand, with a magistrate based at Tsolo overseeing Mditshwa's territory.[2] Just as Cape rule began to unfold, however, the rebellion of 1880–81 redefined the terms of colonial engagement with local peoples and environments. Restoring order and resettling populations in the war's aftermath forced colonial authorities to refine their political and environmental strategies. More intensive schemes to restructure boundaries of local political authority and assume control over essential natural resources worked hand in hand.

Besides its anticolonial dimensions, the rebellion in the early 1880s brought to a head escalating tensions between African groups with various political motives, cultural affiliations, and economic positions.[3] In Thembuland, Kwa-Matiwane populations in the two magistracies of Umtata and Engcobo were predominately comprised of Thembu clans, Hala and Jumba, as well as ama-Qwathi under Chief Dalasile, all of whom had formally recognized Ngangelizwe's authority as paramount.[4] Yet the paramount's courting of Cape ties in the prerebellion period exacerbated strained relations with Dalasile, who resented both Ngangelizwe's use of authority and the expansion of colonial power in the Qwathi chiefdom.[5] Another source of tension in both Thembuland and Mditshwa's territory was the accelerating influx of Mfengu and other immigrant populations from the Eastern Cape and other parts of the Transkei.[6] Many of these newcomers, with much longer and closer interaction with colonial society than the majority of their new African neighbors, had already incorporated "progressive" economic ideals and Christian beliefs and practices into their culture and thus were viewed much more favorably by the colonial government. Various missionary groups—Free Church of Scotland, Anglican, Wesleyan Methodist, and Moravian—also expanded their local activities in the late nineteenth century, deepening lines of cultural and economic differentiation within and between many African communities.[7]

The combination of such competing interests and allegiances exploded into open conflict in 1880. Those Mpondomise, Qwathi, and Thembu groups who rebelled, led by such prominent chiefs as Mhlontlo, Mditshwa and Dalasile, directly opposed the colonial government's imposition of taxation, threats of disarmament, and broader exercise of power over their chiefdoms.[8] The war itself was relatively short-lived, however. By April 1881, colonial forces had broken the backbone of the resistance, and officials began the difficult task of stabilizing colonial control among the region's now incredibly fractured and war-torn populations.

In restructuring the social landscape of KwaMatiwane in the rebellion's aftermath, colonial officials combined three immediate and interrelated goals: the establishment of social control and political stability in the Territories, the interspersing of officially designated "rebel" and "loyalist" African communities across the region as a means to prevent future outbreaks, and the opening up of particular lands for the expansion of white settlement from the Cape Colony. In defensively redesigning the region, government leaders resuscitated a well-worn tradition in colonial expansion on the Eastern Cape frontier: the creation of human "buffer zones."[9] Following the war, colonial authorities strategically located African groups that had allied themselves with the Cape government as "buffers" between former rebel groups or independent African polities, on the one hand, and European settlements along the widening Cape Colony border. European and African populations in much of the Engcobo, Umtata, and Tsolo districts, as well as populations in the adjoining Qumbu and Maclear districts, were reassigned specific territorial boundaries. Nodes of European settlement and colonial administration at Tsolo, Umtata, and Engcobo, as well as the expanding European farming populations in the Elliot and Maclear districts, would be "protected" by corridors of loyalist Mfengu, Thembu, and Mpondomise groups; former rebels would be relocated on the other side of these "buffer zones."[10]

An essential component of this resettlement mosaic was the partitioning of forest access. Throughout the rebellion, as in previous wars, many local people had exploited nearby forests as strategic sites of refuge, hiding themselves and their livestock from enemy forces, relying on wild foods for daily subsistence, and utilizing plants and trees for the manufacture of weapons and ammunition.[11] Even after the fighting ended in 1881, many fugitives from either side of the Umtata-Tsolo district border and in the Engcobo District continued to hold out in local mountain forests.[12] Over the next few months, thousands of individuals, driven by hunger and desperation, eventually moved down into the nearby valleys to seek refuge with loyalist Thembu and Mfengu communities and peace with the colonial government.[13] Yet the threat of rebellion still seemed very real to many African and colonial authorities, each of which sensed serious risks in resettling in their midst former rebels who might at any moment take up arms or flee to military strongholds in the forested mountains.[14] In order to preempt any such unpredictable maneuvers, officials ensured that the relocation of rebel communities involved distancing them from advantageous natural areas.[15]

First, colonial authorities sent troops to root out any rebels still hiding in the mountains, capture any livestock secreted there, and regain military control over the forest areas.[16] Then the complex relocation game began, balancing the need to separate European settlements from former rebel groups with the desire to obstruct the latter's direct access to less easily controlled forest and mountain areas. Officials devised a creative solution to the problem: loyalist "buffer zones" would be used for both purposes.[17] In the Engcobo District, Walter Stanford made it a policy to "cautiously" locate only those headmen and people "of tried Loyalty" in close proximity to any natural strongholds in the nearby forested mountains: the rebel amaQwathi were settled between an artificial line below the Gulandoda Range and a circle of loyalist Mfengu and Thembu communities surrounding the Engcobo magistracy; lands "vacated" by rebels between the Gulandoda Range and the more distant Drakensberg foothills were likewise reserved for settlement by loyalist "natives of good character and inclined to adopt civilized habits," a move favored by Stanford "from a military point of view."[18] In the Umtata District, RM A. H. Stanford established locations under loyalist Mfengu and Thembu headmen along the Tsolo District border; loyalist Mfengu and Mpondomise headmen were similarly put in charge of locations on the Tsolo side of the line. Together, these moves created buffers between Mpondomise rebels settled in the interior of the Tsolo District and the European center at Umtata, as well as between former rebel communities and the bulk of the mountain forests in the adjoining districts. Likewise, toward the northern boundary of the Tsolo District, a mixed location of loyalist Mfengu communities was established, hemming in Mpondomise rebels in the district from the nearby Drakensberg Mountains.[19]

Initiating this strategic mapping of local populations and resource areas at the close of the war was one thing; making it run smoothly would be another matter altogether. As colonial authorities attempted to restore stability and regularize social and ecological control in African communities, local tensions exacerbated by war and resettlement would complicate official plans. Management of people and resources was soon mired in Africans' multiple, conflicting experiences of change under expanding colonial rule in the postwar period.

POSTWAR STRUGGLES OVER RESETTLEMENT, AUTHORITY, AND RESOURCE CONTROL

Alongside postwar resettlement schemes, settling the parameters of political and environmental authority in different magisterial districts and subdistrict wards was an essential ingredient in official strategies to stabilize colonial control. In the early 1880s, responding primarily to mounting conflicts over natural resources among and between African communities and European settlers, colonial authorities instituted several commissions to determine political and natural boundaries within Thembuland and East Griqualand, and between these territories and the expanding settler communities in what would become the Elliot and Maclear districts.[20] More persistent resource disputes accompanied the restructuring of authority within and between subdistrict locations. Conflicts immediately developed in many locations in the Umtata, Tsolo, and Engcobo districts over access to productive resources, interwoven with deeper contests over power and authority in newly constituted locations. As different groups competed for soils, grasses, and forests, they brought their disparate economic situations and cultural affiliations to bear on the local politics of resource control.[21] Officially appointed headmen were often at the center of these contests, exacerbating ongoing tensions in their designated wards by exploiting their unique influence over resource access.[22]

Some of the immediate sources of tension were the economic disparities between former rebels and loyalists. The displacement of war and the losses in livestock, land, and crops incurred by colonial forces left former rebel communities destitute. An extreme drought in 1881 only compounded their woes. While these communities often struggled to stave off famine in the war's aftermath, magistrates in all three districts concerned noted the relative prosperity of farmers who had allied with the government and had been able to maintain their crops and livestock during the disturbances.[23] In many areas, such disparities generated conflicts over critical agricultural and grazing resources. In mid-1881, for instance, RM Walter Stanford described the intense "enmity" existing between Mfengu residents and rebels being relocated among them in the Engcobo District, warning that land disputes between the groups had the potential "to start the war afresh." A local headman reported that "the surrendered rebels, are burning all the grass . . . & consequently depriving the Fingoes of grazing ground for their cattle. . . . [I]t appears that the destruction of the grass, is an act of spite on the part of the rebels, they, having no cattle of their own to feed."[24] While this type of economic retaliation died down in the months after the rebellion, economic disparities intensified by the war continued to cause tensions in many communities in subsequent years.[25]

Such disputes were exacerbated by the government's installation of appointed headmen to oversee newly aggregated populations of differing cultural, political, and economic situations. In KwaMatiwane and beyond, colonial incorporation significantly reconfigured local political landscapes.[26] Prior to annexation, the headmanship was essentially a political appointment under local chiefly control: while the position was often inherited within prominent families, chiefs held the power to dismiss and replace any disloyal or unsatisfactory appointees. Upon annexation, the government instituted a policy of direct rule in each magistracy, subdividing districts into multiple locations and independently appointing salaried headmen to supervise them. In practice, the creation of such headmanships became a bit more complicated. Paramounts and lesser chiefs, for instance, stripped of much of their earlier status, were given special salaries and became "headmen" of locations in their own right. In many areas in the early years of colonial rule, chiefs were also able to maintain the right to nominate individuals, invariably their allies, for other headmanships.[27] Moreover, the position of headman often became an informal form of hereditary office, passed from father to son in many instances. In all cases, however, the magistrate of a district assumed ultimate authority to appoint or dismiss any headman for uncooperative behavior or poor performance of his duties, regardless of chiefs' opinions on the matter. Magistrates often selected men from commoner lineages as headmen precisely because they were less tied to royal lineages and thus were more likely to serve colonial interests. By becoming headmen, then, many commoner men were now able to pursue new sources of authority, wealth, and privilege independent of, and even in competition with, chiefs' drastically diminished sphere of influence.[28] In their locations, variously situated African leaders and residents jockeyed for position in these rapidly changing configurations of power.

For areas directly affected by the rebellion, the combination of changes to local economic and political landscapes in the postwar years produced perpetual conflicts over authority and control over productive resources. As William Beinart has described for the Qumbu District, loyalist headmen, officially bestowed with great powers in their locations in the postrebellion period, often used their advantageous positions to favor their followers and marginalize their enemies. At a time of population increase and intensified competition over resources, headmen could and did exercise environmental control as a political weapon, aiding allies and fomenting chronic resource disputes.[29] In KwaMatiwane perpetual accusations of favoritism and corruption similarly surrounded headmen's activities, revealing sharp lines of social and economic difference in local communities.[30]

In the Umtata District, in the locations along the Umtata River, relations were particularly hostile between Mfengu headmen and Thembu residents. From the early 1880s on, disputes over lands and political allegiances arose between the Mfengu headman Fodo at the Ncise location and abaThembu of the Jumba clan placed under his authority.[31] In one particularly vivid example in 1899, one elderly Thembu resident threw sand in Fodo's face and threatened him. As Fodo explained in court: "[H]e abused me said you are a wizard you brought charms from Pondoland. Although you are killing us don't you know that you are only a Fingo & a dog & that I am an old Tembu of Gangelizwe [the former paramount] & you will eventually come under my feet."[32] In the nearby Kambi area, the Mfengu headman and active Wesleyan Methodist Paul Nkala was regularly accused by former Thembu rebels placed under his authority of favoring his Christian and Mfengu allies and punishing non-Christian abaThembu when allocating land. Adding his own partisan views on the matter, Nkala explained to the magistrate: "I am the Head-man of the School through the word of the missionaries, & the rebels were handed to me by the magistrate. . . . These people I understand them as I lived with them a long time. They do not like the Christians or the Professors Fingoes & White people."[33] In other KwaMatiwane locations, the divisive realignment of groups of varying economic status and social affiliation in the postwar period also nurtured hostile struggles over land access and political authority, even well into the 1890s, 1900s, and beyond.[34]

Local tensions surrounding headmen's authority and resource control were further complicated by the expansion of colonial forest conservation efforts in the region, in which appointed headmen played an integral role. Prior to the establishment of the Forest Department's Transkeian Conservancy in the late 1880s, magistrates were solely responsible for regulating popular forest access in their districts, and they heavily relied upon African political appointees in the process. As with land allocation, struggles quickly surrounded headmen's exercise of forest control in their locations, as they often exploited their hand in forest access for their own personal benefit, favoring political allies while imposing fees on and settling political scores with others.[35] One particularly well-documented case of such local politics and the problems they generated is the tumultuous history of headman Thomas Ngudle. Ngudle was part of a migration of Mfengu peasants from Fingoland to the Tsolo area in the initial period of colonial expansion in the region. His father served as headman in the Mjika area in the early colonial administration of Tsolo District in the 1870s, and Thomas succeeded him prior to the 1880 rebellion. During the war Ngudle and his followers fled his rebelling neighbors to find refuge with nearby loyalist communities across the Umtata river, returning home after the fighting had ceased.[36] Finding Ngudle a tried and true colonial ally, officials retained him as headman of a large location in the newly engineered postwar Tsolo District, comprising a mix of former loyal Mfengu and rebel Mpondomise populations. In addition, Ngudle resumed a position he had held since 1879, when he was appointed as the first forest ranger in the area.[37] This unique combination of Ngudle's formal positions was at the center of heated conflicts throughout the 1880s.

Even before the war Ngudle was a particularly unpopular figure. His dominance over local political and natural resources brought to a head some of the strains between different social and economic groups in the area, often expressed in ethnic terms, that influenced the rebellion itself. As he himself later recounted to the local magistrate: "[A]s a Fingoe Headman and Forest Ranger in this District, I was especially marked out, hemmed in, and I and my people were plundered of almost everything we had, by the Pondomise upon the outbreak of the late War."

In the years following the rebellion, as in other locations under Mfengu headmen in the region, the mutual animosity between Ngudle and his rivals continued to erupt in struggles over political and economic power.[38] As impoverished Mpondomise rebels surrendered and were placed under Ngudle's authority, Mfengu and non-Mfengu factions perpetually clashed over how political authority was exercised and how local resources were allocated. Political rivals regularly obstructed his commands, while Ngudle himself routinely used his power to marginalize and punish Mpondomise and other non-Mfengu factions, relegating them to inferior residential and farming sites while granting Mfengu allies privileged access to choice lands. In one memorable instance, Ngudle even assaulted an influential Hlubi rival with sticks after he failed to follow the headman's orders concerning which lands to plow.[39] Ngudle's position as forest ranger further entangled him in perpetual disputes. Conflicts surrounding the contravention of forest regulations were often opportunities for Ngudle and suspected forest offenders to fight deeper political battles and settle personal scores.[40] By the late 1880s, as magistrates and foresters caught on to Ngudle's abuses of power and as local resource politics became increasingly divisive, colonial officials decided to take action. Officials reorganized the location, reassigning one of the chief sources of animosity—Ngudle's Mpondomise rivals—to a new location under their own headman, and limiting Ngudle's authority to local Mfengu communities, before finally dismissing him.[41]

Although Ngudle's relations with location residents was complicated by his dual role as headman and forest ranger, the types of conflicts that beleaguered his career were by no means unique. Across the region, disputes over headmen's control over resource access continued to be embedded in deeper political, cultural, and economic cleavages in local communities, ones which were aggravated by the colonial restructuring of locations in the postwar years.[42] Headmen's environmental control was at the epicenter of fractious jockeying for political and economic power as differently situated groups negotiated the wider realignment of authority in this emerging colonial order. For officials aiming to exert greater governmental control over local political ecologies, African authorities' involvement in resource management would thus pose serious challenges.

COLONIAL RESOURCE CONTROL AND THE
UNCERTAIN ROLE OF AFRICAN AUTHORITY

While headmen's influence in resource access inspired political struggles within newly designated wards, African authorities' ambiguous participation in the emerging institutions of colonial resource control also proved problematic for government officials. In KwaMatiwane and across the newly annexed Territories, developing a colonial system of resource management became intertwined with the wider challenges of incorporating African political authority and environmental "custom" into an evolving framework of colonial law and administration. In complex dialogues with officials, chiefs and headmen sometimes found sufficient space to insert themselves and their own ways of imagining their environmental "traditions" and prerogatives into colonial practice.[43] More direct official efforts to involve African authorities in regulating popular forest access often stumbled against chiefs and headmen actively protecting their power to interpret and exercise forest control as they saw fit, whether for personal gain or to shield local residents' environmental practices from colonial intrusion. Although their capacity to derive benefits from colonial resource control was uneven and partial in this era of political transition, their ability to negotiate the terms of their own colonial engagement significantly shaped the local meaning of official resource policies.

Following annexation, many chiefs in the Transkei sorely felt and contested the cumulative effect of political reorganization under direct colonial rule — the chopping up of their jurisdictions, their personal demotion, and the elevation of individuals of lesser rank and status. The atomizing of chiefdoms into magisterial districts and headmen's locations particularly undermined chiefs' environmental control and their ability to legitimate their political authority through symbolic ownership of their territories' lands.[44] After Thembuland's annexation in 1885, for instance, Paramount Chief Dalindyebo immediately contested the new magisterial boundaries undermining the territorial integrity of his authority in the chiefdom. Dalindyebo was particularly alarmed by magisterial interference in chiefly decisions over forest use throughout Thembuland: "We also want to know for what reason our forests are no longer under our control. Thus if we want to cut wattles or wood for any other purpose we have first to obtain the permission from Magistrates. . . . Formerly the forests were under the control of the headman appointed by [Paramount Chief] Ngangelizwe to whom the people had to apply for permission to cut but now this is taken out of the hands of the headman & all alike have to apply to the Magistrate."[45]

Colonial domination of chiefly power also brought significant changes to the operation of legal institutions. Upon annexation, magistrates assumed authority over civil and criminal cases, intentionally undermining the preexisting chiefly court system. In practice, however, chiefs continued to preside over a range of civil and even minor criminal cases in their local areas, although now without the formal power to enforce their decisions and with magistrates presiding as superior appellate judges in all such cases. As officials grappled with new and unexpected African realities in the Transkei, they also often relied upon African chiefs in articulating those "native laws and customs" which could be selectively incorporated into the colonial legal system.[46] Thus, even though chiefs operated from increasingly subordinate positions, they were able to help shape the fluid legal boundaries of their authority, particularly concerning resources.

A useful example of the nature and ambiguous outcome of chiefs' negotiation strategies in the postannexation period is Thembuland. Paramount Dalindyebo repeatedly inserted himself into various resource conflicts in the chiefdom in the late 1880s, and his success in deriving political mileage from new colonial institutions was decidedly mixed. In 1889 Dalindyebo tried to claim unlimited royal access to certain forest resources in the Thembuland district of Mqanduli, a move which both upset local political rivals and went against magistrates' fledgling attempts to restrict forest access. When the paramount sent subordinates to cut wood for him in the district, a man living in the vicinity, Ngxishe, seized the wood, rebuffed the paramount's demands to explain himself, and asserted that only a rival chief in the district, Sipendu, had legitimate authority over resources in the area. The Mqanduli magistrate hearing the case decided in Ngxishe's favor, finding that Dalindyebo's position as Thembu paramount no longer brought him any privileged authority over resources in this magisterial district.[47] Exploiting this opportunity, Sipendu continued to ignore the paramount's authority over local forests: when the latter sent men to hunt or cut wood in the area, Sipendu drove them out of the forest, seized their wagons and implements, and even threatened to kill them.[48] This situation infuriated Dalindyebo to no end, and he and his councillors repeatedly, and unsuccessfully, brought their concerns to the chief magistrate in Umtata, complaining that the demarcation of new colonial boundaries in Thembu territory had emboldened the "desire on some of his chiefs not to acknowledge his supremacy." They also criticized the Mqanduli magistrate's intervention in the matter, trying to appeal to the chief magistrate's own sense of supreme authority over lesser appointees: "[The Mqanduli magistrate] is a servant of this office what right has he to interfere with Dalindyebo. [H]e ought not to interfere with Dalindyebos [sic] people he ought to get authority from the C.M."[49]

While such arguments fell on deaf ears in this instance, Dalindyebo and other chiefs were much more successful in negotiating political benefits when they could mesh their claims to "customary" prerogative with official interests in maintaining order and enforcing environmental regulations. For example, the paramount garnered administrative support for his assertion of "customary" power to resolve particular forest disputes in outlying regions of Thembuland. In one disagreement between two lesser chiefs in the Mqanduli District over their respective spheres of influence, Dalindyebo asserted his legitimate chiefly power to preside over and settle such conflicts "as heretofore according to our Custom," albeit with the right of appeal to the chief magistrate.[50] In this and a subsequent case, CM Elliot endorsed the paramount's right to intervene in such disputes, even superseding the authority of the Mqanduli magistrate.[51]

Chiefs were even more adept at extracting personal advantages from their direct participation in colonial environmental restrictions. Prior to annexation, chiefs across the Transkei accumulated wealth from a variety of fines and fees imposed on their subjects for different transgressions, privileges, or occasions.[52] In the late nineteenth century, various Thembu chiefs pursued opportunities in the new colonial legal system to continue to collect certain "customary" environmental dues from their subjects. Shortly after Thembuland's annexation, for instance, Dalindyebo and the Umtata district magistrate struck an agreement: the chief would prohibit people from cutting smaller thorn trees without his express permission. A civil suit before the magistrate in 1889 tested the meaning of this arrangement. After some thirteen individuals were found harvesting thorn trees in the district, Dalindyebo sued the group for damages in the magistrate's court. Finding the commonage resources involved to be under the chief's "customary" authority, the magistrate ruled that Dalindyebo be personally compensated two head of cattle for the crime.[53]

A similar case came before the Engcobo district magistrate in 1894, as Chief Silimela prosecuted an individual for contravening the chief's personal order that certain mimosa trees in his location not be cut. Confused about the particulars of this case, the magistrate relied upon the advice of the assistant chief magistrate, John Scott: "[U]nder our common or statutory law there seems to be no means of dealing with the case referred to. The parties being all natives you are at liberty to apply native law, under which the authorities had, and exercised, the right to reserve growing trees and forbid cutting of wood at certain spots, and disobedience of which orders under the native law is punishable by fine."[54] Scott's ruling here reflects the "hybrid" legal product of European and African authorities' negotiation of environmental authority in the early colonial period.[55] While colonial officials held the ultimate power to decide such cases, they were still feeling their way through a legal and political framework for controlling such communal resources as grass and forests. Such uncertainty and fluidity provided opportunities for some African leaders to assert their own interpretations of the meaning of "customary" resource authority in the new colonial system. In both of the above cases it is not clear where officials derived their insights into the "traditional" workings of environmental authority, yet it seems they relied rather heavily on chiefs' own claims regarding "customary" restrictions and penalties and their confluence with colonial interests in controlling popular resource access. Such cases thus illustrate how magistrates and African authorities each exploited the new colonial system to convert their respective interests in environmental "tradition" into legal precedent.[56]

Many chiefs in KwaMatiwane also found ways to perpetuate "customary" sources of personal income by arranging with officials to serve as informal forest supervisors. Besides the loss of past royal dues and fines, many chiefs in KwaMatiwane and beyond were also deeply upset by the impact of colonial annexation on their ability to impose and collect revenue from sawyers and wood traders migrating to their territories from the Cape and Natal.[57] In KwaMatiwane, some chiefs had accrued handsome profits through such timber licensing arrangements, particularly from the 1870s onward, as a growing number of European and coloured woodcutters took advantage of the region's rich, commercially "untapped" forests and fled the more restrictive regulatory environment of the Cape.[58] Following annexation, in response to complaints by chiefs in the Engcobo District about such lost revenues and to induce African leaders to cooperate with new colonial forest regulations, RM Walter Stanford initiated a tradition of relying on chiefs and headmen, along with other "trustworthy" African men, as "unpaid" forest guards.[59] Whenever these individuals assisted in bringing forest offenders to justice in the magistrate's court, they would be entitled to a substantial percentage of the fines imposed.[60] This informal arrangement became standard practice in the district in subsequent years and was expanded to some areas of East Griqualand after Stanford took over the position of chief magistrate for the territory in the mid-1880s.[61]

As colonial conservation efforts expanded from the late 1880s onward, and as rewards to informers of forest offenses became standard policy, magistrates and newly installed Forest Department officials particularly extended such benefits to cooperative chiefs and headmen.[62] At the same time, many African authorities actively pursued more formal positions for themselves and their allies in forest management. Headman Thomas Ngudle's appointment as ranger in 1879 was an early exception, but more chiefs and headmen successfully secured appointments as salaried forest guards for relatives and allies by the late 1880s and early 1890s, expanding a strategy they already pursued to influence magistrates' selection of location headmen.[63] Through both formal and informal arrangements, African political leaders thus wove their influence into the fabric of colonial resource management in these early years.

This fact proved to be a double-edged sword for officials. With limited finances and personnel to police the vast forest tracts of KwaMatiwane in this period, enlisting the aid of local chiefs and headmen was vital. Yet this dependence on African authorities did not guarantee their dependability. In the absence of regular official supervision of their day-to-day activities, chiefs and headmen were often able to shape the meaning of their environmental authority as they saw fit, much to officials' ire. In many cases African authorities willfully opposed or ignored colonial forest laws in order to shield local residents' livelihoods from governmental interference, or manipulated their role in forest control to suit their own political and personal pursuits.[64] In 1888 and 1889, for instance, the first head of the newly appointed Transkeian Forest Conservancy, Caesar Henkel, grew particularly wary of certain Thembu chiefs nominally serving the colonial administration in the Baziya area of the Umtata District. One such chief and headman, Makaula, continued to manage local forests on his own accord, allowing African men from far and wide to cut timber without the requisite licenses or fees, and bypassing any directives from the magistrate or the Forest Department.[65]

A different threat to colonial forestry came from Nqweniso, a Mfengu headman at Mbolompo, who, together with his appointees, regularly and illegally issued harvesting permits for local forests to many African residents without any reference to officials.[66] In the Tsolo District, Thomas Ngudle exacerbated the frictions surrounding his headmanship by exploiting his position as forest ranger to develop a small wood "empire" in his location—erecting and running a saw mill, cutting business deals with local sawyers in exchange for ready access to desirable forest areas, and actively trading in timber himself.[67] One of the main reasons for his eventual dismissal in 1889, in fact, was this sustained pattern of corrupt business deals and his inconsistent supervision of forest access.[68]

African authorities thus assumed ambiguous and unpredictable positions in the emerging colonial regime of resource management. As they negotiated political and environmental restructuring with government officials, both newly subordinated chiefs and newly appointed headmen asserted their hand in forest control and strove to draw benefits from adapted interpretations of "customary" chiefly prerogative. In the process, African leaders' political maneuvers complicated and altered official designs to install new institutions of forest administration systematically across the Transkei.

STRUGGLES SURROUNDING COLONIAL FOREST STAFF

Negotiations over environmental authority shifted significantly in the 1890s, as colonial forest employees increasingly patrolled local forest access. Following the 1890 enactment of the first comprehensive forest regulations for the Transkei, the colonial government gradually expanded and regularized the deployment of European forest officers and African forest guards to enforce the new measures.[69] One dimension of this strategy was the formal reduction of African headmen's influence in local forest access in the mid-1890s, given the many collisions between colonial forestry ideals and African leaders' actual practice of resource control. Increasingly skeptical of headmen's suitability and penchant for forest supervision, Henkel and other foresters began limiting the appointment of location forest guards to African men whom they deemed "trustworthy" and "independent" of headmen.[70]

In KwaMatiwane African residents experienced and responded to the expanded influence of colonial forest personnel in a multitude of ways. Frustrated by increased government interference with local resource control at particular moments of economic and ecological stress, people often violently lashed out at representatives of colonial environmental authority. Men and women also contested the specific ways in which forest officers and guards exercised their expanding authority over local resource access. Personal disputes between these colonial appointees and African residents often dictated the dynamics of local forest politics, as much as competing interests in resource access. Headmen felt particularly threatened by newly appointed forest guards, often engendering extremely tense relations between these groups over their respective fields of authority. Thus the expansion of colonial environmental control became quickly entangled in the deeper political-economic concerns and relations of local residents.

As KwaMatiwane forest patrols became regularized in the mid-1890s, relations between forest personnel and African residents very soon turned hostile. As officials readily acknowledged, forest officers and guards were primarily engaged in unpopular "police work," equipped with small arsenals of intimidation—horses, guns, and ammunition—to be used in deterring and pursuing forest offenders.[71] Adding fuel to the fire were certain local foresters' particularly intrusive enforcement techniques in the 1890s and 1900s. An especially vivid example is Thomas Adams, based at the Nqadu and then Baziya forest reserves. Adams developed an official reputation as an exceptionally "zealous" enforcer of forest regulations, often generating more revenue and prosecutions in his forest district than any other forester.[72] Yet, during his tenure in the area, local residents repeatedly complained to officials that Adams physically assaulted suspected forest offenders, among other abuses, eventually leading to his relocation to another post.[73] While Adams's actions were particularly oppressive, his example suggests more generally how foresters' new regulation of forest access was personalized on the ground, often shaped by individuals' little-supervised styles of enforcement and coercion.

The notoriety of forest personnel certainly contributed to local residents' own hostile responses to the expansion of colonial forest patrols from the mid-1890s onward. Over the course of 1895 and 1896, forest officers and guards in the Umtata District reported that local people were repeatedly attacking them, destroying their property, and threatening them with death.[74] In early 1897, Forester Adams and a forest guard fought with three Mpondomise men at the Nqadu Forest in the Tsolo District as they attempted to arrest men for illegally cutting reserved trees. Adams was struck several times with a knobkerrie, another man tried to take away the guard's gun, and the group was only dispersed when Adams was able to fire his revolver in the air. Some ten days later, a forest guard scuffled with two men in the same vicinity; the men held him to the ground and threatened to kill him. Then, just a few days later, the chief conservator of forests in Cape Town requested that backup staff be immediately dispatched to the same locale following a struggle between forest officers and a group of African "poachers": "Natives flogged [Forester] Adams horribly one of whom he shot twice on leg. Also flogged new Guard bound him down to open veldt."[75] In the next few months similar verbal and physical fights broke out in the area, leaving some foresters "trembling for the peace of the Territories."[76]

Such popular animosity toward local foresters was in part a response to the many socioeconomic and ecological pressures affecting local communities at this particular moment. Over 1894 and 1895, drought had a devastating impact on harvests throughout the region, reducing the money available for the payment of colonial taxes and increasing household indebtedness to local European traders. As people attempted to recover from this blow over the next couple of years, rinderpest began to rapidly reduce livestock herds, killing an estimated 85 to 90 percent of cattle in many communities in the region.[77]

Compounding these problems were growing pressures on local wood supplies in locations in the mid-1890s. Foresters demarcated reserves in the KwaMatiwane mountains at the same time that more people relied on these valuable wood sources, particularly from outlying regions in Thembuland where trees were a rapidly dwindling resource.[78] As forest officers and guards more extensively patrolled forest access in KwaMatiwane communities from the mid-1890s on, their activities thus became entangled in multiple strains on residents' livelihoods. In a period when people in KwaMatiwane and neighboring areas increasingly expressed their resentment toward multiple colonial interventions amid growing rural insecurity,[79] KwaMatiwane residents responded to the government's intensification of forest reservation and policing by increasingly assaulting those representatives of state authority most directly restricting popular access to vital resources.

Further economic and ecological stresses in the early 1900s continued to poison local residents' relations with foresters. The South African War (1899–1902) led to the closing of mines and job losses, leaving many migrant laborers from the region with less income for purchasing food and meeting their financial obligations to the colonial state. Periodic drought conditions coincided to further undermine rural production and help create a sharp rise in illness and disease in many communities. Even after the war ended and some of these pressures abated, perpetual droughts and invasive locust swarms increasingly took their toll on many local pastures, livestock herds, and food supplies.[80]

Alongside these forces, colonial forestry expanded further in KwaMatiwane. Over the late 1890s and early 1900s, foresters began small timber plantations in several locales, re-demarcated certain areas in the Tsolo District and removed several dozen men and women cultivating and residing within the newly constituted boundaries, and erected beacons and fences around these and other forest lands across the region.[81] In the Tsolo District in particular, the coincidence of drought and the beaconing off of demarcated forest areas created a severe resource crunch, as more people came to restricted forest areas in search of wood sources, cultivable lands, pasturage, and game for food and skins. Over the early 1900s, foresters increasingly complained about their difficulties in policing local forests amid this rapid increase in resource use.[82] In the frustrated words of Conservator A. W. Heywood in 1901, "The Pondomisi natives have shown themselves, so far as forest offences are concerned, to be the most lawless tribe in these Territories."[83]

Heywood's comment reflects foresters' general answer to the growing resource pressures of these years: intensify enforcement. The primary way to curb mounting forest offenses in the area, officials declared, was to bolster the authority and involvement of local forest guards. Thus, in late 1901, guards in the Tsolo District were given the authority to serve summonses for forest offenses, receiving a small payment for every summons leading to a suspect appearing in court. The following year the secretary for native affairs ordered that forest guards in the area be sworn in as police constables so that they could more effectively serve summons, identify offenders, and bring them to justice.[84] Guards' expanded powers and incentives for prosecuting suspected lawbreakers are part of the reason for the higher number of forest convictions in the Tsolo District in these years.[85] In 1901, for instance, the district had by far the highest number of convictions for forest offenses in all of the Territories, some 252 out of a total of 629.

This more intensive surveillance and prosecution of local forest users in the district and in adjoining areas in the Umtata District also helps explain the high incidence of violent conflict between foresters and local residents at this time—particularly attacks on forest officers and guards—which also surpassed that of any other area in the Territories.[86] In a typical case, when a forest guard questioned a man about cutting kraalwood in the Bele Forest, the suspect "refused to shew me the permit drove his oxen on and said 'today your day has come to die.'"[87] In another instance in the nearby Baziya Forest, when a forest guard investigated three suspected offenders, the group apprehended him, forced him to walk six miles, then left him for the night with bound hands and feet.[88]

As the above examples suggest, amid this violent transition of forest authority, KwaMatiwane residents often vented their frustrations most severely at African forest guards. As both members of African communities and representatives of the colonial state, guards, like other African colonial employees, were often condemned for bending to the government's will and abandoning their communities.[89] Thus, as Forest Guard Joseph Macingwane, working in the Jenca area of the Tsolo District, testified in 1900, a man he attempted to arrest for cutting in a demarcated forest had "threatened my life saying that I was identifying myself with the white people."[90] In some cases, residents conveyed the same intimidating message through more symbolic means. While officials strained to contain a wave of popular assaults on foresters and their property in the Umtata District mountain forests in the mid-1890s, for instance, they reported on the prevalence of particular acts of animal mutilation. As Conservator Henkel anxiously wrote to the chief magistrate in 1896: "[African guards] are constantly threatened by the Natives of the various locations around the [Baziya] forests and their horses mutilated. Forest Guard Edward Dungane, a very zealous and trustworthy guard, has had several horses killed, one a special good horse stabbed to death with assegais, others were mutilated by having their ears cut off."[91]

Rather than just a random act of spite, this form of animal mutilation was commonly associated with witchcraft accusations. Many instances of animal stabbings in KwaMatiwane surfaced as criminal cases in the 1880s, 1890s, and 1900s, often implicated in deeper disputes over resources, familial relations, and local politics in rural communities.[92] In many situations, animal stabbings were a means of expressing such disputes through publicly and symbolically identifying the owner as an "evil" public enemy, a manipulator of powers of sorcery for malevolent ends.[93] Periodically in this and other areas of the Transkei, confrontations between rural communities and local representatives of the colonial state were expressed in such violent terms, as people vented their resentment and frustration over government interventions in symbolically powerful ways.[94] Mutilating forest guards' horses similarly stigmatized their owners as "traitors" opposed to the prosperity and interests of community members.

People may have also singled out forest guards for such violence in response to many forest guards' using their positions to settle personal scores. As one headman in the Tsolo District complained about Forest Guard Nkonya in 1904:

> Many time has he brought many of my people to the office there on the slightest provocation, he does this not only in matters relating to the forest department but in many others too, but what troubles me most is the way he looks after the forests. . . . [P]lease let him have his duties far from these forests near about here all the people here are tired of his actions, I do not say they do not trespass; no! That they do, but this man will never be a great help towards surpressing [sic] this for sometimes he will just accuse a person because he has a spite against him.[95]

This last point echoes that made by the Tsolo magistrate several months earlier, reporting on a common grievance in the district: "Several influential Natives . . . have of late represented to me the desirability of employing only Europeans as Forest Guards and Detectives. They state that Natives abuse their official position and make miserable the lives of those who do not happen to be in favour with them. Forest matters are responsible for most of the complaints of the people of this District and if it is possible to remove the cause of complaint by the employment of Europeans as Guards and Detectives I am of opinion it should be done."[96]

Forest guards' lives became closely intertwined with those of local residents, and existing disputes over everything from garden boundaries to romantic relations might intensify both a guard's vigilance in prosecuting particular people and their animosity toward him. In a series of incidents in the Jenca area of the Tsolo District in 1901, a forest guard, Nompintsho, prosecuted a man named Nkumbi and his wife for illegally cutting wattles in a nearby forest. Just a few days later, Nkumbi sued the guard for adultery, claiming that he had discovered his wife meeting secretly with Nompintsho on a small island in the Inxu River. Although both men denied the cases had anything to do with one another, forest offenses disputed in colonial courtrooms often directly involved such personal vendettas.[97]

At the same time, guards were generally poorly paid and little supervised in their day-to-day practices, tempting many individuals to manipulate their resource gatekeeper status in order to reap various personal benefits and only further souring their relations with villagers.[98] One illuminating indication of such practices is the fact that in some areas local African men masqueraded as forest guards to try to extract particular advantages apparently associated with such positions. In 1907 in the Tsolo District, a Mpondomise man extorted money from a woman whom he had seen breaking the forest laws, "representing to her he was a Forest Guard."[99] A local missionary detailed a separate case near the St. Cuthbert's mission station that year:

> Our shepherd . . . has been to me this morning to report that last week some women from his kraal and from two neighbouring kraals were getting wood in the Qudu forest. While they were there a man named Cana Bana of Nogwazitole's location came up, and took one of the women aside and kissed her and tried to induce her to sin with him. He said that he was a forest guard, and he accused her of breaking some rule of the Forest department, and he used that fact as a means of terrifying her into yielding to his proposals. She however refused and got away. He then went to another group of women belonging to the same party, who were getting wood in another part of the forest. There also he picked out one woman and tried to play the same game, and he went so far as to throw her on the ground. . . . It is not the first time that I have heard of similar misdeeds being perpetrated by Native men who claimed to be Forest guards.[100]

To contend with the changing conditions of resource access represented by forest guards and their imitators, local residents employed their own tactics of persuasion beyond violence and threats. Individuals who possessed sufficient "bargaining power"[101] frequently offered livestock, money, or food to buy the confidence of a guard and avoid potential prosecution and fines for breaking forest regulations. As one elderly man who grew up in the Tsolo District described the situation in the 1920s and 1930s, if a guard came looking for poached timber at someone's home, offering sufficient quantities of homemade beer was enough to dilute his curiosity or at least make him "forget" the reason for his visit.[102] Bribery, however, could be a tricky game to play. Guards could extort fees and "gifts" from people or drag individuals to court for attempted bribery charges, whether legitimate or not. Although local residents could make similar accusations of their own, charging guards with demanding illegal payments for forest access, the word of state forestry employees weighed much more heavily in colonial court proceedings than the claims of African villagers. And forest guards could exploit this reality to their advantage, framing people they particularly disliked and deflecting attention away from their own illegal activities.[103]

Through such practices, forest guards further infuriated one particular group in KwaMatiwane: headmen. For many headmen, the expanded activities and authority of forest guards represented threats to their established spheres of local political and ecological control, threats they directly tried to thwart. In 1900, for example, Sam Ndayi, headman at Jenca in the Tsolo District, repeatedly threatened and physically assaulted an African forest guard in revenge for the latter's prosecution of forest offenses by Ndayi and local residents. The headman not only lashed out at the government's interference in resource control in his location but also contested the forest guard's personal infringements on Ndayi's political power. Ndayi publicly beat his adversary at a most symbolic site—alongside the headman's cattle kraal, where public gatherings and appeals occurred before him—in order to emphasize as strongly as possible the headman's supreme authority in the location.[104] In the Ncembu location a few years later, Headman Bikwe Ndlebe of the Ncembu location had a protracted struggle with Forest Guard Nkonya. Bikwe complained to officials that all of his present difficulties originated following a previous court case involving them both; ever since, the forest guard "has acted in a revengeful manner towards me."[105] While Nkonya used his position to retaliate against Bikwe and his followers, Bikwe actively worked to remove Nkonya for both oppressing local residents and undermining the headman's own authority.

As headmen and forest guards negotiated their relative positions in these years, such personal, political, and professional interests were often inseparable.[106] Given these relations, it is not surprising that many headmen were not always very inclined to cooperate with guards in enforcing forest regulations in their wards. Many headmen continued to confound official forestry strategies, refusing to divulge the names of suspected forest offenders and failing to assist guards and officers with prosecutions. While certain headmen might

of the councils, holding ultimate veto power.[109] For this and other reasons, Africans in many districts of the Territories mobilized tremendous opposition to the Glen Grey legislation, particularly the council system, from its inception through the mid-1900s. Particularly in Fingoland and East Griqualand, popular movements slowed down the introduction of the councils, sometimes with the full support of key chiefs and headmen who feared such a change would erode their local political influence. By the mid-1900s, however, such movements had fragmented and dissolved, partly through the government's dismissal and discipline of recalcitrant African leaders and official efforts to secure local leaders' political loyalty through economic enticements.[110] Many chiefs and headmen once actively opposing the councils now became central figures in both district council meetings and in their representation at the Transkeian Territories General Council (TTGC) sessions. After a shaky start and with all of its limitations, the council system became the main organ of African political representation in a growing number of districts, gradually extending to all African populations of the Territories.[111]

The development of the councils and the more enthusiastic participation of African authorities in the system by the late 1900s and early 1910s were partly due to the shifting socioeconomic positions of chiefs and headmen. Many members of senior chiefly lineages, together with senior headmen and other relatively prosperous African men, increasingly comprised the "progressive elite" across the Transkei—exposed to mission education, pursuing agricultural and cultural "advancement," and accepting the broad outlines of colonial administration even if quibbling with its details. As this elite came to dominate council membership, the councils thus became essentially conservative bodies, generally complying with governmental initiatives and recommending only moderate reforms.[112]

At the same time, as they became more embedded in the colonial bureaucracy, prominent chiefs and headmen tried to utilize their council positions to compensate for their dwindling influence and support among the rural masses of their locations. Chiefs and headmen regularly proposed in council sessions that the government enhance their local control in such arenas as land allocation, public gatherings, the organization of labor for communal projects, and the enforcement of various colonial edicts.[113] They likewise attempted to expand their authority in the legal sphere, proposing that they serve as assessors "to assist in determining questions of custom" in civil suits decided by magistrates, and even that they themselves should try many of these and other cases, including forest offenses, according to "native custom."[114] As they played more central roles in the council system, then, chiefs and headmen asserted their dual importance to the administration in ways that reflected some of the ambiguities of the colonial system and their positions within it. While they could facilitate the "progressive" application of colonial governance, they were also uniquely qualified to define, perpetuate, and enforce "customary" social norms.[115] In the realm of resource control, chiefs and headmen likewise strove to impress upon the Transkei administration their ability to control popular forest use in both a "progressive" and "traditional" manner.[116]

The first significant opportunity for council chiefs and headmen to assert their hand in forest access was in 1907, as they helped persuade the government to initiate a short-lived and problematic experiment in limited council control over certain forest areas. At the TTGC meeting in January 1907, councilmen successfully negotiated a promised "concession" from Prime Minister Leander S. Jameson: all "undemarcated" forests in Transkeian districts where councils had already been established would come under district council authority.[117] Much as they had portrayed the launch of the council system itself as a gradual course in "civilization," Native Affairs authorities lauded Jameson's latest endeavor, "allowing the people a voice in the control of the patches of bush and forest on their commonages."[118] Transkei administrators continued to depict the new system of political representation as embodying the democratic "voice" of "the people," a situation from which many chiefs and headmen in the councils eagerly sought benefits.

By contrast, Jameson's concession did not sit well at all with agriculture and forest officials. After several years of sharing forest management duties with magistrates, the Forest Department had just finally assumed complete responsibility for all Transkeian forests in 1903. Relinquishing any forest control to African leaders at this juncture seemed counterproductive to the goal of resource preservation. Senior authorities in these departments lobbied the secretary for native affairs to reconsider the concession, with "confident" predictions that the proposed change would invariably result in "the extinction of undemarcated forests, at an early date."[119] Such displeasure was echoed in the wider settler community in the Transkei, and even the former conservator Henkel felt compelled to come out of retirement to rail about the issue. Native Affairs officials tried to calm these fears by arguing that the new arrangement in no way altered the general course of forest "protection," as all of the "valuable" forests had already been demarcated in the districts affected by the Prime Minister's decision, and resident magistrates, as district council chairmen, would still be able to supervise council management practices. Additionally, the secretary for native affairs agreed to treat this entire exercise as an "experiment": should council forest management prove unsatisfactory, all such undemarcated areas would revert back to Forest Department control.[120]

The new arrangement, officially promulgated as Proclamation 288 of 1908, was troubled from the start. Councillors and magistrates in the TTGC bickered over the desired scope of chief and headmen's influence in location forest control.[121] On the ground, the new rules generated additional disagreement and disarray by allowing district councils, now generally in charge of location forests, to grant authority to local headmen to supervise areas covered by "mimosa, protea, and other bushes commonly known as scrub." Moreover, given councils' limited capacity to station forest guards effectively in all areas, headmen were able to exercise de facto control over location forests in many locales. In some places, headmen further dominated the prosecution of forest offenders as they actively exploited a provision in Proclamation 288 that entitled them to a handsome portion of any fines imposed for such efforts.[122] Within two short years, the combination of these and other realities riddled the council forest scheme with inefficiency and confusion. One magistrate condemned the system before the TTGC in 1910: "Who controlled those bushes—the Council, the Magistrate, or the Headman? . . . The whole thing was a farce,—a troublesome, irritating, farce."[123] By 1910, the majority of African councillors, recognizing defeat, came to the same conclusion. With government backing, the TTGC decided to return all undemarcated forests to the Forest Department.[124]

Foresters and magistrates drew important lessons from this brief, dismal episode in council forest management. The failure of the TTGC to supervise indigenous resources in an "enlightened" way confirmed foresters' prejudgments and sufficiently convinced Native Affairs authorities that African leaders should not be trusted with the complex responsibilities of modern environmental management and statecraft. Thus in 1910 officials began comprehensively redrawing boundaries of authority over Transkeian forests. A commission of local forest officers, the magistrate, and select district council members began inspecting all undemarcated forests in each district, reserving all "valuable" ones worthy of strict preservation to the Forest Department, and placing any remaining smaller forests and "scrub" under the control of location headmen for the less restricted use of location residents. Over the next few years, as the department consolidated its control over Transkeian forests even further, officials rationalized the new "hybrid" forest system taking shape across the Territories. From now on, colonial foresters would be responsible for the largest and "most important" forests in the region; African headmen would manage "their" forests for the benefit of "their people."[125]

As officials dismantled council forest control and created a new "headmen's forest" system, chiefs and headmen in the TTGC saw opportunities to reassert themselves as legitimate caretakers of Africans' "communal" resources. Moiketsi

Lebenya of the Mount Fletcher District reminded the assembled magistrates that prior to the introduction of councils "they (the Natives) had their own regulations" regarding local forest protection.[126] Headman Mkwenkwe of the Mount Frere District took this line of reasoning even further: "[T]he forests had been under the control of the Chiefs before the white people came here and placed white officials in charge of them. At that time the forests were all in good order, but then the white man came and issued orders that the forests were not to be touched unless by permit. That was when the Native people commenced to destroy the forests and steal the wood, because they were being ordered by the white men instead of by their Chief." Mkwenkwe went on to argue strongly for undemarcated forests to "be placed in the hands of the Chiefs and Headmen." Anyone advocating otherwise was "trying to spoil the Native law": "Headmen of locations knew how to control their people, and when certain bushes were allotted to them they would know the best way to distribute the wood among the people."[127]

While headmen speaking before the council often differed in their overall political stances regarding state intervention in community affairs,[128] the comments quoted above reveal some common threads. African leaders' legitimate role in local forest management was based firmly on precolonial precedent. "The Native people" had, prior to colonial takeover, controlled local forests in their own "customary" way, and African political leaders were necessarily at the center of such regulatory systems. It was thus only natural, councillors argued, that devolving forest management to "the people" today should mean placing control in the hands of headmen, the modern, legitimate heirs to precolonial political custom and those who knew best how to handle the people and serve their interests.[129] Moreover, as Mkwenkwe argued, adherence to "customary" authority over forests and "the Native law" was vital if popular destruction of the environment was to be constrained. Embedded in Mkwenkwe and others' remarks was an important theme: headmen were worthy of the government's trust in regulating forests because they could control popular resource use in line with the "progressive" interests of colonial officials.[130] Thus, even as Headman L. W. Masiza from the Tsomo District criticized the idea of additional forest laws, he emphasized Africans leaders' capacity for "responsible" resource management.[131]

As headmen's forests were created in the Transkei, African leaders utilized and combined such "modern" and "customary" logics to influence the course of official policy both inside and outside of council sessions. Headmen were integrally involved in not just the discursive but also the material construction of forest landscapes and their control during these years: several of the councillors involved in the above debates worked alongside foresters and magistrates in the commissions appointed to decide which forests should be demarcated and reserved for government and which should fall strictly under headmen's control.[132] Furthermore, headmen's depictions in council sessions of headmen's forests as "the 'People's' forests" which had been "handed over to the headmen in trust for the people" were clearly compatible with official designs for restructuring forest administration. New regulations in 1912 and 1913 supported such assertions, granting headmen sole authority over the harvesting of all undemarcated trees and bushes growing on location commonages.[133] This new division of forest administration—headmen's authority over "their" forests, Forest Department control in government reserves—would structure forest use across the Transkei until the late 1930s, when state officials would begin dramatically abandoning "trusting" Africans with resource management. Until this point, headmen across the Transkei were able to exploit this remaining niche of African forest control and the space it afforded them to interpret and legitimate their local authority in alternative ways.

In KwaMatiwane the creation of headmen's forests in the early 1910s presented colonial officials with immediate challenges. More formal and exclusive control over location forests offered headmen greater opportunities to appropriate local resources for their personal benefit, either for their own use or for sale to African farmers, European settlers and traders, and European and coloured sawyers.[134] In the Engcobo District, for instance, foresters regularly complained about local headmen's rather broad interpretation of their new range of authority. In one case a European sawyer and his children were apprehended with a concealed wagonload of illegally cut wood from Headman Vetu Gcanga's location. It turns out that the sawyer was operating in Gcanga's location "on the halves" with the headman, publicly trading his own share of whatever he felled. Several other headmen in the district similarly cut wood in location forests—or hired others to do so and split shares—and then sold timber at the Engcobo and Umtata markets. In a typical case, a forester complained that constraining one headman's wood-selling activities was rather difficult since "Sigidi evidently regarded his Forests as being for his particular use & benefit."[135]

In some locales, headmen also allowed "their" forests to benefit the people of their locations to a much greater extent than officials had intended. After officials simultaneously absorbed more forest areas into the state's demarcated reserve system and created headmen's forests in KwaMatiwane in the early 1910s, local residents more intensively exploited the less restricted wooded patches in their locations. With this growing pressure on wood supplies in undemarcated areas, people also placed ever greater demands on their headmen to create loopholes and flexibility in the colonial regime of forest management.[136] Oral informants and contemporary official reports attest to the limited restrictions on location forest access imposed by a number of head-men during these years. Dabulamanzi Gcanga (DG), for instance, recalled that headmen in his area of the Engcobo District were regularly lenient:

DG: And these forests, some of them belonged to the government, they were demarcated for the government, and nobody would go there, except by special permission from the forest man. But there were those forests that were in the hands of the headman. They would ask permission from the headman to go and cut wood. . . .

JT: Did the headman always give permission, or were there some times he said no?

DG: He always gave permission, he always allowed it.[137]

In addition to headmen loosely enforcing forest restrictions, there is ample evidence of certain individuals using their authority to surreptitiously provide men and women ready access to resources in nearby demarcated reserves. Officials began to notice such practices in the few years prior to the formal creation of headmen's forests, when council control enabled headmen to wield great influence over forest use in their locations. In 1910, for instance, a particularly notable conflict of interests erupted after some 93 African men in the central Tsolo District participated in an illegal hunt, moving freely be-tween undemarcated and demarcated areas. After many individuals were prosecuted and fined, the Tsolo magistrate vented his irritation with the sev-eral headmen of the locations involved and asked the CMT that they be rep-rimanded: "The Headmen mentioned declared that they know nothing about the matter and treat it with indifference. To my mind it is impossible for so large a body of men to collect without the knowledge of the Headmen and more especially of . . . [those people] who live near to the forests."[138]

A decade later, officials in the western Engcobo District were particularly incensed about headmen and residents concocting even more elaborate arrangements of illegal forest access. In 1922, the district forest officer in the area expressed his frustration at the "absurdity and dishonesty on the part of the Headmen" in at least ten locations, who were shielding the illicit wood-harvesting activities of their residents through permit fraud schemes:

The Headmen issue permits (written permits are not always given) to natives to cut produce in their forests. . . . The permit holder proceeds to the Headman's Forest and when he is not observed, enters the adjoining Demarcated forest and procures and hauls into the Headman's forest the produce which he could not obtain in the latter. The Forester-on-patrol meets a sledge load of (say) Yellowwood poles on the location area, asks the owner for his permit—the permit is produced and the law is satisfied! The natives are evidently trading on this procedure. . . .[139]

Besides such instances of headmen "screening" forest offenses activities in their wards, other examples point to some headmen using their positions as location forest gatekeepers to extract personal benefits from location residents. Headmen across the Transkei in the early twentieth century, both demoted chiefs and the "incipient class" of commoners occupying headmanships, increasingly demanded special favors and payments to compensate for their meager salaries and limited political and economic mobility in the colonial system.[140] In fact, given the corruption and favoritism practiced by many headmen in the past, officials strove to nip such practices in the bud when passing control of location forests over to headmen's authority in the early 1910s. One of the provisions instituting the forest transfer, for example, explicitly prohibited headmen from charging any fees or rewards for residents' access to forest resources.[141] Despite such precautions, officials had difficulty regulating headmen's daily exercise of resource control once headmen's forests were created. In a typical case in the Engcobo District in 1915, one man complained to the resident magistrate that his headman only granted access to kraalwood to residents who offered him some form of payment.[142] One elder from the Tsolo District recalled similar practices by certain headmen, who expected residents to offer a little "something," such as money or livestock, in exchange for land allotments and forest use in the location.[143]

In pursuing these and other practices during this period, headmen often invoked specific interpretations of their authority over the forests and peoples in their locations. In particular, many headmen increasingly claimed and exercised their "customary" prerogative to regulate women's access to local forests. This local development followed broader efforts by African male leaders and rural patriarchs across the Transkei, along with colonial authorities, to control women, their labor, and their mobility more fully during this period.[144] In this era of intensifying labor migrancy, rural male elders and headmen in local communities generally became more reliant on women's labor power. They were also becoming particularly dependent on women's wood harvesting to secure daily and seasonal fuel needs. On several occasions in the 1910s and 1920s, many African leaders in the TTGC emphasized women's vital economic roles in African households and advanced proposals for greater male control over women's labor and environmental practices.[145] In 1918, for example, one councillor from the Mount Ayliff District, Nota, explained that "[t]he Native people depended very much on the people who collected little pieces of wood, even though those persons were full grown women" and that "it was putting their women in a difficult position if they were to be brought before the Magistrate for committing an offence." To cope with this situation, Nota suggested that headmen "be authorised to punish lightly women and children found committing forest offences in forests under the control of Headmen in preference to prosecuting them in the Magistrate's Court," since, after all, "the people who had been given the forests to control had children and wives who had been trained to work for them."[146] Echoing this sentiment, Charles Pamla, an influential councillor from the Umzimkulu District, argued that male political leaders had the "customary" prerogative to control and punish female forest offenders: headmen had the right to "rebuke their people in the manner suggested *as they had always done.*"[147]

As such interpretations of headmen's "customary" authority and women's status circulated at the higher echelons of Transkeian politics, many headmen in KwaMatiwane put similar ideas into practice in their locations. Accounts of the years following the creation of headmen's forests suggest that women were often treated differently from men when it came to prosecuting forest offenses. More often than not, headmen would conduct their own proceedings in the case of female suspects, although women still possessed and pursued the legal right to appeal to their resident magistrate afterward.[148] And, in overseeing women's cases, local headmen could also invoke their own definitions of "customary" privilege. R. T. S. Mdaka (RM) explained how this worked in the Manzana area of the Engcobo District in the 1920s and 1930s:

JT: Do you remember when you were young, did people ever have any problems with the government about using the forests?

RM: Women had trouble with the government. They went and cut wood without permission.

JT: You mean without talking to the headman first?

RM: Without talking to the headman first.

JT: Okay, so what would happen?

RM: They would be brought to the *inkundla* [court, at the headman's kraal], and their case would be tried. And if found

guilty, . . . [he acts out, like a man making a command]
"Go fetch me utywala!"

VT[149]: They would have to prepare the "kaffir-beer."

RM: For men to go and drink.

JT: For the headman?

RM: For the headman, to go and drink, and his men. . . . It's
the headman first. . . . And if that particular person didn't
want to do what the headman said what she must do, then it
goes—the case was taken—to Engcobo, to the magistrate.[150]

As Mdaka's description suggests, headmen continued to play an ambiguous role in the control of location forests following the restructuring of the late 1900s and early 1910s. In relations with rural men and women, headmen's authority could mean more open forest access in certain cases or greater exploitation in others. Although officials had reduced headmen's territorial sphere of influence, they still had a difficult time regulating headmen's daily exercise of authority, whether used to deflect state intrusions into local people's livelihoods or to appropriate forest resources, wealth, and labor power for their personal benefit.

~

By the time the headmen's forest system was established in the early 1910s, the structures of environmental authority had passed through some fundamental shifts in the Transkei and in KwaMatiwane in particular. Whereas in the 1880s independent chiefs brokered the initial transitions of colonial rule and the reorganization of forest control, a few decades later the government controlled the majority of forests in the region, with state-salaried chiefs and headmen supervising only smaller undemarcated tracts in their compartmentalized locations. Giving specific meaning to such broader structural shifts were complex negotiations and contestations in particular settings over both the control of popular resource access and much deeper changes in local power relations through these years.

Throughout this era, African authorities' unpredictable and ambiguous participation in colonial resource management presented ongoing challenges for officials. Particularly in the earliest, most fluid and permeable moments of colonial administration, KwaMatiwane chiefs and headmen exerted their influence and at times could infuse their interests and self-representations into the new colonial legal and political institutions governing resource access.

Their ongoing capacity to exploit their authority over local resources in ways they determined themselves and beyond official desires and control—whether for their personal profit, to protect or deny the interests of their constituents, or in competition with newly installed forest staff—would continually frustrate colonial schemes for managing and tempering local forest use.

Local men and women's experiences of and responses to shifts in environmental authority and the larger patterns of colonial restructuring in which they were embedded further complicated the picture. Colonial strategies of reorganizing KwaMatiwane populations and environments in the 1880s exacerbated existing social and economic rifts within and between various communities of the region. Conflicts erupted over differently situated groups' experiences of local transformations in environmental control and power relations, often manifested in political disputes surrounding headmen's authority. As both headmen and forest personnel played increasingly pivotal resource gatekeeper roles over the next few decades, local residents responded to their wider social, economic, and ecological situations as they negotiated the everyday politics of forest access—bribes and extortion, legal disputes before magistrates and headmen, and intimidation and physical violence in their interactions with forest guards. While people expressed resentment toward the extension of colonial environmental authority at the turn of the century, their particular personal relationships with foresters and headmen also shaped their collective interactions. As with the other examples described in this chapter, such negotiations involved more than just the interplay of competing interests in forest resources themselves. They reflected varying stakes in localized experiences of shifting social, economic, and environmental relations in the early colonial period.

2 ⌁ Environmental Entitlements in the New Colonial Order, 1888–c. 1905

ALONGSIDE THE POLITICAL MACHINATIONS surrounding the authority structures of resource management, the implementation of colonial forest restrictions entailed complex negotiations within and between African communities and government circles over environmental entitlements. Forest officials, magistrates, and differently situated African men and women all responded to shifts in popular rights by bringing into play their particular interests in both resource access and wider changes of the early colonial period. In the late 1880s and 1890s, when officials began inscribing popular forest restrictions into colonial law, Africans across the Transkei immediately criticized the impact of such constraints and directly questioned the legitimacy of a government that intentionally criminalized its subjects. Up to the early 1900s, foresters and magistrates negotiated amongst themselves, often viciously, not only the acceptable scope of African forest rights but also the relative power of their departments in managing these rights. When officials tried to implement new institutions of wood access during these years, African men and women expressed and asserted their particular interests in the wider entitlements of colonial engagement and in the gendered transformation of the Transkei into a migrant labor reserve.

Such interactions and responses also helped shape the nature and limits of colonial environmental interventions in this early period. Officials' different perspectives on how far to restrict popular forest use, how best to manage Africans' environmental behavior, and how much to tie forest rights to gendered economic restructuring all affected how willing they were to work with each other, and how successful they could be, in regulating resource access.

Africans also took advantage of fractures in the colonial administration in these years, at times altering the trajectory of certain forest policies through protest and slipping through the cracks and discrepancies of forest enforcement in their locales. Rural men and women were more consistently successful in manipulating new rights institutions through their everyday practices, as they negotiated their specific socioeconomic and ecological predicaments. Men staked claims that hut-tax payments entitled them to free forest access and women asserted their rights to cut and sell fuelwood beyond the confining colonial definitions of their "customary" rights and "subsistence" practices, as differently placed individuals gave particular meanings to the colonial institutions now impinging on their livelihoods.

FOREST LAWS, POPULAR DISCONTENT, AND JOCKEYING FOR COLONIAL POWER

Complications surrounded the articulation and implementation of colonial resource restrictions in the early years of forest conservation. In response to newly imposed limitations on forest access in the late 1880s and 1890s, Africans across the Transkei directly questioned the state's legitimacy in constraining their environmental practices and livelihoods. In this fluid period of colonial development, such protests at times exerted sufficient pressure to help modify the course of colonial interventions. At the same time, popular discontent with the expanding environmental reach of colonial power became an important factor in officials' own internal debates and power struggles. For much of the 1890s, Native Affairs and Forest Department officials pointed to the ongoing problems of forest administration not just to promote competing schemes for resource management but also to criticize their opposing colleagues and assert their own departmental influence in the emerging colonial order. Only by the turn of the century would officials settle their major squabbles and more cohesively approach the question of managing popular forest use.

When the Transkeian Conservancy was launched in the late 1880s and forest officials began sharing responsibilities with magistrates for environmental management in African communities, deep fissures soon developed between the different departments. While they might have shared broad visions of colonial expansion and settler progress, officials often engaged in heated disputes over the proper policy course to attain these goals. Foresters generally prioritized securing the financial and industrial potential of Transkeian forests. Through proper "scientific" management by well-trained "experts," forest resources could accrue significant revenue to the state, provide raw materials for white settler agriculture and industry, and help conserve environmental supplies for future generations. Yet, from its beginnings, the Forest Department was hampered by its secondary status in the eyes of Transkei administrators, subordinated to other ministries in the Cape government, and chronically beleaguered by funding and personnel shortages.[1] Native Affairs officials had other priorities: facilitating the smooth navigation of "native affairs," the gradual "civilization" of the African population, and the uninterrupted flow of African migrant labor to white-owned farms and mines in the Cape. Moreover, having only recently ended a series of wars in the region and begun incorporating African communities into the colonial fold, magistrates were reluctant to introduce into their districts what they viewed as potential sources of popular resentment and social instability.

Such differences surrounded the initial implementation of resource restrictions in the Transkei. After Forest Act 28 of 1888 was introduced in the Cape Colony and authorities debated its extension to the Territories, forest officials resisted devolving any ounce of state forest control to local African communities. The newly appointed Conservator Henkel even invoked the "precedent" of precolonial chiefs' environmental "conservationism" in the Transkei to buttress his claims. Since African leaders, such as the Gcaleka Xhosa Chief Sarhili, had previously used their comprehensive authority to restrict popular resource use and thereby protect the environment, the new colonial government's centralized control over forests would merely modernize a preexisting "tradition."[2] Diluting this control by granting Africans legal rights to forest resources, forest officials asserted, would only hamper the ability of the government to preserve forests, limit resource "abuse," and collect forest revenues.[3]

Leading authorities in Native Affairs viewed the situation quite differently. Many officials had already spent the past decade debating and devising unique regulations for Africans in the new colonial system to meet the "special" conditions of "native life."[4] In a similar vein, magistrates lobbied to adjust Forest Act 28 to what they perceived to be the unique social conditions and policy needs in the Transkei. Influencing their sense of caution were experiences with sporadic popular protests against the forest restrictions previously attempted by some magistrates.[5] Yet magistrates also invoked the need for sensitivity to popular resource interests and the specter of popular discontent to legitimate their assumed positions as new "chiefs" benevolently protecting "native rights."[6] With the prime minister's backing this logic prevailed. In September 1890 the administration introduced a new system of forest regulations for the Territories (Proclamations 209 and 308), formally imposing restrictions on popular forest access, yet, much to forest officials' chagrin, allowing particular African resource rights to continue.[7] Even as they installed new institutions of forest access, officials marked their own battle lines around the nature and scope of popular entitlements.

As soon as foresters began implementing the new regulations, Africans across the Eastern Cape and Transkei threw their own oppositional voices into the fray. The leading Xhosa-language newspaper in the region, *Imvo Zabantsundu*, responded with a series of scathing editorials.[8] When Act 28 was first introduced in the Cape Colony, the editor sharply commented, "From its rigid severity, we confess we did not expect that the Ministry were in earnest in promulgating the new Forest Act. That enactment cannot be regarded by practical minds as anything else than a Crimes Manufacturing Act. . . . we have no sympathy with a law that can only have the effect of converting a whole community into a set of criminals."[9] After Proclamations 209 and 308 were enacted in the Territories, and as prosecutions under the new laws became commonplace in the Eastern Cape and Transkei, *Imvo* amplified its critique with an editorial entitled "Is a tree worth more than a Native?"

> The Government, through its Forest Officers, is dealing out very hard measures to our people on this Frontier. Nearly every piece of ground that can be taken is proclaimed under the Forest Law, and almost every plant that grows is declared too sacred to be cut with an axe. For houses, and kraals, and house-fires, our people must have wood, and that they have for generations cut from the bush or forests close to their doors. Now suddenly, they find these forests filled with policemen in strange green clothes, who pounce down on the unfortunate man who cuts a wattle, and hasten him to the magistrate, who speedily sends him to the common gaol.[10]

These prominent critiques scrutinized the fundamental principles underlying colonial environmental expansion, particularly the apparent criminalization of the majority population. And officials closely tracked such commentaries,[11] for disquieting stirrings of popular resentment were rapidly spreading across the Transkei. In meeting after meeting with government representatives, communities repeatedly questioned the legitimacy of the government's domination of forest access and its potential impact on people's ability to freely secure resources for their everyday livelihood needs. In Willowvale District in early 1889 the magistrate reported that the impending introduction of expanded forest restrictions "created some stir amongst the natives, as they felt that they would be suddenly deprived of certain privileges which they had enjoyed for a number of years."[12] A year later in Umtata, Thembu communities rushed to voice their displeasure with leading Native Affairs and Forest authorities, surprising the three chief magistrates of the Territories with the volume of popular discontent with the new forest laws: "So great is the interest evinced on this subject by the Tembu nation that although very short notice was given to [Paramount Chief] Dalindyebo, and only himself and a limited number of Councillors invited to be present fully six hundred men arrived with their chief, amongst whom were observed the principal Chiefs and Councillors living within a radius of many miles from Umtata."[13] The magistrates further acknowledged "the vast political importance" of settling this issue for colonial stability in Thembuland and beyond, for the outcome of this session in forest diplomacy "would soon be known by all other tribes" as word spread to African communities throughout the Territories.[14]

As the government implemented Proclamations 209 and 308 throughout the region in the early 1890s, discontent deepened across the countryside. Chiefs, headmen, and community representatives openly questioned the government's right to proscribe the use of wood for making kraals and to limit access to particular "reserved" tree species, particularly emphasizing the burdens these latter restrictions created amid growing localized problems of wood scarcity.[15] In 1895, before the chief magistrate and several hundred Thembu chiefs and followers, Dalindyebo's spokesman went straight to the point: "The Chief has come here first about the forests. The number of reserved trees covers nearly every tree suitable for hut building purposes, which shows that the Tembus have nothing to cut or utilize for their buildings, and means their destruction. It is not possible to build huts without the use of some of these reserved trees, our huts fall down if we build without them. If we are not allowed to cut any of these trees how are we to live?"[16]

By the mid-1890s government leaders could no longer ignore this popular chorus of forest grievances. During Prime Minister Rhodes' tour of the Territories in 1894 he encountered frustration with the new forest restrictions at nearly every public meeting with African communities.[17] Chief Makaula of the Mount Frere District made a particularly strong plea as he presented the complaints of Bhaca communities at a large public gathering before the premier. He suggested that the government accept a "compromise" with African communities across the Territories: the Forest Department should only exercise its constricting regulations in specified reserve areas, allowing Africans to use other forests in their neighborhoods freely, albeit under the supervision of resident magistrates. Recognizing that African discontent with colonial forestry "was assuming serious features," Rhodes took Makaula's cue and promised that this compromise would be enacted, eventually resulting in Proclamation 388 of 1896.[18] Although government leaders would later take sole credit for inspiring this policy, its origins lay more in protracted negotiations between African communities and officials surrounding forest restrictions.[19]

This realignment of colonial forest management only aggravated an increasingly stormy relationship between forest officials and Native Affairs staff and their contending visions for regulating Africans' resource access.[20] In the preceding years, interdepartmental debates over how to define popular forest rights had become increasingly entangled in disputes over which arm of the colonial administration could best manage both popular resource use and popular discontent. As they jockeyed for departmental influence, each group thus attempted to legitimize their cause by blaming the other side for the problems of forest administration and asserting their own unique professional ability to resolve these issues, particularly their capacity to handle "native sentiment."

Forest officials regularly condemned magistrates for the sad state of forest affairs in the Transkei. A common trope in forestry reports was the tragic narrative of popular resource "abuse" and environmental decline following magistrates' takeover of annexed African territories. The abrupt ending of precolonial African political relations, once independent chiefs were deposed or absorbed into the colonial system, had left a damaging power vacuum, granting African residents open resource access unbridled by magisterial supervision.[21] A particularly prominent and outspoken district forest officer in the Transkeian Conservancy, Knut Carlson, articulated this narrative most forcefully in an 1896 article in the popular *Cape Agricultural Journal.* Immediately after colonial annexation and "[t]he power of the Chiefs was quelled," Carlson argued, "some of the Magistrates, who took personal interest in the preservation of the forests, revived the old custom of the Chiefs and told the people not to cut in certain forests. This nominal protection, however, could not be expected to have the same effect as the autocratic power of the Chiefs, and for a long time there was really nothing to stop the rapid destruction then taking place."[22] That is, of course, until foresters arrived on the scene to steer resource management in a proper direction.

Foresters thus portrayed themselves as the professional class uniquely qualified to tackle the Transkei's environmental challenges. According to Henkel and others, foresters not only possessed the relevant intellectual expertise but the necessary local knowledge for the job. While magistrates sat behind a desk every day, foresters "are constantly in the saddle and are familiar with every path, and the name and value of every tree"; they were, some even claimed, the "natural protectors" of local flora and fauna.[23] To share forest management with magistrates was thus inappropriate and inefficient, a situation "anomalous" in the world of state forestry.[24] "Dual control"—the overlapping spheres of the Native Affairs and Forest departments—directly resulted in unnecessary confusion and ineffective forest protection on the ground. Moreover, foresters asserted, this "hybrid" form of governance was a central source of African discontent, as it was much too difficult for the limited "native mind" to understand anything departing from the "traditional" style of centralized African government. As Henkel asserted in 1893, "Dual control leads to confusion and *does* bewilder the Natives." When Africans faced different arms of the government administering local forests, "[T]hey naturally say: 'How many different Magistrates have we?'"[25]

Magistrates, by contrast, often sought to temper the agenda of foresters in these years, adapting the goals of resource conservation to the greater interests of social stability and to the special requirements of "native conditions." Foresters were seen as lacking the proper understanding of how Africans functioned and what style of governance was needed to handle and gradually modify their ways. Moreover, this lack of wisdom and tact had caused unnecessary confusion and discontent in African localities. Henkel was often particularly singled out for his promotion of fiery conservationism to the detriment of Native Affairs protocol.[26] Underlying such criticism of Transkei forest administration was an argument for centralized magisterial control over "native affairs." Magistrates, like forest officials, condemned the evils of "dual control" to buttress their department's claims to managing African resource use. In 1894, for instance, James Rose Innes, the under-secretary for native affairs, blamed ongoing forest management problems squarely on the Forest Department. In the days when magistrates alone controlled resources in recently colonized districts, he argued, forests were "better secured, and discontent was not known, both results owing to the influence of the Magistrate over his people, and to the difficulty of getting the Native to understand any other authority, as representing the Government but the Magistrate, and the natural result of his disregarding all orders which do not come from his Magistrate."[27]

Such claims again relied particularly on images of Africans' seemingly innate understanding of centralized authority. The subsequent division of colonial governance in the Transkei into separate departments was too much for African minds to bear. As leading magistrates often tried to convince their superiors, "Many masters perplex and confuse the Natives."[28] Moreover, unlike the brutish narrow-mindedness of foresters, magistrates possessed the requisite tact, patience, and experience to assume the role of Africans' new chief and father.[29] Such arguments increasingly appealed to the highest Cape authorities by the middle of the decade, as reflected in the 1896 handover of unreserved forests to magisterial control.[30]

These contending interests in resource management and control widened into more open and hostile crossfire the following year. Pushing interdepartmental relations to the brink was a period of rural crisis confronting officials with the fears and realities of forces moving beyond the orbit of government control. The growing swell of popular discontent with governmental forest restrictions only increased in this tense climate and contributed to official fears of social instability.[31] This combination of pressures forced officials to articulate more clearly their competing visions of how best to manage and control Africans and their resources. Seizing the opportunity to champion the cause of centralized magisterial control, in mid-1897 Walter Stanford, CM of East Griqualand, argued that a consolidation of government was in order during these pressing times, and that Native Affairs should therefore take over control of the Forest Department. When he assumed the position of superintendent of native affairs later that year, Stanford asserted that "much unnecessary irritation" among Africans had been caused by the administrative tactics of the Forest Department and that the advent of foresters on the scene had brought "greater destruction in forests" than when magistrates alone held sway. He then launched into a comprehensive argument for the supreme authority of chief magistrates in the Territories.

> They are Paramount Chiefs displacing the old Native order. . . .
> One of the commonest forms of Native expression addressed to the
> Chief Magistrates at a public meeting is "We do not see the Government: we see only you." The Natives despise diversity of control.
> It is undermining the Chief Magistrate's position to place by his side
> officers working on lines entirely apart from him. The duties of forest officers, sheep inspectors and the like cannot be dissociated from
> the general welfare and the political feelings and interests of the
> Native people.[32]

Such comments naturalized not only the special position of magisterial authority in the "native mind," but also magistrates' unique ability to maintain popular confidence in and obedience to colonial rule.

Raising his voice in strong opposition was William Hammond-Tooke, the secretary for agriculture. In lengthy tirades in late 1897, Hammond-Tooke criticized the impact of accepted "native policy" on both African populations and their natural resource base. While the early years of colonial takeover might have required magistrates to adopt the role of Africans' "Baba," or father, these changing times required more "enlightened" uses of authority.[33] The Transkei, he urged, needed to be wisely managed by a collection of "expert" departments: the Native Affairs Department should concern itself solely with "all tribal matters," while the uniquely trained Agriculture Department staff would best "preserve and exploit" local resources and ensure the African's "future and his way of salvation."[34] More pointedly, Hammond-Tooke condemned magistrates' depiction of Africans' sense of justice as merely a smokescreen to promote their own favored mode of "benevolent despotism." Magistrates claimed that "natives prefer an autocrat" primarily to defend their monopoly of power and prestige within the colonial administration. According to the claims of Stanford and others, Hammond-Tooke sarcastically elaborated, the typical African supposedly "goes away rejoicing, blessing the Government as the author and giver of all good" when Native Affairs imposes forest restrictions, yet, when the same is done by foresters, "he straightway [sic] denounces the Forest Department as a robber that has deprived him of his hereditary right."[35]

Such charged comments between colonial authorities reveal how deeply official debates over regulating African resource rights were embedded in disputes over departmental influence in the emerging colonial order. After escalating so dramatically amid an administrative crisis in 1897, however, such open interdepartmental hostility would be relatively short-lived, as changing conditions in the Transkei helped shift official attitudes toward more conciliatory approaches to sharing the reins of environmental management. The mounting duties and difficulties of "native administration," particularly during the rinderpest epidemic of the late 1890s and the onset of the South African War, led Stanford and other Native Affairs authorities to devote magistrates' attention to more immediately destabilizing matters than forest management. Moreover, many magistrates at the turn of the century increasingly recognized that forest management was failing under their watch in the face of such other pressing responsibilities, and that a more strategic way to manage Africans' resource rights and curb popular "destruction" of forests was to hand over their management to a separate conservation staff.[36] Capitalizing on such sentiment was the new conservator A. W. Heywood, who succeeded the much more fiery and controversial Henkel after the latter's retirement in 1898. Over the course of 1899 and early 1900, Heywood busily worked to patch up interdepartmental relations, convening with high-ranking administrators, promoting the importance of the Forest Department, and gradually convincing Native Affairs authorities to loosen their grip on environmental control in the Transkei. Through Heywood's initiative, Prime Minister William P. Schreiner and SNA Stanford eventually accepted a new scheme for managing Transkeian forests, whereby all forest areas, both demarcated and undemarcated, would be controlled by the Forest Department and a comprehensive new forest act would be proclaimed to regulate them. Interdepartmental negotiations, bureaucracy, and the war all slowed down the transition process, but by 1903 the transfer of power was complete.[37]

The extension of colonial control over African forest access was thus in an uneasy infancy by the early 1900s. Africans across the Transkei responded with intense criticism of colonial environmental expansion, and at times their protests reverberated profoundly through a still developing and fractured colonial administration.[38] Foresters and magistrates brought their own power struggles to the table, using the emerging problems of managing popular forest use as a battleground for their own administrative aspirations before finally working toward bureaucratic compromise. Officials and Africans alike, albeit on increasingly unequal terms, negotiated the initial terms of state environmental intervention and their meaning within a novel, unpredictable, and rapidly changing colonial context.

THE CHANGING CONTOURS OF "FREE" RESOURCE ENTITLEMENTS

While policy protests and debates influenced the broader path of environmental schemes in the Transkei, daily negotiations of particular colonial institutions shaped the deeper meaning of resource rights on the ground. One of the central components of the new Transkei-wide forest regulations in 1890, building from practices previously introduced in some magisterial districts, was the "free permit" system, guaranteeing Africans the right to access particular forest resources without payment. According to section 4 of Proclamation 209 of 1890, resident magistrates would now issue permits within their respective districts "to natives being occupiers of land under the tribal or communal system," for cutting and removing wood from government forests for building materials, agricultural implements, and other items "ordinarily used by natives," free of charge; license requirements and tariffs applied to all other types of wood use. As officials implemented the free permit system across the Territories, however, it was immediately entangled in a web of negotiations among African residents and state actors over the meaning of popular resource entitlements in a rapidly changing colonial context. Africans asserted alternative perspectives on the scope of their rights while magistrates and foresters negotiated their contending policy interests, together reshaping the nature of the permit system in actual practice.

For much of the 1890s, the free permit system poured salt on the wounds suffered in ongoing "dual control" battles between the Native Affairs and Forest departments, undermining the efficacy of enforcing forest regulations on the ground. Desiring administrative support for limiting Africans' wood rights and inducing them to pay the colonial government for harvested wood, forest officials regularly decried the new system's financial and ecological toll from its very start, particularly arguing for a quantifiable cap on Africans' rights to "free" forest resources. "Natives take advantage of these Permits to built [sic] huts and cattle kraals more than absolutely necessary," Henkel claimed in 1891, citing instances of African men dramatically overstating the number of livestock they possessed when applying for permits in order to inflate their apparent kraalwood needs.[39] Subsequent reports quantified and lamented the tens of thousands of pounds' worth of forest produce given away to Africans through permits rather than accruing to state coffers. Officials were described as being besieged by Africans' "exorbitant" requests for permits, sometimes numbering in the hundreds per day.[40] Foresters often blamed magistrates for this sad state of affairs, accusing them of generally "coddling" Africans and indiscriminately distributing permits to African men without knowing how many wives they actually had, and, thus, without determining for how many huts wood was "legitimately" required.[41] According to Henkel and other foresters, permits and tax payments were supposed to confine popular forest use to a quantifiable level of "subsistence" rather than entitle Africans to unlimited and unregulated resource access.[42]

In 1893 and early 1894, as foresters restructured the permitting process in response to complaints from African leaders and various officials, they simultaneously moved to enshrine this more limited definition of African entitlements.[43] Rather than having Africans regularly travel to the resident magistrate for permits, and to ease magistrates' workload, Henkel suggested that the conservator annually issue permits to headmen via their magistrates, a practice already operating in the Eastern Conservancy. For Henkel this maneuver was also an opportunity to redefine the legal and territorial boundaries of forest use under permits. The idea centered on the influential role of headmen in controlling the quantity of forest produce removed through free permits. Since a headman was directly involved in the registration of huts and tax collection, he would know exactly how many new huts were recently built, registered, and "paid" for in his location. On this basis he could accurately assess the actual wood needs for huts and kraals in his community, a numerical figure that could be inscribed onto the annual permit he received from the magistrate on his constituents' behalf; any right to cut produce in excess of this amount would require special permission from colonial authorities. Making headmen the locus of the permit system further implied a critical shift in the territorial boundaries of forest rights: permits would now allow access only to undemarcated forests in headmen's locations and not demarcated government reserves. Over the course of 1894 and 1895 the new measures won the approval of senior authorities in the Native Affairs and Agriculture departments and took effect in 1896, with the handing over of location forests to magisterial control.[44]

This administrative switch brought very mixed results for the operation of the free permit system, however. Many magistrates continued to be more tolerant of Africans' wood rights and thus more lax in enforcing permit restrictions than forest officials, particularly when they felt that social stability in local communities was at stake.[45] Foresters regularly complained that forest regulations were "completely ignored" by some magistrates; in 1894, the Willowvale magistrate simply decided not to follow the permit restrictions, returning the permit books untouched to the presiding forest officer.[46] Following the handover in 1896, a number of magistrates continued to bypass the complexities of the permit system in order to avoid perceived sources of "native irritation." Others continued to endorse receipts from individual African taxpayers when they applied for permits at the district office, rather than issuing permits to headmen for their locations.[47]

Exacerbating such problems of magisterial "leniency" were the difficulties forest staff faced in policing popular use of permits, as residents in many locales employed various strategies of evasion and deception.[48] In many Kwa-Matiwane forests in these years men and women regularly evaded patrolling foresters by closely monitoring their movements and work shifts. Forester H. L. Caplen, in charge of the forests of upper Thembuland, proposed in the mid-1890s that the conservancy hire undercover African detectives, since local people regularly tracked the movements of conspicuous uniformed European officers; Caplen himself in fact regularly traveled "unexpected" routes to avoid detection by potential forest offenders.[49] Informants recalling their own similar strategies in the 1920s and 1930s suggest some of the ways men and women of an earlier generation contended with increasing surveillance of their wood harvesting practices. One eighty-four-year-old woman who grew up at Qelana in the Tsolo District boasted how she and others routinely monitored forest guards when illegally collecting firewood: "We had no problems, because we used to time the guards, and then we ran to the forest so that the guard would not catch us." Another elder from the same area described how people "would identify the horses belonging to the guards." If they spotted these horses in a particular forest area, they would hide until the coast was clear.[50] More commonly people waited until the cover of nightfall, especially clear nights with plenty of moonlight, to stealthily exploit forest resources and avoid detection.[51]

Even when foresters were able to chase down and apprehend suspected forest offenders in a particular forest, many such individuals falsely identified themselves, their headmen, and their locations, making court summons and prosecution difficult if not impossible.[52] And then there was the problem of insufficient evidence for prosecution. Concealing harvested wood was so common that foresters routinely went to suspects' homes to inspect their huts and livestock kraals for recently chopped and hidden trees and bushes.[53] In one typical instance in the Tsolo District in 1897, while officers searched the home of a man who had cut reserved trees to construct a sledge, his wife scrambled to hide the poles and branches in a different part of the kraal.[54] Another local case involved two women who illegally collected bundles of mimosa bush to use for firewood. The forester searched their homestead, discovered the wood, left temporarily, and then returned to find that the women had hidden the bundles he hoped to confiscate as evidence.[55]

Other popular responses in these years further hobbled official designs for the permit system's operation. Permit access to wood might be free from tariffs, but the specific temporal, spatial, and biophysical restrictions imposed on wood harvesting often made following the rules highly impractical. Travel and transport conditions drove some individuals to disregard the conditions of their permits, or at least to employ them as justifications when defending their actions in a magistrate's courtroom. Thus, one defendant complained to the Tsolo magistrate in 1897 that it was impossible for him to comply with the three-day limitation of his permit, since "[t]he Rivers were flooded & my cattle sick & that is why I did not use my licence sooner."[56] In a separate case another man testified, "I cut this load of firewood long ago and hid it because I could not ride it away on account of my cattle dying."[57] Other cases reveal that some men attempted to bypass regulatory impositions altogether by using their own, borrowed, or even expired permits and licenses as cover to harvest certain types or amounts of wood beyond the conditions printed on these documents.[58]

In numerous KwaMatiwane locales, many residents were further frustrated by the apparent advantages the new wood access system offered to their more prosperous African neighbors. The constraints imposed by permits forced a growing number of men to forego this "free" access and instead pay for wood at government reserves, a situation which seemed to unfairly advantage individuals with the required wagons, animal transport, and cash. At a large public gathering at Engcobo in 1892 the Thembu chief Mgudhlwa explained to CM Elliot that men in various locales had asked their leaders to protest the forest regulations "because some can afford to buy yokes, and others cannot."[59] The following year Henkel likewise described a general effect of new resource institutions on African class relations in many local communities:

"The majority of Aborigines [who] have no wagons to proceed from their locations say from 36 to 50 miles to procure wattles for hut building, are therefore dependent upon their more enterprising neighbours who own wagons."[60] Indeed, for a small group of male African wood riders and traders operating in KwaMatiwane forests in the 1890s, the free permit system generated an expanded African client base. Moreover, these entrepreneurs further confounded the policing of wood access by illegally using their free permit rights to trees to secure wood for potential sale to local African and European buyers.[61]

Another significant factor shaping people's negotiation of free permits was their alternative understanding of the very meaning of this new institution and the entitlements it bestowed. The basic mechanics of the permit system and their location within wider colonial policies were critical in this sense. One of the central features of the new system was making Africans' access to free permits conditional upon married men's regular payment of hut tax.[62] As in other interventions into rural life, officials here placed responsibility for household production decisions squarely in the hands of men.[63] As they applied for forest access permits, men could thus determine how different household wood needs and labor requirements should be prioritized, despite the regular contributions of other family members to the various economic activities necessary to generate income for paying the requisite hut tax. This close relationship between hut tax payment and free permit access was directly expressed through the permitting process itself. Men were required to present their latest hut tax receipt to a local authority who then endorsed the particulars of the permit—including time, place, and wood type restrictions—directly onto the back of the receipt.[64] This procedure contributed to very different understandings of the meaning of both this crucial piece of paper and the nature of African participation in the colonial system. While government officials viewed tax payment as a useful way to control and manage local forest use more effectively, African men often interpreted and invoked their regular compliance with hut taxes as a "contract" with the state, enabling them to access local resources without government interference.[65]

As Sean Redding has described, colonial authorities in the Transkei instituted hut tax to serve multiple purposes, including discouraging polygyny, enshrining a particular model of rural patriarchy, accruing needed revenue, and encouraging a steady supply of African men into the migrant labor force. Yet for Africans, she has argued, the significance of hut tax in its early years had both material and symbolic dimensions. Regular tax payment ensured access to magistrates and their courts, enabled men to acquire land allotments through their wives, offered public affirmation of a man's ritual passage to adulthood, and protected families against the possibilities of prosecution or appropriation of property for noncompliance and more generally against the state's perceived supernatural power.[66] Redding's analysis begins to ask important questions about colonial and African perspectives on the meaning of tax collection and payment, yet is also generally limited to exploring the historical functions of these state rituals. Left unexamined is how the nature of the hut tax itself—what the tax was "on" and what Africans were actually "paying for"—was also implicated in colonial negotiations over meaning. Africans not only felt, viewed, and used the functional roles of taxation in alternate ways; they also questioned the basic definition and purpose of the hut tax as envisioned by state actors.

One of the central and often-stated motives for imposing the tax was to draw revenue from African communities for their occupation of land formally under colonial control.[67] In an 1883 meeting with the Tsolo magistrate, for example, some headmen explained that members of the Tola clan, only recently settled in the district, felt they should be exempt from paying the newly instituted tax: "[They say] We have no fixed place of residence. How can we pay Hut Tax?" The magistrate then criticized this logic: "[T]he Tax is really 'Land Tax' and means that all persons who have land or gardens have to pay 'Hut Tax.' It is a Tax on the Land."[68] This example reflects both some of the deeper misunderstandings surrounding taxation's introduction and the close connections local officials often reinforced between such payments to the state and popular resource access.

Africans across the Transkei also associated environmental entitlements with hut taxes, but in much broader terms than officials anticipated. In 1884 Chief Magistrate Henry Elliot commented on a source of popular discontent in many areas: "The Native view of hut tax is 'We agreed to pay it, and in doing so did a foolish thing, and we are justified in evading it, if we can. Government did not build our huts or provide the material, why should it tax us upon our own labour, and the natural products of our own land?"[69] Elliot's words suggest that many Africans perceived and publicly invoked connections between tax payment to the state and their access to a wide array of natural resources, even as they evaded it. In many other cases, as people complied with taxation, they turned such associations into expectations of government obligation. As one local official described popular sentiment in the Matatiele District in the late 1870s: "[T]he natives look upon their 'hut-tax receipt' as a title to their homestead and garden, and also as a means of claiming protection as British subjects." While this and other official statements on popular acceptance of the hut tax strove to legitimize colonial policy, they also suggest that many Africans interpreted and asserted particular resource claims through their tax payments.[70]

Redding has noted that African men in the Transkei often viewed hut tax payment as affirmation of their "adult" status in the colonial system and all of the material benefits that position conveyed.[71] Yet taxpayers' expectations and claims to environmental entitlements also extended well beyond colonial definitions. Through marriage and hut taxes, a man expected all of the entitlements of full adult village membership—not just residential sites for himself and his household and arable field allotments through his wives, as recognized in colonial hut tax laws, but also the right to graze livestock on the location commonage and for everyone in his household to be able to utilize the location's other environmental resources, such as wood and water, for their everyday livelihood needs without hindrance. This point was even acknowledged by Conservator Henkel himself in 1895, as he reflected on his decade of experience with the hut tax system and popular discontent with state forest interventions. Africans, he noted, generally felt that "[b]y paying 'Hut tax' for one or more huts, according to the size of their Families, they are relieved from all other taxes. 'Hut tax' to them, means: a piece of arable land, grazing for their cattle, goats, sheep and horses, forest produce such as wattles, poles, kraalbush, sledges, hoe and axe handles, sticks for assegais &ca."[72]

Such interpretations of taxes and entitlements often complicated how African taxpayers responded to new forest policies and asserted "free" wood rights in the early colonial period. In 1888 in the Umzimkulu District, for instance, local men failed to heed newly introduced forest regulations, based on their reinterpretation of colonial entitlements. Despite the district magistrate's attempts to explain the actual wording of the forest laws, Henkel reported, the residents "still persist in cutting wattles and brushwood, as it was for these, they say, 'hut tax' was paid."[73] Some twenty years later, headmen and residents of the Kentani District similarly appealed to the government to excuse them from restrictions on harvesting mimosa and other scrub trees growing on their locations, given the fact that they regularly paid taxes to the government.[74] These examples suggest that throughout the era of the free permit system, whereby the state deliberately tied forest access to popular taxation, Africans often perceived the relationship between these institutions and asserted claims to the resource entitlements they conveyed in ways that disrupted colonial economic and environmental policies. When forest officials repeatedly complained in the 1890s and 1900s about Africans flooding them with free permit applications and popular "abuse" of permit "privileges," they were in part observing—and misreading—the material effects of wider popular understandings of permit access and the nature of the resource "benefits package" accompanying tax compliance.

From the late 1880s to the early 1900s, then, negotiations among magistrates, foresters, and African residents over the nature and scope of popular resource rights shaped the course of the free permit system on the ground. Africans' claims to more expansive entitlements, along with their strategies of evasion and deception, complicated colonial resource management at a time when officials themselves disagreed on how best to balance environmental and social management priorities in the Transkei. By the early 1900s, frustrated foresters faced the reality that Africans' alternative approaches to resource entitlements, their "abuse" of permit specifications, and the uneven efforts of magistrates all made the permit system and the attempt to impose a strict cap on popular wood use "almost a dead letter" in many localities.[75] Together such forces gave real and unanticipated meanings to the new resource access institutions being formally inscribed into colonial law.

EN-GENDERING AND CUSTOM-IZING "SUBSISTENCE" FOREST RIGHTS

Complicating the problems of confining Africans to "subsistence" wood use were particular negotiations over the gendered orientation of colonial forest policies amid broader gendered changes in rural economies in the early colonial period. While conflicting colonial priorities weakened the permit system, Forest and Native Affairs officials generally found common interests in using resource access to help transform the Transkei into a "labor reserve" economy, based on the out-migration of African men. From the late 1880s to the early 1900s new forest laws particularly restructured men and women's relative access to fuelwood through restrictions on their "customary" wood rights. KwaMatiwane residents coped with these new constraints alongside local economic and environmental changes, and although people regularly questioned the contradictions and inequities fuelwood regulations posed for their livelihoods, officials were not inclined to alter the wider, gendered restructuring of wood access. More effective in this early period were women's particular efforts to overcome imposed limits on their "customary" and "subsistence" practices and maintain some control over their own environmental behavior.

Despite their other disagreements, colonial officials shared desires to transform the Transkei in gendered directions that could directly benefit the Cape economy. Magisterial authorities regularly applied financial pressures—hut taxes, dipping fees, fines, etc.—to induce African men to meet their growing cash needs through wage employment at white-owned mines or farms.[76] In the early years of the Transkeian Conservancy, forest officials similarly strove to constrain the economic activities of African men by specifically restricting male environmental practices and imposing tariffs on certain forms of wood use. In 1890, for example, Henkel explained that restricting African men's forest rights and imposing wood tariffs would benefit the Cape economy, "as the Natives would be forced to seek work among the farmers, on the railway lines under construction, and at the gold-fields, if they had to pay for forest produce."[77] Although magistrates in the 1890s were much less inclined than foresters to undermine African men's rights comprehensively and force them to pay for all wood access, Native Affairs officials similarly envisioned environmental restructuring as a fruitful means of reshaping rural Africans' livelihoods in gendered ways. Determining the scope and territorial extent of African rights to specific resources in the Transkei would limit Africans to "fixed" sites of residence and economic activity, forcing men to migrate for remunerative labor in white-owned farms and industries and return to the "native reserve" only when their contracts expired, and forcing women to remain in the countryside bearing the brunt of labor burdens at "home."[78] In the early years of conservation, there was thus general agreement among officials that resource policies should serve larger efforts to direct African men's "productive" and women's "reproductive" lives.

This consensus was reflected most clearly in the development of policies regarding fuelwood access at the turn of the century. Officials narrowly construed Africans' "customary" rights to fuelwood to aid their goals of limiting popular wood consumption and reshaping rural gender relations in ways that benefited the colonial economy. From their beginnings in the late 1880s, fuelwood regulations limited popular fuelwood "privileges" to "domestic use" only, and further reserved such entitlements for women alone. In constraining the ability of men to access fuelwood without payment while "recognizing" women's gathering rights, successive regulations simultaneously placed cash pressures on men and positioned women's heavy labor burden as the default option.[79]

At the turn of the century African women in KwaMatiwane were indeed primarily responsible for collecting, with the aid of children, fuelwood for their families—traveling by foot, gathering by hand (known as *teza* in isiXhosa), and carrying materials on their heads back to their homes. Yet, contrary to official depictions of "customary" labor practices, this was not the only way that households harvested fuelwood.[80] Men and boys, generally responsible for tending livestock, often utilized ox-drawn wagons, carts, and sledges to aid in the collection of fuelwood, especially when larger logs or heavier pieces of wood needed to be retrieved. Take, for example, one account of agricultural work in Thembu communities in the Baziya area in the early 1870s. Rev. R. Baur, a Moravian missionary who had established a station there a few years earlier, emphasized how the sexual division of labor was far from absolute: "When the seed-time comes, I always see the men working with their hoes together with the women. And in harvest-time it is pretty much the same. The women generally are the drawers of water, and are obliged to fetch firewood from the forests, often in heavy loads upon their heads; but even this is being ameliorated, as many of the Kaffir men go to the forest with a pair of oxen, and fasten a load together for home consumption."[81] While such descriptions of firewood collection practices are unfortunately rare for the early colonial period, accounts of similar practices still being pursued in the 1920s and 1930s in various KwaMatiwane communities suggest that Baur was describing a pattern applicable to many different parts of the region.[82]

The new fuelwood laws of the late nineteenth century ignored such complexities and restricted free access rights merely to women's "customary" teza practices. Walter Stanford first instituted regulations affecting fuelwood access in the region in the mid-1880s, first as Engcobo district magistrate and then as CM in Kokstad. His East Griqualand regulations in 1885 saddled African men with tariffs and licensing requirements for cutting and removing firewood by the sledge- or wagonload; tariff- and permit-free collection of fuelwood by hand was allowed, but limited only to wood that was dry, fallen, and from unreserved tree species.[83] These regulations became the basis for subsequent Transkei-wide fuelwood laws in the 1890s and early 1900s.[84] Colonial officials thus delimited a narrow range of environmental practices which they deemed "customary" and therefore a legitimate basis for fuelwood entitlements.[85] Africans could collect fuelwood freely only if they transported it bodily, normally by the headload, the common practice of women and children. Men's regular use of alternative forms of transport would be charged.

When colonial authorities began enforcing these regulations in the late 1880s in KwaMatiwane, many men and women recognized and resented the ways in which fuelwood access was being reorganized along a gender divide. Although the new laws did not explicitly refer to male versus female forest rights, officials commonly intended and enforced such distinctions, and they were widely understood and felt by local residents.[86] In an 1888 criminal case in the Tsolo District, for instance, one woman from the Umgna area testified that local people generally understood forest restrictions to mean that "no man might cut there, but we were not prevented from 'Teza'ing." A man who had been assigned by his headman to watch over local forests in the area offered a similar explanation of his duties: "[P]revent them from being used for

Figure 2.1. Woman carrying a headload of fuelwood, with Gulandoda mountain in the background, c. 1959. *Photograph courtesy of University of Fort Hare Library, Piper Collection*

any other purpose except as firewood & that only could be gathered by women alone."[87] This uniform policy of restricting men from accessing fuelwood freely and employing labor-saving animal transport was particularly burdensome for communities farther away from forest areas and wood sources, where the laws often translated into women having to travel greater distances by foot, carry heavier headloads, or make more frequent trips to procure regular fuelwood needs. Soon after the new regulations were implemented in the Tsolo District in 1886, many residents submitted questions and complaints to the magistrate's office, wanting to know "[w]hether people may use sledges to carry their firewood from the forests where they have no waggons and it is too much labour for the women to carry it on their heads—free of charge."[88] A few years later, at a large public meeting with the resident magistrate, headmen and other male residents again voiced similar concerns and again to little avail: "We want wood. Women are allowed to get wood. We live far from the Forest."[89]

Despite such complaints, local residents increasingly felt the gendered constraints of colonial resource rights, as greater state interference in Kwa-Matiwane environments combined with major stresses to local economies and ecologies over the following decade. Official demarcation of local forests in the early 1890s and the subsequent confinement of free permit use to the

Figure 2.2. Bringing the harvest home in a sledge, c. 1954. *Photograph courtesy of University of Fort Hare Library, Piper Collection*

much smaller and less accessible undemarcated forests in 1896 effectively reduced the area available for men to utilize their permit rights and led to more intensive exploitation of and competition for the resources of these tariff-free areas. The arrival of rinderpest in the area in 1896 and 1897 added additional pressures. Scarce livestock meant men were much less able to transport fuelwood by wagon or sledge, whether accessed through free permits in undemarcated areas or through licenses in demarcated reserves, for domestic use or for sale to African neighbors. The disease's tremendous toll on local livestock further translated into a substantial loss in dung locally available for fuel.[90] Together, such strains on local livelihoods and resource availability increased local demand for cheap fuelwood sources and thus placed greater pressure on women to exploit more intensively their tariff-free teza rights in the relatively abundant demarcated forests. Given the headloads-only rule imposed on teza access, women were left to bear the physical brunt of securing scarce fuelwood sources themselves, without the use of more efficient and labor-saving forms of transport.[91]

This situation was compounded by additional strains in the early 1900s. Cycles of drought, poor harvests, and debt, particularly following the South African War, increasingly drove more local men away from their homesteads and into the migrant labor force in search of income, leaving rural women with greater economic burdens.[92] Additional forest demarcations in the early 1900s also further limited male access to undemarcated forests and intensified use of these dwindling resources, adding pressures on women and their teza access.[93] In 1905, officials then expanded their assault on male participation in fuel collection by dramatically raising the price of firewood wagonloads available for purchase at government reserves.[94] Even if certain families were capable of devoting resources to purchasing and carting fuelwood loads from demarcated forests, such conditions made it an increasingly difficult option to pursue. By the mid-1900s, then, socioeconomic and ecological constraints in KwaMatiwane communities had increasingly transformed fuelwood collection into *de facto* "women's work."

Foresters and magistrates alike were generally quite pleased by this development. Confining popular wood rights in gendered ways seemed to be successfully serving the twin colonial goals of tying women to the rural areas and pulling cash-hungry men into labor migrancy. Normalizing and ensuring women's default position as household "subsistence" providers were key components of this equation. For example, in the mid-1900s, when officials debated raising firewood tariffs in demarcated forests and their potential impact on rural Africans, SNA Walter Stanford reassured his colleagues with the following logic: "As it is only proposed to increase the tariff of firewood per wagon load, the ordinary native whose supply of fuel is obtained in bundles by the women will not be affected." Since the "ordinary native"—inevitably defined as male in official discourse—could always rely on women to collect fuel themselves, as was "customarily" done, imposing stiffer tariffs on firewood accessible to African men could not be considered preferential or unjust. After reducing African fuelwood rights to a limited and rigid definition of customary women's work, officials then relied on this interpretation to justify and normalize further gendered constraints on African wood access. Such logic undergirded official fuelwood policies for years to come, ignoring Africans' protests. For example, when members of the Engcobo District council proposed in the TTGC a few years later that the government broaden men's fuelwood access rights, the CMT and the SNA immediately rejected the idea and repeated the justification that even "the poor man" could always get "his women" to carry wood from the forest, "as he had always done."[95]

While people in KwaMatiwane contended with officials' clear commitment to such broader gendered changes in access rights, local women further struggled against the narrow interpretations of "custom" and "subsistence" imposed on their everyday wood-harvesting practices. In particular, women consistently contested colonial efforts to limit teza rights to only the supposedly "customary" collection of fallen wood by hand. Soon after assuming his post in the late 1880s, Conservator Henkel launched his own personal crusade to enforce teza restrictions in several of the region's mountain forests in such rigid terms. In May 1889, Henkel garnered local magistrates' support to prohibit African men's tree-felling in the highly prized Kambi Forest yet allow women to collect firewood by the headload there as long as they did not bring axes into the woods.[96] That same month, Henkel instructed the CM Tembuland that similar policies applied to Thembu districts: women were entitled to "headloads," "but they should not enter forests with axe or hatchet as they are in the habit of chopping pieces of wood off good blocks laying on the ground and also steal wattles, when provided with a hatchet."[97] By November, the magistrate at Engcobo had received numerous complaints from residents in one locale that a local forest guard, claiming to be following Henkel's instructions, regularly arrested many women "for entering forests with axes." Both the magistrate and local women were taken aback and confused by the action, "as the people have not been informed that any alteration has been made in the rights and privileges they have always enjoyed, and I am not aware of any regulation prohibiting the women from taking axes into the forests, the only places in the district where firewood can be obtained."[98]

There were quite practical reasons why women strove to maintain their rights to utilize axes when harvesting wood. An axe was often a necessity when dry, fallen pieces of wood were too large, too hard, or too difficult to bend and break by hand; chopping the wood into a workable headload was the only feasible way for women to transport the fuel home.[99] Women also had legitimate legal cause to be surprised by these prosecutions. Although Henkel strongly desired to keep African women from cutting green timber in the KwaMatiwane mountains, only incomplete regulations officially existed on paper. Even when the first proclamations reserving teza rights were enacted in 1890 and 1896, the laws merely prohibited using standing and living trees for firewood: nothing explicitly forbade women from carrying axes into forest areas and using them to cut dry wood or dead trees.[100] Nevertheless, unofficial interpretations of teza restrictions were routinely enforced in local forests. It was already a fairly common practice to confiscate axes from African men caught cutting reserved trees or carrying axes in a demarcated area without a legitimate license. Forest patrollers simply extended this practice to women found using or simply carrying axes in any forest, even when they were not breaking any official regulations and were merely using these tools to dig out a tree stump or cut dead wood.[101] In an 1899 court case before the Tsolo magistrate, for example, a local forest watcher explained that seizing axes from suspected female forest offenders was standard practice and the explicit directive of presiding forest officers.[102]

After a new regulation in 1903 finally expressly prohibited the use of axes for teza collection in demarcated forests, women in KwaMatiwane continued to struggle with a particularly difficult blend of official and unofficial policing of their livelihood practices.[103] In the Baziya forests, local forest staff and the Umtata magistrate aggressively enforced the new teza restrictions. Forester Thomas Adams praised the magistrate for his determination to strictly prosecute teza offenses as a means of sending a stern message to local communities about forest law enforcement. The magistrate's recent conviction of a group of women for cutting dry firewood with axes had produced the desired effect: the Baziya Forest "looks safe now," Adams wrote, because people in the area were "very frightened of the RM of Umtata," and "I never see a women [sic] come or go to the Forest with an ax."[104] During the next several months, Adams' own rigid enforcement of teza restrictions drove leading men in four locations neighboring the Baziya and Mbolompo forests to petition the magistrate and complain that "women are not allowed to enter the undemarcated forests or small patches of bush along the ravines with their small axes to cut some dry sticks with for fuel." Even according to the new teza rules of 1903, there were no restrictions on women using axes to cut dry wood in undemarcated forests. When Conservator Heywood investigated the matter, Adams reassured him that the complaint had no basis, even though there had been other occasions where guards had confused the regulations and unnecessarily seized axes from local women.[105] Such confusion, in fact, had led a group of thirty women in the Tsolo District to travel down to Umtata to personally lodge their complaint at the CMT's office. It seems that some forest personnel had interpreted the expanded teza constraints of 1903 as a green light to completely prohibit women's access to certain forests and vital sources of dry firewood, even when they were without axes. In the Assistant CMT's words, these women were thus left with "no means of gathering fuel."[106]

As local women grappled with such narrow interpretations of "customary" teza practices at the turn of the century, they also negotiated the confines of their "subsistence" wood rights. Just as forest officials felt that Africans should be limited to a quantifiable level of "necessary" wood use in their exercise of free permits, the new forest laws regarding teza rights from the late 1880s onward only allowed for Africans' "bona fide use" of forest resources for "subsistence" and not for commercial profit, since official interpretations of women's "customary" practices did not include selling harvested resources.[107] In trying to tie African women to the domestic sphere of the reserves, colonial authorities particularly desired to constrain teza rights so that women could not easily exploit their resource access for the accumulation of independent income.

Yet many KwaMatiwane women were not content with their resource entitlements being limited to imposed definitions of "subsistence." Although evidence for local women's economic activities in the early colonial period is relatively scant, sources suggest that women had actively traded the resources they personally gathered and most directly controlled—such as thatching grass and fuelwood—prior to the institution of colonial forest regulations, and that they strove to protect their commercial interests at the turn of the century. From the earliest days of European expansion into the region, newcomers commented on local women exchanging fuelwood for desired colonial commodities. For example, when two Moravian missionaries first explored upper Thembuland in the 1860s they described how women in the Baziya area actively traded fuelwood in exchange for tobacco.[108] In subsequent years, European travelers engaged in similar fuelwood barters with abaThembu in the region, and women presumably continued to take part in such exchanges.[109] As European settlement increased in the region in the 1870s and 1880s, women were able to expand their economic networks and find lucrative markets for the sale of these commodities, particularly local European trading shops. Recalling his earliest days in upper Thembuland in the late 1870s, CM Elliot noted how he had been appalled to see, even in the less forested flatlands, women clearing "every tree" and "digging out the roots of the trees and selling them to the traders as fuel."[110] In the late 1880s, Conservator Henkel complained about the very same situation in the Umtata District.[111]

From the late 1880s onward, officials made concerted efforts—through teza regulations and their enforcement—to limit such activities, yet many KwaMatiwane women continued to pursue opportunities to sell and barter wood, even given the risk of prosecution and fines. Women were regularly brought before magistrates for firewood-cutting offenses in the 1890s and early 1900s, often caught on their way to hawk wood at markets in nearby European towns. Conspicuously selling wood at public markets became increasingly difficult and impractical, however, for women would likely be inspected, found to be without a sales license, and fined. Facing such restrictions, some women continued to sell and exchange wood at traders' shops in various locales, a more convenient practice that could more easily avoid official detection.[112] In 1908, for instance, one European trader in the Tsolo District described how women had been his sole source of fuelwood, as he pleaded with the magistrate to ease new forest restrictions: "[Headman] Goniwe tells me that the Ntywenka Forests are closed to the girls getting dry wood out, to sell. I have been depending on them bringing me wood & have not a bit of wood on the place. [C]ould I get a licence for the girls to carry me a load on their heads[?]"[113]

These intermittent insights into women's wood trading suggest that many local women were not satisfied with wasting economic opportunity in order to comply with official definitions of their legitimate "subsistence" practices. Nor were they content with having their power to choose how they harvested wood stifled by officials' selective reading of "customary" practices. Local residents might not have been able to alter the overall path and gendered orientation of fuelwood policies affecting them in these years, but women in everyday contexts regularly pushed the imposed limits of their resource entitlements.[114]

⌐

The initial colonial restructuring of popular resource rights from the late 1880s to the early 1900s was thus very much shaped by multiple layers of negotiation and adaptation among and between Africans and officials. There was certainly nothing automatic or inevitable about the process or the final product. Popular discontent with the government's stripping of Africans' forest entitlements across the Transkei helped modify the nature of colonial policies at a time when officials themselves debated the appropriate course to follow. Divergent official perspectives on the propriety of forest restrictions in turn muddied enforcement on the ground, a situation many men and women happily exploited through their own evasive strategies. The free permit system and teza access were both reshaped by local people's daily responses to their fluid situations in a rapidly changing colonial landscape.

All of these examples further reveal how changes in and negotiations over resource rights were about much more than just environmental access. Foresters and magistrates inserted into these negotiations their concerns with popular instability and their relative claims to bureaucratic power. By asserting alternative perspectives on the meaning of forest reservation and free permit entitlements, many people in the Transkei brought into the equation questions about the broader livelihood benefits the government owed to taxpaying African subjects in the new colonial order. And in the negotiations over fuelwood access, both officials and African men and women recognized the deeper transformations in rural gender relations and livelihoods implicated in restrictive teza policies. Over the course of the early twentieth century, as colonial resource control intensified and built upon its early foundations, Africans would continue to respond to changes in forest access from their particular locations in a wider colonial landscape.

3 ↜ Shifting Terrains of Wood Access in the Early Twentieth Century, 1903–1930s

BEGINNING IN THE 1900S, particular changes in the scope and nature of colonial forest control more extensively shaped Africans' experiences of wood access across the Transkei. Finally enjoying a more secure position in the colonial bureaucracy, the Forest Department garnered stronger administrative support to exert greater pressure on Africans to reduce their reliance on indigenous forests and instead purchase their wood needs from the government, primarily through the slowly expanding fleet of state tree plantations being established across the Transkei. A central part of this strategy was the expanded absorption of forests into the state's domain from the early 1900s to the early 1910s, effectively reducing the territorial extent in which African men in particular could exercise their wood access rights.

Such developments grated against the historical experiences and environmental perspectives of differently positioned rural men and women. With dwindling access to trees for building materials, more and more men indeed turned to purchasing wood from state plantations in the first few decades of the twentieth century, but often only after exhausting less costly options. As they increasingly transferred scarce financial resources to the state amid growing problems of poverty, many people also recognized and resented the colonial domination of their environmental rights and livelihoods that these exchanges represented. In KwaMatiwane people not only criticized the government's general control over the distribution of wood, but further lamented the quality and nature of the tree species being reserved or "made available" to them. Even as they grappled with particular changes in the availability of indigenous and exotic species, men and women continued to assert their own sense of the economic value of different tree types.

Women's experiences and negotiation of shifts in wood access particularly centered on the forest resource most crucial in their livelihoods: fuelwood. Across the Transkei, growing numbers of women responded to their mounting burdens of resource scarcity and male out-migration by increasingly purchasing fuelwood from state plantations in the 1910s and 1920s. But women also held a dubious distinction in these years, following the colonial government's consolidation of forest control: they possessed the last remaining popular rights to access wood in state demarcated forests, through teza. As they increasingly capitalized on this access in the late 1920s and early 1930s, especially in response to economic depression and scarce financial resources, their adaptations clashed with shifting state preoccupations. Almost in proportion to women's growing labor pressures and problems of fuelwood scarcity, officials more consistently and aggressively targeted women's environmental practices in a broader campaign to combat the "crisis" of environmental decline in the "reserves." In KwaMatiwane residents adapted their environmental behavior in turn, manipulating the resources they harvested and reorganizing the labor of teza. As in their other negotiation strategies in the early twentieth century, even if African men and women could not reverse their broader and deepening subordination in the colonial political ecology, they found ways to reshape the particular meaning of shifts in wood access in their everyday lives.

"WEANING THE NATIVE"

The rising administrative status of the Forest Department in the mid-1900s emboldened forest officials to expand their efforts to restrict popular wood rights over the next decade. Following the department's consolidated control over the majority of forest areas after 1903, the creation of an independent forestry bureaucracy and the permanent appointment of its first chief conservator in 1906 further ensconced forest conservation more securely in the Cape colonial administration.[1] Exploiting this supportive environment, forest officials more vigorously campaigned among Transkei authorities to increase the extraction of forest use payments from African subjects. One product of this campaign was the expansion of government tree plantations in the 1900s, originally begun on a small scale in the late 1890s, whereby Africans could purchase exotic species of wood for their everyday needs and thus "spare" the indigenous forests. Interwoven with this initiative was a more comprehensive effort, backed by a growing number of magistrates, to reduce the territorial extent of Africans' "free" wood rights and thus force Africans to pay for wood. By the late 1900s and early 1910s a conducive administrative environment enabled forest officials to finally end the free permit system and successfully limit the majority of popular forest rights to a progressively decreasing base of undemarcated "headmen's forests" in the Territories.

By the turn of the century, forest officials increasingly supported a strategy to develop state-run plantations to meet Africans' wood needs and thereby, as they often repeated, "wean the native" from the indigenous forests.[2] Transkeian foresters had experimented with planting quick-growing acacia and eucalyptus species at limited sites in the early 1890s, and by the late 1890s a handful of plantations were operational at certain key towns and in the Fingoland districts, where new district councils were involved in their management. Following A. W. Heywood's appointment as conservator in 1898, the department more aggressively pursued plantation work in many areas of the Transkei.[3]

The condescending notion of "weaning" Africans from forests aptly conveyed foresters' perspectives on popular wood use: Africans were immaturely addicted to consuming indigenous trees and could not control themselves; they therefore required the foresight and guidance of forestry experts to regulate their forest use, provide for their wants, and see to their long-term environmental interests. Embedded in this discourse were two other important themes, both with far-reaching implications: Africans' interests in wood were simple and indiscriminate, so satisfying Africans' wood needs was merely a question of supply; and the government was justified in charging Africans for wood. Africans simply needed wood, regardless of the source or type. If their wood demands could be supplied from government tree plantations, then the masses had no legitimate need to access the economically valuable timber species in indigenous forests.[4] And, foresters argued, once one accepted that the government should not only take over supplying wood to Africans but also charge them for these resources, there was really no reason to continue free wood access through existing popular "privileges."

The Forest Department's consolidation of forest control in the early 1900s began to put such ideas into practice. By expanding the domain of demarcated government reserves in the Territories by some 62,000 acres, the department effectively limited the extent and quality of remaining undemarcated forests Africans could access with free permits.[5] With this expanded sphere of control and a plantation system well under way, forest officials then more confidently and vociferously worked to convince Native Affairs authorities that Africans should pay the government for access to everyday wood needs.[6] Leading foresters in the Eastern Cape and Transkei expanded their assaults on free permit access, rabidly denouncing both Africans' uncontrolled "annihilation" of indigenous forests and the severe financial toll made possible by the permit system. In early 1905, both Heywood and Eastern Cape Conservator Joseph Lister proposed legislation to abolish free permits altogether and instead charge Africans for cutting indigenous trees.[7] In the Eastern Cape districts, local magistrates responded to the proposal with grave apprehension. Given the already bad reputation of the Forest Department and its regulations in many communities, magistrates sensed the legislation posed serious threats to the local maintenance of social control, and SNA Stanford suspended the plan for the Eastern Cape.[8]

In the Transkei, despite signs of popular discontent in several districts, official resistance to the proposal was much more muted. Some magistrates wholly endorsed the tariff program, while several others were forced to contend with rural Africans' public complaints about the burdens of existing forest regulations and their fears of proposed tariff changes.[9] Although cognizant of these ill feelings and despite the misgivings of some local officials, high-ranking Native Affairs authorities forged ahead with the plan. Once an outspoken critic of such intrusions into Africans' "birthright," SNA Stanford now argued the move was necessary given the expansion of local populations and their "ruinous" impact on indigenous forests.[10] With the blessing of the secretary of agriculture and other officials in his department, Proclamation 43 was enacted on February 1, 1906. Just a week later, however, alarming news arrived from neighboring Natal: an armed group of Zulu men had killed two white police officers in the first of a series of popular disturbances surrounding the imposition of a new poll tax.[11] The Natal crisis sent shock waves through the Cape colonial administration. Fears of antiwhite and anticolonial sentiment spreading to the Transkei communities abounded in official circles, and colonial authorities immediately toned down further state interventions until the situation stabilized. The planned introduction of wood tariffs came to a grinding halt. By the middle of June, Cape authorities opted to revoke the proclamation altogether, a decision many Africans greeted "with loud acclamation," according to one local magistrate.[12]

Despite the ultimate failure of the tariff proposal, the episode reveals an important shift in the balance of colonial priorities. Although some magistrates invoked the issue of popular rights as they conveyed African communities' and their own discontent with forest policy changes, the defense of "native interests" had much less influence on the trajectory of government decision making than the specter of instability of colonial control in African locations.[13] Whereas a decade earlier leading figures in the Transkeian administration had argued aggressively for the protection of "indigenous" rights in their bouts with Forest and Agriculture officials, by 1906 high-ranking Native

Affairs authorities increasingly shared common interests with their colleagues and viewed resource conservation and restrictions on popular forest access as important parts of their management of the Transkei.

Following the botched plan to end permits in 1906, forest officials took advantage of this growing administrative support to push more aggressively for the end of Africans' permit rights. Foresters' condemnation and "documentation" of African environmental wastefulness intensified in their published literature. The call for greater control over African wood use was perhaps nowhere more emphatic than in one widely circulated photograph of two Mpondo men constructing a hut made out of saplings (see figure 3.1), appearing in a 1907 forestry annual report but distributed much more broadly through the publication that same year of Thomas Sim's *The Forests and Forest Flora of the Colony of the Cape of Good Hope* in Aberdeen, Scotland.[14] In both texts the representation of Africans' "primitive" and "uneconomical" building practices were underlined by a tragic caption: "This is what becomes of millions of young trees annually." The photograph's presence in Sim's book—a comprehensive guide to forestry and forest environments in the Cape—is especially revealing. Placing this full-page picture in a volume with very few photographs, the author showcased a visual metaphor of the obstacles facing scientific forestry's progress in African communities across the Kei. The photograph was also particularly placed at the beginning of a chapter entitled "Economic Composition of the Forests," serving as its foil. As Sim painstakingly described the economic uses of various trees in the region, he was sure to spotlight Africans as backward squanderers of valuable forest resources. Such resources could be more productively exploited in the Cape settler economy, the text and image suggested, if only "native rights" to wood were more severely curtailed.

Forest officials also increasingly deployed statistical data to argue for such constraints. Much as state planners in the 1940s and 1950s would employ notions of "livestock units" and "carrying capacity" to calculate both Africans' resource needs and the "optimal" management of local grazing lands, foresters in the late 1900s and early 1910s repeatedly asserted that Africans should only be entitled to quantifiable "subsistence" levels of wood use.[15] By narrowly defining popular wood needs in numerical terms, officials could then justify finite limits on Africans' "legitimate" claims to state-controlled forest resources.[16] Africans not only "abused" the environment but they did so "extravagantly," unnecessarily wasting wood supplies which could serve both the colonial economy and the future needs of Africans themselves.[17] In one 1907 report Chief Conservator Lister even claimed that Africans' "actual" material demands for wood comprised a mere 15 percent of what they routinely harvested

Figure 3.1. Mpondo hut-building, c. 1907. Source: Thomas R. Sim, *The Forests and Forest Flora of the Cape Colony of Good Hope* (Aberdeen: Taylor and Henderson, 1907)

from local forests and that charging for forest produce would favorably restrict the masses to these true levels of need.[18] Added to such rationales was a by now familiar refrain in state forestry discourse: Africans' wasting of wood was unnecessarily robbing the government of revenue. In 1909, for instance, Lister argued the case in terms colonial administrators could easily digest, quantifying the "Value of known removals of Forest Produce by Natives under Free Permits and which should *really* appear under Revenue Returns."[19]

Over the course of the late 1900s and early 1910s such arguments, data, and imagery swayed the minds of a growing number of colonial administrators. Foresters successfully lobbied to redraw and expand demarcated forest areas in the late 1900s and early 1910s, absorbing thousands of acres of undemarcated tracts and bringing significant changes to popular wood access. Once officials inspected location forests and determined which additional areas should be brought under state control, the permit system became locally obsolete. Outside of limited teza privileges, all other African wood rights were now relegated to the smaller, remaining headmen's forests.[20]

This change reflected in microcosm South African state officials' broader campaign in the years following Union to circumscribe African resource rights territorially and consolidate white settler control over land and other resources, manifested most noticeably in the 1910 Union, the 1913 Natives Land Act, and the first Union-wide Forest Act in 1913. This association between state forest consolidation and wider patterns of settler domination in these years was not lost on Africans experiencing such changes. As one informant commented on the period, "The land was taken by the whites. Also the forests were part and parcel of the land, so it was taken by the whites. After the establishment of the four provinces in 1910 . . . the whole land was taken, including the forests. That was 87 percent for the whites and the blacks were left with only 13 percent of their land."[21] Such connections were reinforced in official practice. Just as the Natives Land Act legally confined African populations to racialized reserves, Transkeian magistrates and foresters instituted a territorial division of forest areas that "indigenized" and thereby limited popular resource rights. By the early 1910s, Africans were now restricted to using "native" resources to meet the bulk of their "subsistence" needs under the aegis of "native" authority in "native" areas. This articulation of indigenous rights thus naturalized and rationalized the deeper structural inequality of state resource control.

By roughly 1915, then, forest officials in the Transkei finally were able to institute a system that placed the majority of forest control and management in state hands. While not as restrictive as some might have liked, the expansion of demarcated holdings and the confinement of popular forest claims served as a workable compromise of foresters' and magistrates' interests. Reconstituting popular wood rights would reduce the extent of Africans' "drain" on forest resources, officials hoped, and further pressure African households to purchase their wood needs from government sources and to send off more of their men to provide for their growing financial obligations. For rural men and women, however, such developments would prove to be much more problematic.

POPULAR PERSPECTIVES ON THE COSTS OF "WEANING"

Colonial forestry policies during this period generated contradictions and difficulties in many people's lives that belie the simple optimism of official reports. The Forest Department's strategy of territorially constraining popular forest rights particularly pushed more and more men to purchase their wood needs from state-run plantations in a period of growing economic impoverishment and ecological scarcity. Despite this overall trend, many men turned to buying wood from the government only after first pursuing other available tariff-free options—but as they increasingly paid for wood in the early twentieth century, many people critically viewed such exchanges as signs of their diminishing control over their environments and livelihoods.

Foresters' dreams of "weaning" Africans through the supply of exotic plantation wood got off to a slow and rocky start. In the early years of colonial plantations, supplies were regularly inadequate and unpredictable in many areas. Officials' lack of familiarity with local soils and climates and their choice of inappropriate exotic species for cultivation often contributed to poor plantation growth and yields throughout the late 1890s and 1900s, in turn affecting the amount and quality of wood available to African buyers.[22] A more serious problem was the insufficient number of plantation trees and sites to serve the needs of the massive population of the Territories. A 1906 forest report reveals the significant gap between the area of the few dozen sites the colonial state had claimed and set aside for plantation development over the previous decade and the acreage of existing "actual plantation." Of some nearly 11,500 acres of land reserved for plantations in the Transkei, only about 3,300 acres contained trees in reality. As this latter figure included areas where tree seeds and seedlings had only recently been planted, the percentage of plantations with actual trees available for sale to Africans was considerably smaller.[23] Throughout these years, and continuing into the early 1910s, foresters occasionally reported that such limitations undermined the capacity of plantations to meet African men's wood demands in many localities.[24]

These problems exacerbated many Africans' experience of the colonial remapping of indigenous forest access and its drastic impact on wood accessibility. In Kentani District in the mid-1900s, for instance, numerous African leaders complained to officials that demarcation was making hut- and kraal-building materials incredibly scarce. As one local spokesman summarized the problem in 1905, "We have no forests now of our own which we can use as we like."[25] As men and women leveled similar criticisms in other districts, they also explicitly cited growing wood scarcity as the net result of colonial forestry strategies.[26] In 1906 the Tabankulu magistrate described a major source of popular animosity toward recent forest reservations in the district: "This demarcation of forests took nearly every available bush away from the Natives, and as the Forest Department has no advanced wattle plantations from which to sell poles and wattles, the Natives have absolutely no means of providing for their legitimate wants."[27]

Such sporadic criticisms of imposed wood scarcity expanded across the Transkei by the early 1910s, as growing demographic and ecological pressures intensified the impact of state wood management strategies on local livelihoods. Colonial census figures, incomplete and limited as they are, suggest an overall surge in African populations over the period. From just 1891 to 1911, the total number of Africans in the Territories grew from about 640,000 to 872,000, roughly a 36 percent increase.[28] Local populations were actually much larger than these and later official estimates, since censuses recorded only people who were physically present in the various districts on census days and not the growing numbers of men away on migrant labor contracts.[29] It has been estimated, for instance, that in 1912 nearly 97,000 Africans, overwhelmingly male, left the Territories in search of work[30]—men who would still require wood for their rural homes and kraals even if not physically present throughout the year. While demographic patterns varied across the region, such numbers do suggest more generally how growing populations heightened demand for wood in these years.

As a symptom of these mounting pressures, African political leaders regularly submitted complaints to officials from the 1910s onward requesting the government to open up more forest areas to undemarcated status and release additional species, commonly used as building materials, from restrictive reserved lists.[31] Members of the Pondoland General Council in 1916 were particularly expressive in their frustration and bewilderment at authorities' lack of concern for Africans' resource rights and the growing problem of wood scarcity. When Pondoland forests were first demarcated, Chief Mangala of the Libode District explained, "all forests were taken away from the Natives, not a stick was left." The recent establishment of headmen's forests, ostensibly "returning" some forest areas to African control, were "only sops. What actually happened was that huge portions of the forest areas went to Government, but only very small portions went to the Native people, who wondered at the way their father treated them."[32] Councillor Nongauza identified the imposed problems of wood access more precisely: "Those of his own generation remembered that when they were born they had the forests and fuel for their fires and wood to build their cattle kraals. Since then their forests had been taken away from them. It was very clear that soon the Native would have no place to even sit down on. . . . No people could live without forests. Wood was required to keep out the cold in winter, to cook food and to build cattle kraals. Today there was not sufficient wood for any of these purposes."[33]

It is important to stress here that such problems of wood availability were not automatic or experienced in uniform ways across the Transkei. In some areas people ameliorated constraints on forest access through the production and use of wood from their own or others' homestead woodlots. The most widespread, long-standing woodlot tradition exists in areas of Pondoland, where climate and soil conditions have been most favorable for the rapid growth of planted trees for domestic consumption. From at least the 1890s, many homesteads added planting exotic wattle species for the production of fuel and building materials to the practice of transplanting indigenous trees around or near their residential sites, combining wood production with multiple other tree benefits.[34] Increasingly over the early twentieth century, some African men in other areas also created their own home woodlots, taking advantage of wattle seeds provided by the government to tax- and council rate-payers or directly harvesting and sowing wattle seeds themselves.[35]

The impact of such domestic tree production on the broader picture of wood access in the Transkei was highly uneven and slow in development, however. While woodlots in some areas of Pondoland could supply a good deal of their homesteads' domestic wood needs by the 1930s, woodlots in the rest of the Territories were still insignificant overall in supplying wood to most people even well into the following decade.[36] Moreover, such woodlots did not always serve men and women's wood needs equally. In KwaMatiwane, for instance, many home plantations were oriented more toward male wood priorities and building activities than women's fuel needs.[37] Looking back at the situation in the 1900s and 1910s, when colonial "weaning" strategies were initiated, the impact of domestic woodlots across the Transkei was also markedly different. The noted preponderance of Africans' complaints about wood scarcity during this earlier period, even across Pondoland, suggests that, on the whole, people through much of the Transkei indeed felt pressures from the colonial state's expansion of both forest reservation and constraints on popular wood rights. The growing number of Africans purchasing their basic wood needs at government plantations in the early twentieth century further indicates a more general response to localized problems of wood scarcity.

Data on African men's wood-buying, for instance, reveals the increasing importance of wood from state-run plantations in many Africans' livelihoods. Table 3.1 specifically tracks the monetary value of men's hut wattle purchases from government plantations through the 1920s, one of the most complete and consistent data sets on African wood-buying during this period. In the period up through the mid-1910s, with wattle prices remaining relatively stable, wood purchases increased steadily. Whereas men collectively spent only about £16 at the turn of the century, by the beginning of 1917 the annual total reached nearly £580. To put such figures in perspective, government plantations reported selling only about 600 bundles of wattles in 1900, with each bundle comprising about 25 to 30 saplings, or a total of 15,000 to 18,000 trees.[38] Based on the available information, it can be roughly estimated that by 1917 African men bought somewhere between 865,000 and 1,039,000 wattles annually from government stations, a substantial increase from the early 1900s.[39]

Men's purchasing of hut-building material also noticeably accelerated at key moments during this period. In the late 1890s, many men exploited their

free permit rights to available wood in local undemarcated forests and avoided buying wood from the government.[40] In the early and mid-1900s, in the wake of new forest demarcations and the consolidation of Forest Department control, wattle purchases increased substantially, as fewer forest areas were then open to free permit access. When undemarcated forests were first handed over to district councils' control in 1908, wattle-buying declined; between 1910 and 1911, when these areas were returned to Forest Department control, such purchases rose immediately and dramatically. While wattle-purchasing temporarily declined in the early 1910s, as drought and East Coast Fever debilitated men's economic resources and thus their ability to pay for wood,[41] plantation sales rose to high levels during the rest of the 1910s and throughout the 1920s.[42]

It appears, then, that African men exploited tariff-free access to undemarcated wood sources whenever they could, resorting to plantations only when such options dwindled, particularly as forest demarcations expanded and the free permit system was terminated. Officials regularly noted that Africans seemed particularly reluctant to spend scarce financial resources on wood, especially during periods of economic distress, and viewed wood-buying only as a last resort, when indigenous wood supplies were "absolutely unobtainable" in particular locales.[43] In fact, some plantation managers in the early years offered temporary incentives to "wean" African men and interest them in purchasing their hut wood, such as thinning out overstocked wattle crops and selling them at reduced rates, especially during the hut-building season.[44]

Evidence further suggests that a growing percentage of men were frustrated by their increasing payments to the government for regular wood needs.[45] Many Africans in the Transkei experienced greater pressures to spend scarce money on wood at a time of intensifying economic stress. African households' deepening impoverishment and mounting need for cash income were reflected in the swelling numbers of men entering the migrant labor force by the 1920s. In 1912, 70,000 more men left the Transkei in search of work than just twenty years earlier.[46] William Beinart has estimated that the percentage of male absentees from the total Transkeian population rose from about 13.4 to 17 percent between 1896 and 1921, with some districts, particularly in Pondoland, experiencing an increase as sharp as 11 percent during the same period.[47]

While people coped with the indebtedness and poverty driving such changes, they also contended with the state's dominance over indigenous wood sources and the inflated prices government plantations charged by the 1920s. Officials themselves often recognized the strain plantation prices placed on the majority of African households. In 1922, the Engcobo magistrate urged that the government reduce its wood tariffs for very pragmatic reasons:

"The Native must have building material, and if we place it beyond his reach, he will steal to obtain it. Every wattle pole sold has the effect of saving a yellow wood."[48] A few years later, R. D. Barry, RM Mt. Frere District, acknowledged that "prices for plantation produce in these Territories were higher than they were almost anywhere else . . . and the result was that the Natives, especially those in poor circumstances, were almost compelled to steal the produce of the indigenous forests."[49] This problem was also taken up by African leaders in the Transkeian Territories General Council. In a 1926 session, Councillor Sakwe from the Idutywa District placed wood prices in the historical context of many men's daily lives: "Before they could put up a roof they had to pay £1, whereas in former days they used to get poles from Manubi and other forests. They had not to pay anything then, but that was stopped now. . . . [I]t would appear that the plantations were a money-making business."[50]

Sakwe's comments reflected a growing resentment in these years against not only the deepening financial burden of plantation purchases but also the very principle of paying the state for wood access. Such sentiments colored the memories of elder informants in KwaMatiwane, who similarly invoked the notion that colonial control of essential wood supplies subverted the tree entitlements to which people were previously accustomed. Informants particularly stressed the absence of economic sanctions for harvesting wood before the government took over indigenous forests and created plantations. Samuel Qina asserted that prior to colonial forestry people could simply access wood "without paying anything, because there were no foresters those days."[51] According to Anderson Joyi, "There was no buying and selling . . . before the arrival of the white man."[52] "We were using firewood from the forest, we were not paying for them," Tsikitsiki Nodwayi added. "A woman would simply go and fetch a bundle and put it on her head."[53] Various informants then moved from such statements to portray the government business of owning and selling wood as inherently "unnatural" and illegitimate. For Samuel Qina it was this concern, rather than the plantation prices themselves, which deeply troubled residents in the Tabase area: "Although, I may say, those days they [government plantation wattles] were cheap, they were cheaper than today, but they [people] were reluctant to pay, because they said 'no, these trees were created by God, why should we buy them?'"[54] Another explained, "[W]hen the government came, the white government came, they took the best forests for their own use, and they used to sell everything there. If I want to get some wood there, then I'll pay."[55] Wele Boyana even more emphatically depicted the ulterior motives of colonial forest schemes: "The forest was ours. It was ours. This forest was disturbed by the whites, who planted the gum tree, the pine tree, the wattle tree, and the whites took the forest from us. You [referring to me] have taken it, you have taken it. . . . The whites came and planted the trees, so that we can buy the trees from them."[56]

Such bitter reflections on the colonial past cannot be interpreted uncritically or devoid of context.[57] For many local men and women interviewed in the late 1990s, paying the state for regular household wood supplies had long been a fact of life, particularly for those individuals physically removed from areas officially designated as plantation sites during the apartheid era. These traumatic experiences and, in some cases, related and ongoing plantation land-claim disputes, understandably shape people's perceptions and representations of the long-term meaning of state wood ownership in their lives.[58] Moreover, informants' criticism of selling wood and extracting revenue from popular tree use was focused exclusively on the government, even though some of these same individuals either had lived or continued to live in homesteads with small woodlots, or had at some point in their lives paid for access to wood from other such homesteads.

Yet there are also continuities, between the critical statements of Africans in the 1920s and these memories of a much later generation, which enhance our understanding of the widespread frustrations with colonial "weaning" strategies that were developing in the colonial era. Even as men in KwaMatiwane and across the Transkei increasingly turned to plantation wood for building materials in the early twentieth century, this shift was neither automatic nor uncontested. While people negotiated their limited options for securing wood supplies, many also resented and questioned the government's restructuring of wood control and access and its commitment to "money-making" at Africans' apparent expense. For a growing number of people, it seems, their reliance on plantations also represented their increasingly subordinate position in the broader colonial political ecology.

STRUGGLES OVER TREES' ECONOMIC VALUE

Every forest where there was sneezewood was turned into a demarcated forest. Why was their father, the Government not sympathetic. Surely the Native should be valued higher than sneezewood. Had the Natives not some rights in their own country.

—*Councillor Nongauza (1916)*[59]

[People] cursed the government for not allowing them to get their indigenous forest trees. They cursed the government for that.

—*Samuel Qina (1998)*[60]

In addition to broader resentment of the restructuring of wood access, many people more specifically criticized government constraints on accessing particular tree species. Efforts to "wean the native" were fundamentally founded on colonial evaluations of natural resources and rural African needs. As long as African subsistence requirements were met, colonial officials argued, the state was justified in redistributing and limiting access to wood as it saw fit. Officials thus routinely reserved what they viewed as the most valuable and sound stands of indigenous timber, leaving the remainder for the exercise of African rights.[61] As one magistrate asserted in 1912, if higher quality timber species and forests were regularly opened to Africans they would not be put to their "best use."[62] Moreover, the quality of wood left "available" for popular use was relatively poor. Colonial authorities commonly deemed lower "scrub" or "rugged and inaccessible parts of the district" to be suitable for "native requirements"—or even "best adapted to the primitive uses of the Native"—and thus designated as less regulated, undemarcated areas.[63] Some officials even asserted that it was unnecessary to make the most desirable timber species in the Transkei readily available to Africans, since these trees were "of no value in the household economy of the native."[64]

Despite their differences, foresters and magistrates thus collectively supported a fundamental principle: Africans could make do with whatever type of wood the government chose to make available. Yet Africans were not so easily "weaned" from their environmental preferences. When people formally protested the government's distribution of wood, they also specifically criticized the quality of available wood supplies and the restrictions imposed on access to certain species. In pursuing their livelihoods and negotiating colonial constraints, men and women also often exploited forest resources quite selectively and asserted competing evaluations of the worth of both indigenous and exotic tree types.

Throughout the formative years of colonial forestry, officials regularly reported on African men publicly complaining about, or more quietly circumventing, restrictions on certain reserved species which people preferred for making huts, kraal fences, hoe handles, yokes, sledges, and many other implements.[65] Facing this reality, officials often took rather elaborate steps to contend with tree preferences in particular locales. In KwaMatiwane, for instance, officials and local men in the early 1900s competed for dwindling supplies of *umthathi* (sneezewood, *Ptaeroxylon obliquum*).[66] Due to umthathi's durability and strength as fencing poles, telegraph posts, construction materials, and railway sleepers, officials had regularly restricted its cutting in both demarcated and undemarcated forests across the Territories and parts of the Eastern Cape.[67] Men in KwaMatiwane likewise valued the species' unique qualities, finding it well-suited for manufacturing hoe and spade handles or hut and kraal poles. One informant in the Tabase area recalled how desirable umthathi was for many farmers in the early 1900s: "A tree that would never rot. Harder than iron poles. . . . My old home, . . . we had poles of sneeze-wood that were there when I was . . . married already, but they are still there. Never rots, never rots that tree, never rots."[68] Recognizing the tremendous popularity of the species, some foresters even attempted to hide umthathi saplings in denser forest areas when planting them, rather than place them out in the open where local residents would more easily detect and harvest them for construction materials.[69]

More protracted struggles developed over preferences for kraal fencing materials in local mountain forests. Officials in KwaMatiwane and the Trans-kei as a whole had long attempted to convince African men, even through special tariff rates and permit restrictions, to forsake the "uncivilized" and "de-structive" use of indigenous wood for kraal-making and instead utilize sod or stone. Such constraints intensified in the early 1900s, despite the scarcity of sod or stone in many locales and some men's complaints that "sods are of no use for cattle kraals."[70] In KwaMatiwane a particular development in local for-est environments further influenced kraalwood negotiations: the growing dominance of the lemonwood tree (*umvete* or *uvete, Xymalos monospora*).[71] By many colonial accounts, lemonwood was a "worthless" species—it had lit-tle value as timber for most industrial and agricultural purposes, "being gen-erally crooked and ill-grown," did not make good fuel, and soon rotted when used for fencing posts.[72] The tree's local dominance in the early 1900s thus ob-structed one of the goals of scientific forestry: to reserve the basic elements of forest growth—sun, soil nutrients, and water—for the maximum yield of com-mercially desirable species.[73] Conservator Heywood clearly depicted the per-ceived lemonwood crisis in 1903: "With increased growing space Lemonwood crowns have spread out, and bid fair to monopolise an undue proportion of certain forest areas in the Tsolo district. . . . Regeneration of good species or indeed of any species of timber under Lemonwood seems to be quite impos-sible, and nearly all Yellowwood and Sneezewood having been removed years ago, large areas at Nqadu have been entirely taken possession of by the com-paratively useless Lemonwood."[74]

In devising a remedy, Heywood strove to combine the department's in-terest in maximizing the economic output of local forests with the existing framework of forest regulation. Foresters in the Cape and Transkei regularly attempted to hinder the growth and thin out "inferior" or "rubbish" species, often invoking images of Darwinian battles being waged and won through ag-gressive forest management.[75] To combat lemonwood Heywood took this

Figure 3.2. Man building a sledge with saplings, c. 1930s. Photograph courtesy of Cape Town Archives Repository, Jeffreys Collection

strategy one step further. Allowing and even encouraging Africans in the region to harvest this "inferior" tree would enable the "best species" to dominate local forests.[76] Officials were so excited by this proposition, in fact, that they envisioned an opportunity here to ameliorate the kraalwood "problem" as well. To kill both birds it was first necessary to reinvent the value of lemonwood in official discourse. Worthless in all past colonial assessments, lemonwood now became "eminently suited" for African wood needs, particularly for kraal-building.[77] Beginning in 1904, the Forest Department took concrete measures not only to facilitate but to *ensure* African men's exploitation of lemonwood for kraal fences. Aiming particularly at KwaMatiwane and other inland mountain forests, Heywood directed foresters across the Transkei to restrict kraalwood licenses issued to African men solely to wagon- and sledge-loads of lemonwood. Despite years of concerted official efforts to undermine Africans' kraalwood use, now, when kraalwood harvesting could conveniently serve other state forestry interests, the practice was encouraged within the confines of new species restrictions.

Many men in KwaMatiwane immediately criticized these impositions on species access. In a petition to their magistrate in June 1904, spokesmen from four locations abutting mountain forests in the Umtata District complained about the new regulations, including the order "that no other kind of kraal-wood [is] allowed besides the wild lemon." Conservator Heywood immediately rejected their plea.[78] Over the next few years, such official determination to harness and channel African men's demand for kraalwood continued to drive forest management in KwaMatiwane and beyond. In 1906, Chief Conservator Lister reiterated the urgent need to get Africans to utilize lemonwood for their needs, particularly in the inland mountain forests, to ensure "the healthy development" of desired timber species. As the presence of lemonwood increasingly took on pathological proportions in colonial discourse, officials more resolutely ignored questions of species preferences in local communities.[79]

In KwaMatiwane officials were never very successful in controlling either lemonwood growth or kraalwood use over the long term. Evidence suggests that many African men in the region circumvented the new licensing restrictions and, more significantly, continued to harvest and use trees other than lemonwood for kraalwood.[80] Many individuals purposely avoided the government's forced choice of lemonwood and asserted their alternative species preferences for kraal-making and other purposes. When asked which tree species were preferred most by men and women for construction materials and firewood in the 1910s, 1920s, and 1930s, not one person I interviewed included lemonwood on their list. Moreover, it seems likely that many men in the area, when looking for kraal-making materials and when able to choose between and access different species, avoided lemonwood for many of the same reasons that officials did: it was not as strong, durable, or easily workable as other tree types. Although dwindling supplies of kraalwood in local headmen's forests during these decades drove many people to abandon species preferences in favor of whatever types of wood were most accessible and least burdensome to procure free, it also appears that many still strove to balance these priorities for years to come. Throughout the 1920s and 1930s foresters continued to complain about the preponderance of lemonwood "and other less valuable trees crowding the better ones" in local mountain forests.[81] The persistence of lemonwood, despite its tariff-free accessibility and amid general problems of wood scarcity, suggests that many men continued to harvest what they viewed as more valuable species instead, complicating official environmental strategies.

Similar challenges beleaguered colonial plans to harness popular demand for exotic plantation trees. In the early years of plantations, many African men in the Eastern Cape and Transkei rejected the black wattle (*Acacia mearnsii*)—the tree planted most extensively during this period—as unsuitable for construction purposes, due to its susceptibility to different insect pests. As one conservator reported, "Bark beetles attack the Wattles after they have been cut and built into a hut and do considerable damage. This the Natives know and hence their prejudice."[82] Officials also repeatedly recorded and discounted local men's concerns about the strength and durability of wattles for hut poles and kraal fencing. In a rare correspondence on the subject, magistrates and foresters in various Eastern Cape districts described in 1905 how African men were avoiding purchasing black wattle trees from state plantations, "as the natives declare that the saplings rot too quickly."[83] Even as men in the Transkei became regular purchasers of plantation wood in later years, authorities still noted buyers' selectivity when choosing different exotic species to suit their needs.[84]

In KwaMatiwane people especially questioned the suitability of plantation trees for firewood. Although forest officials in the early twentieth century assumed that Africans would accept any type of wood for fuel, men and women in KwaMatiwane were quite selective when in positions to choose between different fuelwood sources. In particular, women's species preferences and practical experience with making fires to serve specific tasks entangled colonial efforts to "wean" African women from indigenous forests. In fact, many women found the exotic species made "available" by the government to be poor substitutes for the known, desired qualities of specific reserved indigenous trees. As one informant summarized this sentiment, "The government had seeds for its own forests, but it didn't have seeds for God's forests."[85] In interviews, elders recalled that eucalyptus species were clearly less desirable than indigenous trees for fire-making, echoing women's opinions in the wider Transkei and in other parts of southern Africa.[86] Wattle species were also problematic. For one thing, as Nozolile Kholwane explained, the wattle was challenging to convert to fuelwood: "[I]t's difficult to cut into pieces—it's too hard."[87] Several others emphasized the different way particular species burned. Dabulamanzi Gcanga offered a typical description: "The wood of the plantations is only good for building houses and for building cattle kraals or sheep pens. It is the natural forest that will make a good fire."[88] When I asked one elder, originally from the Qelana Mountains, how people historically felt about using exotic species, he elaborated on this theme: "No, they never like it, because the gums and the pines have no good fire like the natural forests. Even though it burns, it has got no coal. If it burns out, it just burns out. But the tree from a natural forest, if you burn it you get coals. Even if the fire is not there, you get coal. During the harvest time, we barbecue mealies, and then it's good to do it on good coal, and you get good coal from the natural trees. But you won't get coals with the ones that have been planted, so it's difficult to barbecue mealies."[89]

People not only preferred indigenous trees for such purposes, but also sought particular species when harvesting firewood. Informants recall a number of diverse species and their unique attributes as fuel. For Nozolile Kholwane (NK), the best around was *ugqonci* (*Trichocladus ellipticus*).

> NK: We were using ugqonci, whether you are baking bread or barbecuing mealies, it was the one with good coals.
>
> JT: Were there any other trees that you were using to cook anything else?
>
> NK: Even the ugqonci was okay when it is rotten, when you are cooking utywala [homemade beer]. It was good when you are cooking utywala like that.[90]

Along with ugqonci, some of the most prized species for firewood were the very same trees which conservators most vigorously "protected" in the Kwa-Matiwane area, as well as in all Cape and Transkeian forests, at the turn of the century: the yellowwoods. In official accounts of the colonial period, and even in more recent years, the value of such species as firewood has generally been dismissed, reflecting state actors' own interests in promoting plantation substitutes and reserving these indigenous trees for timber production. In a 1985 publication by the former Transkei government's Department of Agriculture and Forestry, for instance, *umkhoba* (common yellowwood, *Podocarpus falcatus*) and *umceya* (real yellowwood, *Podocarpus latifolius*) are defined as "very good" for timber yet "poor" for firewood purposes.[91] Yet when answering questions about their firewood preferences in their youth, elders often placed umkhoba and umceya at the top of their lists, because of their long-burning coals, useful for preparing meals or brewing utywala.[92]

Such tree preferences, flying in the face of official comments on Africans' environmental needs, continued to be exerted throughout the early colonial period, even as state authorities increasingly implemented more comprehensive forest restrictions in local communities.[93] Despite government efforts to "wean" them, men and women still often resisted state attempts to impose its definition of economic worth on tree types and continued to harvest the reserved indigenous species which best served their economic interests.[94] As people calculated how and where to obtain wood amid changing colonial conditions, they assessed and pursued their interests in the particular qualities of different species, alongside other livelihood costs and benefits.

DEALING WITH FUELWOOD SCARCITY AND THE "PERILS" OF TEZA

[T]he Native people were like persons in a desert exposed to the storms and like bucks in the open flats without shelter from the hunter's bullets. Wood had become so scarce through the Forest Department's regulations that their women had to go and steal wood for their fires and when they were caught they were taken to court and fined.

—*Councillor Jiyajiya* (1916)[95]

The unfortunate and injurious servitude of dry firewood collection in headloads by native women is one of the greatest curses of exploited forests in the Transkei.

—*Conservator of Forests, Transkeian Conservancy* (1926)[96]

As the negotiations over wood preferences suggest, the problems of official "weaning" strategies bore particular implications for differently situated men and women in specific locales. Women's livelihoods and environmental practices continued to be most directly affected by changes in fuelwood access. By the early 1910s, localized experiences of the constraints and impracticalities of teza policies became more systemic and widespread, as problems of fuelwood scarcity and pressures on women's labor became more acute in many areas of the Transkei. Women responded to these shifts by increasingly turning to government plantations as a wood source over the following two decades. But as the physical limitations and financial costs of plantation wood persisted amid a worsening economic situation for many African families by the early 1930s, many women also more heavily exploited their remaining teza rights in demarcated government forests. This response rather tragically coincided with a growing preoccupation by government authorities with a perceived crisis of environmental decline in the "reserves." Officials interpreted women's heightened use of indigenous forests as a serious ecological threat necessitating greater state control and were now much more united in their view that suspending Africans' fuelwood rights and more forcibly "weaning" African women, in particular, were justified, no matter the impact on local people's lives.

By the early 1910s, the toll of recent state forest demarcations, ongoing teza restrictions, and mounting resource demands made fuelwood increasingly scarce and women's fuelwood collection more problematic in many parts of the Transkei. African leaders in the Transkeian Territories General Council submitted popular complaints about the pressures felt from wood access policies in the face of declining fuelwood sources, particularly in often meager headmen's forests, in many locales.[97] Moreover, drastic ecological changes in the early 1910s, including drought, bad harvests, and the tremendous loss of livestock due to East Coast Fever, exacerbated women's labor load in securing fuel supplies, at a time when more men entered the migrant labor force on a regular basis and women felt the general burdens of domestic work more intensely.[98] Reduced herds meant that women had to rely more on their own energies to transport fuelwood from forests. Meanwhile, reduced livestock and crop yields resulted in less dung and crop residue available for local fuel use, forcing more and more women to search for and exploit wood sources further afield, exacerbating problems of scarcity.

Given the combined pressures of the early 1910s, many women increasingly resorted to an alternative officials eagerly welcomed: purchasing fuelwood from government wattle plantations. For growing numbers of women, the savings in labor and time spent searching for and harvesting fuel materials often outweighed the financial costs of buying ready-cut plantation wood.[99] To purchase an *inyanda* or headload bundle, for instance, women would usually trek to a forester's station, pay a few pennies, "and then they're given a permit to collect the dry firewood. . . . It was already cut by the foresters there and left there to dry."[100] Official reports suggest that Africans across the Transkei increasingly pursued such options from the early 1910s onward. Records of fuelwood sales from departmental plantations, including both headloads and wood transported by sledges and wagons, reveal that a new pattern of fuel access began to take shape during these years (table 3.2).[101]

TABLE 3.2.

Annual Fuelwood Sales, 1910–30

	Demarcated Forests		Government Plantations	
	VOLUME (CU. FT.)	VALUE (£)	VOLUME (CU. FT.)	VALUE (£)
1910	354,363	1,120	109,298	895
1911	269,115	834	156,920	1,221
1913	161,707	525	177,210	1,689
1917	129,292	445	333,484	3,113
1918	95,709	354	453,182	3,918
1919	112,436	397	471,312	4,291
1920	101,126	378	473,374	4,274
1924	100,664	537	446,267	4,380
1927	115,308	687	478,919	4,531
1928	101,136	642	567,187	4,962
1929	108,818	665	590,697	5,412
1930	104,826	610	801,696	6,572

Source: This table is compiled from Forest Department annual reports for the years cited. All figures are rounded to the nearest pound sterling.

Africans' purchasing of fuelwood from government plantations rose steadily and jumped most dramatically at particular moments of social and ecological pressure, such as the mid-1910s. In contrast, during these two decades the amount of fuelwood sold from government forest reserves declined overall.[102] By 1930, compared to the amount sold at plantations, fuelwood sales at forest reserves were relatively small. The general trend during these years, then, was that Africans were now buying their fuelwood in ever greater numbers, with government plantations increasingly becoming a regular fuel source. It was a situation that foresters happily embraced. By the mid-1920s, the Transkeian conservator gloated over the local demand for plantation firewood and its contribution to forest preservation: "[T]he indigenous forests were thus being less molested."[103]

Despite official pleasure at this turn of events, major structural constraints on fuelwood access persisted. Even when Africans chose to buy firewood from government plantations, supplies were much more limited than for hut- and kraal-building materials in these years. For one thing, the scheme of plantation wood supply itself had been gendered. The early period of plantation development in the Transkei resembled a pattern of state forestry experienced in most areas of southern Africa in the twentieth century, subordinating women's environmental priorities to the resource needs and interests of men.[104] Foresters had designed plantations principally to meet the demand for hut- and kraal-building materials, following the broader interest in getting African men to earn cash to pay for produce and reflecting the general official perception that only men's work was real and legitimate. As a result, officials considered supplying firewood at government plantations secondarily, if at all. Since women could always get wood in their "customary" fashion, there was no need to prioritize meeting popular fuelwood needs. The annual reports of the Forest Department, in fact, regularly failed to mention firewood when detailing yearly plantation produce sales in these years: plantations were often explicitly described as production centers of sticks and poles for male building activities.[105] Thus, this gendered orientation contributed to the increasing scarcity of fuelwood available to African women. Only in the early 1910s, as more extensive fuelwood cutting and scarcity across the Transkei began to cause greater official worry, did foresters seriously begin to consider supplying firewood more systematically.

Adding to problems of plantation supply were difficulties of physical access and availability.[106] Even when women pursued plantation fuel in greater numbers during the 1910s and 1920s, it still required traveling to often remote sites by foot and transporting heavy loads back home, particularly if stock-powered transport was not available or was too expensive to hire. Such burdens were exacerbated by the fact that government plantations continued to be inconveniently located for the vast majority of communities in the region. On more than one occasion in the mid-1920s, Transkeian Territories General Council councillors noted this chronic problem as they petitioned the government to rethink its wood supply policies at both state-run plantations and forest reserves. In the 1924 session, councillors proposed that the government "open all the demarcated forests for people to buy all forest produce from the nearest forest to the buyer. . . . The motion asked that they should be allowed to get wood from all the Government forests, so that they could buy wood close to their homes"; a similar motion was passed the following year.[107] Leading magistrates and the chief conservator of forests immediately rejected both proposals.[108]

By the early 1930s the situation was no better. In 1931 the Forest Department reported that only some 61 plantations with a total planted area of less than 15,000 acres served the entire population of the Territories, which by

Figure 3.3. Women carrying heavy poles in the Engcobo district, c. 1950. *Photograph courtesy of University of Fort Hare Library, Piper Collection*

now surpassed one million.[109] At the same time, intensive exploitation of wood sources in undemarcated forests had left paltry wood supplies. In 1932, one official critically assessed the state's plantation policies of the past, noting that plantations were generally great distances from many communities, while in districts with the densest populations the undemarcated forests "contain only a few large trees which stand as ghosts to upbraid us! To-day there are ab-solutely *no* wattles and *no* poles and practically no firewood or kraalwood in the vast majority of Headmen's forests."[110] Throughout the following decade magistrates and conservators often recorded such problems and their impact on fuelwood access. For many households, the nearest plantations that could serve regular fuelwood needs were still as far away as 18 miles.[111]

For women shouldering the heaviest burdens of fuelwood collection, eco-nomic and ecological pressures in the early 1930s only exacerbated such prob-lems of plantation wood access. Droughts and economic depression severely reduced the flow of money into rural households. People thus found it more difficult to pay for forest produce at government plantations along with their other tax and debt burdens.[112] In addition, the severe droughts took their toll on tree crops at tree plantations, reducing the supplies available for purchase and forcing more women to utilize any and all other available fuel sources more heavily.[113] Now drastic problems of wood scarcity in local undemarcated headmen's forests further drove many women to exploit their remaining har-vesting rights in demarcated forests more intensively and creatively. An addi-tional benefit here was that women could use their unique access rights to procure other increasingly scarce household resources from state reserves.

Women and girls in many locales thus regularly employed their "dry fire-wood" rights to surreptitiously cut and remove saplings for hut sticks and kraal wood, to enable small livestock to feed on forest grasses, shrubs and trees, and to cut green trees for fuel.[114]

Such activities increasingly alarmed forest officials across the Transkei. African women not only abused their "customary" rights, officials argued, they also undermined attempts to "wean" the masses to government plantations, a system established for their long-term benefit. The secretary of agriculture and forestry, for instance, complained in 1934 that women in the densely populated Tina Valley of the Qumbu District, where dry firewood supplies had been exhausted, were "breaking down green branches and will not proceed 1 mile further to the Etwa Plantation."[115] The department's report for that year noted that officials had been trying to lure more women to plantation firewood by reducing its cost, but in many areas women still "seemingly prefer to go through the plantations to thieve in the forests in the mountains beyond."[116] Rather than addressing such practices as symptoms of historical problems of resource availability and species access, however, officials were inclined to constrain women's environmental practices even further.

Colonial responses to teza access in these years grew out of broader official preoccupations with environmental "crisis" in the African "reserves." South African authorities as a whole became increasingly concerned about the failure of reserve ecologies to sustain their mounting African populations and the threats such developments posed to racial segregation, social stability, and white economic development.[117] In the Transkei, foresters increasingly criticized the "excessive" trampling of livestock and people through forest areas for compacting local soils and fostering conditions for soil erosion and tree-cover loss. Officials gathered mounting "evidence" of women's intensified use of trees and dung for fuel, with its destructive impact on soil fertility, reinforcing arguments for greater state environmental intervention and control over African women's mobility.[118]

Government reports and evidence brought before the Native Economic Commission in the early 1930s brought even greater official attention not only to the perils of "overstocking" in African communities in the Eastern Cape and Transkei, but also to the deleterious effects of local forest use on soil fertility. With mounting evidence of this "crisis," and with growing state confidence in "technical" solutions to social and environmental problems, South African authorities argued for the necessity of greater state intervention into African livelihoods and resource practices.[119] The onset of economic depression and consequent pressures on government finances further persuaded officials to take more aggressive action in the Transkei. Forest officials launched concerted campaigns to curtail local Africans' fuelwood rights and their ability to avoid paying for forest produce. In 1932, for instance, the inspector of forestry for the combined Natal-Transkei Area, B. R. Simmons, asserted:

> The plantations in the Transkei are annually robbed of a considerable amount of revenue by the practice known as "teza." . . . This practice should certainly now be discontinued in all districts which are amply served by Bunga and Departmental plantations. Large quantities of plantation produce are *lying to waste* in many plantations bordering these forests. These plantations provide material suitable for all native requirements and have been established at great expense for this express purpose, and in order to save the forests from further destruction.[120]

The following year, officials seized an opportunity to comprehensively restructure and reduce teza access. In the 1933 session of the United Transkeian Territories General Council, a council select committee resolved that the collection of dry firewood and other minor forest produce not be limited to African residents of the particular district in which a forest was located. The purpose of the motion was to ease the restrictions on African forest rights during a time of increased wood scarcity in African communities, enabling people to gather materials for their daily needs free from complicated boundary rules. The Transkei's conservator of forests, director of forestry, and chief magistrate fully supported the move and took the necessary steps for the change in late 1933 and early 1934.[121] Yet there was more to this sudden "concession" to African leaders in the Bunga than this picture reveals. Freeing teza regulations from district boundaries also now essentially freed state officials to more efficiently overhaul the parameters of teza access across the Transkei. As Director J. D. Keet asserted, given the fact that so many forests "had been ruined through over-tezaing," this was an opportune time for the government to consider periodically closing certain forests to teza rights in a systematic way, an opinion shared by the secretary for native affairs.[122]

Following the Bunga's proposal and Keet's report, officials quickly took steps to concretize the state's ability to control teza practices. In early 1934 the secretary of agriculture and forestry, W. M. du Plessis, submitted a set of schedules for the temporary closing of particular forest areas to all teza privileges for a period of three years, beginning in 1935; after that period the forests would be inspected and reported on, and officials would decide either to continue to keep the forest areas closed or to close others instead.[123] Du Plessis' report focused on the drastic environmental consequences of Africans' "free" exercise of teza, their criminal abuse of this privilege, and their stubborn backwardness in not utilizing government-supplied plantation produce to meet their needs rather than indigenous trees. Teza restrictions were the linchpin for proper forest management in the Transkei: "The teza servitude is of course a convenient loop-hole for the natives to possess themselves of their other requirements free of charge, and has been taken full advantage of. The forests have been denuded of young indigenous saplings, much prized for hut-building, and consequently the safety of future crops is seriously jeopardised. . . . It is therefore in the interests of the Natives themselves that the forests referred to in the schedules should be closed to the 'teza' servitude for purposes of recovery."

Suggestive margin notes, part of the "technical" descriptions of different forest reserves in the enclosed schedules, reinforced these themes. "Native women hack and destroy trees, tramp over forest and damage seedlings." "Prosecutions for using axes and other implements are very frequent." "Headloads of dry firewood in abundance can be obtained from the Government Plantations." "The Proximity of Plantations is stressed." All such claims were repeated verbatim in many of the schedules, regardless of the particular social and ecological situations pertaining to these diverse settings.[124] Such systematic repetition served a dual purpose: it naturalized assumptions about African women's unnecessary and illogical destruction of local forests, given the ready availability of plantation supplies, while legitimizing necessary interventions by the state. Since Africans were shown to lack the proper reason to alter their behavior for their own long-term benefit, du Plessis argued, more effective state control was essential.

While the tone of du Plessis' report was typical of many commentaries by both conservators and magistrates at this time, not all Transkeian officials responded to his proposals enthusiastically. The magistrates of both Butterworth and Engcobo districts questioned the validity of particular technical "facts" cited as supporting evidence in du Plessis' planning report: local prosecutions for women using axes were very rare, and the nearest plantations that could serve people's daily fuel needs were much farther away and more inconvenient than the report suggested, in some cases a distance of eighteen miles or more.[125] Such criticisms, however, constituted a minor dissent in official thinking. While individual magistrates might dispute the necessity of teza restrictions in specific localities and acknowledge their negative impact on local livelihoods, the overwhelming majority of Native Affairs officers justified the changes with blanket statements on the regular "abuse" of teza "privileges," the availability of sufficient fuelwood supplies to meet local needs, and the extremely "reasonable" prices at which plantation wood was provided. Based on these abstractions, many officials assured their superiors that the teza closings "should not prove a burden to the natives" and that "no undue hardship will be entailed."[126]

Official perspectives had thus changed considerably since the initial restrictions on African resource rights had been introduced in the late nineteenth century, when magistrates and foresters fought fiercely over the nature of African rights and the role of government in restricting them. Now officials were much more unified in their approach and agendas. Very few magistrates rose to defend the integrity of Africans' resource rights by the mid-1930s, and even these opinions generally followed rather than impeded the direction of state policy. In 1935 and 1936 new proclamations firmly established the policy of indefinitely suspending teza rights, authorizing the government to close particular forests to such "privileges" for an initial period of three years, with the option of renewal.[127] Within a few years, forest officials would dispassionately report, the "most easily accessible forests" in many areas were closed to teza rights so that women "will of course have to make other arrangements to obtain fuel."[128] When in 1939 the chief magistrate of the Transkeian Territories sent a circular to district magistrates regarding the desirability of continuing existing teza prohibitions and extending them to additional forests, not one magistrate responded critically. Teza closings on a renewable basis thus became a routine, unquestioned feature of state policy. As many women in the Transkei contended with more intense poverty, resource scarcity, and workloads in the early twentieth century, official responses reflected a deeper consensus among magistrates and foresters that women's resource rights and livelihood concerns should be sacrificed in the name of environmental conservation and social engineering.[129]

EVERYDAY NEGOTIATIONS OVER FUELWOOD ACCESS

Facing such growing problems of fuelwood scarcity and restrictions on teza access rights in KwaMatiwane, local men and women employed multiple strategies to lessen the weight of state forest domination and "make do" in their shifting predicaments.[130] While official policies altered the broader structures of indigenous fuelwood access, individuals experienced and responded to the localized, particular ways that colonial environmental constraints touched their daily lives, exploiting and manipulating weaknesses in the state's regulatory system whenever possible. Given the unique impositions on their environmental practices, individual women in various communities were forced to devise their own ways of creatively negotiating the colonial restructuring of their forest rights, adapting their physical manipulation of resources and their organization of labor, the very things the government's teza restrictions attempted to control. Such varied and individualized strategies helped shape the local significance of forest restrictions and their impact on specific people's lives.[131]

In the 1920s and 1930s local men and women contended directly with some of the gender-specific implications of colonial fuelwood policies. Many men continued to circumvent official attempts to push Africans into a rigid "customary" sexual division of labor through various licensing and teza regulations. Several informants described how young men in these years still regularly utilized animal power to transport firewood for domestic use, particularly when larger logs were needed for special feasts and cultural occasions, or during the autumn months when households were busy collecting enough wood to keep in storage for the coming winter season. Often without purchasing the requisite licenses beforehand, men would fell trees, drag them out of the forest with a span of oxen, then cut them up into smaller pieces back home.[132] R. T. S. Mdaka recalled the situation in Manzana in the 1920s: "[W]hen you were a young man . . . if there was going to be a wedding, you wanted trees, you want the wood to chop and cook for the people. And in that location, not only the home that was going to have a wedding, but even the surrounding kraals had to go, all the young men, it was their duty to go to the forest and help when there was going to be a wedding."[133] Another informant, from St. Cuthbert's in the Tsolo District, recalled how some men still occasionally assisted women with firewood collection in the 1930s by breaking branches off trees and giving them to the women to carry as headloads.[134] Although men periodically engaged in such activities, the potential cost of male involvement in fuelwood collection (through licensing fees or the threat of prosecution and fines if caught without a permit or otherwise breaking the law), together with their increasingly frequent absence from the Transkei, constrained their regular participation in such work.

State restrictions on fuelwood collection of course placed the greatest direct pressure on women. Presented with a narrow range of officially acceptable "customary" practices, local women employed various means to work creatively within the strict letter of the law, or circumvent the law altogether, as they continued to pursue the most labor- and time-efficient harvesting techniques possible. For example, many women continued to assert their rights to use axes for harvesting firewood, despite the risk of being arrested and having their axes confiscated for merely carrying them into a state forest.[135] Or, as Noheke Rangana recalled, some women in the Engcobo area instead manufactured their own makeshift hooks to pull down desired green branches for firewood, since colonial regulations did not explicitly prohibit these tools.[136] As in other parts of the Transkei, women also often physically manipulated trees to avoid prosecution. If sufficient dry wood was not available in a forest, particularly from preferred reserved tree species, then women "dried" the wood, either by cutting down branches with axes or by twisting and breaking off parts of green trees by hand—at night or when forest guards were not around—and then leaving the wood to dry in the forest for teza collection at some later point. Such tactics made prosecution quite difficult, since women could claim they had simply found the wood this way.[137] Some women generated dry wood by setting fire to or barking live trees or by partially breaking branches and leaving them hanging, which would gradually kill parts of the tree and produce a crop of dry branches. Given the number of "accidental" forest fires and incidents of tree "hacking" forest officials attributed to local African women in the early twentieth century, it may be that women continued to employ similar strategies over the years in order to maximize their ecological benefits and disguise their manipulation of local environments and the law.[138]

Women also adapted the ways in which they organized the labor of fuelwood collection. As in other areas of colonial southern Africa, the expanding out-migration of African men from rural areas by the early 1900s sometimes resulted in women and girls taking over formerly "male" labor practices in order to keep their households afloat. Local preexisting prohibitions on female use of livestock, for instance, often eroded in the face of rural families' growing economic pressures.[139] Olga Tyekela, for example, recalled various ways she and others collected fuelwood in the Qelana area: "[W]e go to the forest to fetch firewood, then make small headloads, then put them on the heads, and we arrive at home. And sometimes we used the cattle to go and tow sledges. . . . [Sledges were] cut down in the forest by men and there were girls then who knew how to inspan the oxen [for pulling sledges]."[140]

Another way many women in the KwaMatiwane area coped with their increased labor burdens during these years was to organize work parties, or *amalima*.[141] As in other parts of the Transkei, men and women in the region often used kin- and neighborhood-based networks to organize different amalima to assist in a variety of daily and seasonal labor tasks, such as cultivating, weeding, and harvesting crops, constructing and plastering huts, or cutting and transporting building materials from local forests. Women in particular employed specific types of amalima called *izitshongo* (sing. *isitshongo*) to aid in collecting wood and thatch grass from natural resource areas.[142] Izitshongo were especially useful when a large quantity of natural materials needed to be gathered, such as during the autumn, or on such special occasions as circumcision or wedding feasts, when more firewood was required for cooking meat and brewing beer. If and when sufficient resources were available, women would communicate among their extended families and neighboring homesteads that they desired an isitshongo for firewood gathering on a particular date. Word would spread, women and girls would assemble to work on the appointed day, and the host would prepare food and drink for their efforts.[143] Nozolile Kholwane recalled engaging in izitshongo in the Qelana area of the Tsolo District in the 1930s: "Yes, we used to invite izitshongo, we used to take firewood to other people. . . . We were not paid, but we were cooked for—things like bread, tea, stamped mealies, and *amarhewu*."[144] In the Manzana area, Alice Gcanga also remembered women regularly employing such work parties to assist in firewood harvesting: "They cooked food, and sometimes there would be meat."[145]

Amid greater demands on their labor, deepening restrictions on their teza rights, and ever more scarce or physically and financially inaccessible fuelwood sources, many women in KwaMatiwane increasingly relied on such adaptations to their environmental pursuits. Together, the strategies described above offer glimpses of how local residents negotiated changes in fuelwood access in their daily lives under increasingly difficult conditions in the early twentieth century. Women in particular responded to their combined livelihood pressures and constraints by altering their labor routines and thereby bypassing and ameliorating official regulation of their "customary" practices.

⤸

In 1920, during a discussion of colonial forest management in the Transkeian Territories General Council, Walter Carmichael, RM Tsolo, offered comments that reflect some common sentiments among Transkeian officials of this era.

> There was just one thing he would say about forests. When a woman came before him on the bench, charged with a breach of regulations, he always pointed out that the Forest Department was not merely thinking of the present conditions, it was thinking of the future. If the people had their own way they would leave nothing for the future generations. In breaking the regulations they were like a woman with a child on her breast who took the breast from the child's lips and put it to her own. There was perhaps not very much food for all, but the future generations had their claims as well as the present.[146]

This quote is illustrative in two significant respects. It suggests how far magistrates' commitment to the Forest Department's program of regulating popular resource rights had traveled since the early days of conservation. Moreover, it reflects how thoroughly and with what confidence colonial paternalistic discourse regarding how Africans should be "weaned" and disciplined had permeated official conservation practice, in this case to berate Africans for "starving" their own descendants. Magistrates as well as foresters now almost unanimously legitimated colonial forest management by condemning Africans for their selfish, irresponsible, and insatiable appetite for trees. In particular, official concern increasingly focused on restraining women's environmental practices and suspending their forest rights.

While official environmental sentiment and policy became increasingly constrictive in the early twentieth century, Africans experienced, perceived, and negotiated changes in forest access within the shifting colonial conditions of "reserve" life. People across the Transkei contended with changing definitions of forest entitlements and their dwindling control over an essential resource in their livelihoods—wood—in a period of general rural decline, increased resource scarcity, and expanding male out-migration. As colonial "weaning" strategies and government plantations increasingly dominated Africans' wood access, people critically interpreted and reacted to these changes within the broader context of the colonial government's subordination of Africans' economic and environmental interests. In daily practice, differently situated men and women in KwaMatiwane dealt with gendered wood access constraints in ways particular to and most beneficial in their individual situations, whether by continuing to harvest preferred indigenous species, creating "dry" wood for teza collection, or organizing izitshongo. Grappling with the multiple, mounting burdens of rural life in the Transkei and a government that grew less and less responsive to their local resource needs and predicaments, individuals negotiated changes to their socioeconomic and environmental lives through whatever means they could still most effectively control.

Part II

4 ⌒ Remapping Historical Landscapes

Forest Species and the Contours of
Social and Cultural Life

THERE WERE MANY REALITIES shaping Africans' experiences of and responses to resource access in the late nineteenth and early twentieth centuries outside of the colonial changes to political economies and ecologies described thus far. People participated in and adjusted to complex social and cultural dynamics in their lives, and forest resources were often deeply implicated in these experiences. Understanding how people viewed and negotiated changes in resource access thus requires an exploration of the meaning of different forest species in people's negotiation of their changing social and cultural worlds.

More often than not, such dimensions of forest use existed beyond the full attention or knowledge of foresters reporting from the field. Yet colonial officials also relied upon simplifications of Africans' resource practices and interests to legitimize their management agendas. By emphasizing that Africans' resource needs were merely economic and "simple"—sufficient wood for huts, kraals, and fires—authorities could then further justify restructuring forest access and limiting popular forest rights to these minimal "livelihood" requirements.

Such narrative frames disguised the extent to which differently situated Africans regularly sought access to forest trees and plants as crucial social and cultural resources. In colonial KwaMatiwane, many men and women relied upon a wide range of forest species for use in male socialization and initiation practices, as sources of physical and spiritual healing and protection, and as charms for mediating everyday social tensions. These resource interests were very specific, in terms of both the particular species utilized in different locales and the particular social and cultural needs of different individuals—of vary-ing social age,[1] gender, and status—at certain moments of their lives. As the colonial state imposed its own evaluations of forest species and reserved ac-cess to them accordingly, its policies interfered and contended with local peo-ple's very different, diverse, selective, and complex estimations of trees and plants. While people responded to the shifting needs of their social and cul-tural lives, they brought such realities to their understanding and negotiation of local changes in species access.

ACCESSING A SOCIAL RESOURCE: FORESTS, MALE INITIATION, AND STICK-FIGHTS

Both written and oral sources offer especially illuminating insights into the role of resource access in one local group's social practices in particular: male youths. As they attained the age for initiation into manhood, young males in-tensively sought the specific tree types that most directly contributed to the various ritual and socialization practices of this process. From the early 1890s onward, as the presence of forest officers and guards expanded in KwaMati-wane, they regularly contended with young men during the initiation season as the latter harvested wood for their ritual seclusion huts. Male youths also scoured local forests during these periods to secure the most desirable weapons for organized stick-fighting. Facing social pressures to be impressive fighters, youths employed diverse strategies to maintain their access to resources they viewed as vital to their entry into manhood.

An important component in the annual series of rituals associated with the initiation process in turn-of-the-century KwaMatiwane was the initiates' hut, the *isuthu* (pl. *amasuthu*).[2] The hut's framework was typically built out of wood and required dozens of long sticks or tree saplings for its construc-tion, which *abakhwetha* (initiates, sing. *umkhwetha*) usually cut and gathered themselves from nearby forests. As one elder in the Tabase region, Samuel Qina (SQ), recalled:

> SQ: There were indigenous trees in the forest there. I don't know then what type of trees were they called, but they used to, to make this *ungquphantsi*,[3] this like the hut for the abakhwetha.
>
> JT: The hut for the initiates. So, they used a certain type of wood for that?
>
> SQ: Yeah, they used it for making roofs.[4]

Abakhwetha had additional interests in accessing forests during the initiation process. Fumanekile Sithelo recollected local practices in the Qelana area: "Things that would heal us, we were doing them in the forest. There are secret things that we used to do there in the forest. . . . If an umkhwetha was to be made a man, they used to sing certain songs and go to the forest very early in the morning at 2 A.M., and they'd stay up in the forest. There are things that exist there which will heal the initiates, they are there in the forest."[5]

From the earliest days of state forest patrolling in the region, such popular demands for the forest resources associated with male initiation ceremonies brought African males into direct conflict with locally stationed officers and guards. At the turn of the century, the news of an upcoming circumcision in a local village set off an alarm for KwaMatiwane area foresters: it was a time to be on the alert in their patrols for intensified forest "thieving" and "trespassing." In late March of 1894, for instance, Forest Officer A. L. Raymond, stationed at Manina in the Engcobo District, wrote to Henkel to alert him to the opening of the circumcision "season" in the area and the predictable difficulties the occasion again posed for forest control. He described how local men were flooding the magistrate's office with "free permit" requests to get wood for amasuthu: "The wattle cutting nuseance [sic] is beginning again since the 'Abakweta' business has set in at the Magistracy they are handing the free permits out . . . scribbled over on any piece of paper that comes handy evidently just to get rid of the applicants."[6] Several years later, the RM Engcobo cited the activities of a group of "abakweta boys" as a typical example of the deplorable "vandalism of natives" in a letter to the chief magistrate: they had recently cut down 117 pieces of wood in a nearby forest, "many of them probably living saplings."[7]

Aware of the "toll" abakhwetha could take on local forests, foresters and other officials in the region kept a keen eye on any and all activities surrounding circumcision rituals in local communities. Evidence suggests that many foresters regularly inspected circumcision activities in subsequent decades. In the early 1920s, for instance, a forest guard, Charlie Lengisi, dropped in on a circumcision ceremony in the Ntywenka location of the Tsolo District and then followed up by trailing and prosecuting a group of local men for cutting reserved wood for the isuthu in a nearby forest. The district forest officer later explained to his superiors in Umtata that Lengisi had been hired as an additional forest watcher only recently, for the express purpose of maintaining closer surveillance during the heightened cutting activities of the annual circumcision season.[8] As one elderly man explained to me in the Tsolo District, discussing his youth there in the 1930s, when the time for circumcision arrived in local villages forest guards would not only be on the lookout for increased illegal wood cutting but would even occasionally make surprise visits to the homesteads where abakhwetha lived, to keep a closer watch on the situation.[9]

These few examples reveal that foresters' intrusions into male initiation activities drew struggles over local forest access directly into youths' social and cultural practices. Such struggles were exacerbated by young males' additional pursuit of certain tree types for use as fighting sticks. In her recent study of youth organization and masculine identity-formation in the Ciskei and Transkei during the late 1940s and 1950s, Anne Mager has located stick-fighting as a central ritual in teaching African boys "the rules of manliness." In preparation for the process whereby older uncircumcised youths were initiated into manhood, organized stick-fights between rival groups were the violent and often dangerous ritual arenas in which male identities were most powerfully forged.[10] The historical significance of stick-fighting in youths' lives in Kwa-Matiwane was repeatedly emphasized in my interviews with local men. As soon as the topic of forest access came up in our discussions, most men very quickly, and enthusiastically, moved into a personal reflection on sticks and stick-fights. R. T. S. Mdaka, who grew up at Manzana in the 1910s and 1920s, described the significance of stick-fighting in this way: "You had to know sticks. . . . If you did not know how to shield yourself, and how to hit hard, then you are reduced to, you are a younger boy, even if you are old. Even if you are young, you become an old boy if you know how to hit and how to prevent, how to defend yourself."[11] Other informants similarly reinforced one of the central points conveyed in Mager's article, that stick-fighting was a vital turning point in readying boys for manhood.[12]

Youth fights, often involving many boys and resulting in severe injuries, were a regular feature of life in the early colonial period. Organized battles between boys at dances, weddings, trader's shops, and natural landmarks as well as spontaneous attacks and counterattacks among different social and economic groups within locations, between locations, and across district boundaries were fairly common.[13] Rather than simply reflecting what colonial officials neatly categorized as "faction fights" between distinct "tribes" or levels of "civilization"—Mfengu versus Mpondomise or Thembu, "red" versus "school" boys, etc.—these confrontations involved a complex combination of changing cultural identities and economic situations.[14] "School" boys might be targeted by neighboring "red" boys because they were seen as economically and socially privileged, not just because they were culturally different.[15] And despite colonial stereotypes and the discouragement of missionaries and some Christian parents, "school" boys were often actively engaged in such organized fights. Cyprian Mvambo, who grew up in the 1930s at St. Cuthbert's, an Anglican

Figure 4.1. Boys demonstrating stick-fighting in the Engcobo district, c. 1971. *Photograph courtesy of University of Fort Hare Library, Piper Collection*

mission station in the Tsolo District, emphasized in our interview that, contrary to their reputation as "soft" or merely "Bible-carriers," mission boys were quite regular and aggressive stick-fighters.[16] R. T. S. Mdaka made a similar point about his experiences in the Manzana area in the late 1910s and early 1920s. Although his and other boys' parents in the location were strict and devout church members, they were also sensitive to the importance of stick-fighting in defending oneself and preparing for manhood: "So, they scolded you for fighting. 'You're a Christian, you shouldn't fight.' Red boys beat us. Have we got to be running all the time, when you go to the fields to look after cattle? They understood that we had to fight, we had to know sticks. They too had to fight, they had to know sticks. So, they didn't like people who could not fight, cry and go and report at home, like a woman. Got to fight."[17]

Besides its masculinizing functions, being prepared for stick-fighting also had its practical dimensions. Such fights periodically resulted in severe injuries and even deaths in the late nineteenth and early twentieth centuries. An older boy needed to know how to fight both to prove his masculinity and to protect himself physically.[18] Moreover, stick-fights organized around defending a given "turf" often placed serious obstacles in the way of daily economic activities. As boys herded livestock in disputed grazing areas or attempted to access particular types of grass, wood, or other natural resources only available in a specific locality, they were liable to be targeted by a rival group.[19] Trader's shops were another common site of youth "turf" wars. Again, Mdaka's recollections provide some interesting insights into the inner dynamics of stick-fights.

> This location is divided into parts. There's Umgwali; there is this part, this side. Those boys, Umgwali boys, didn't allow us to go to the shop on the other side of the school. They said that it was their shop, and if they saw us there they beat us like anything. We had nowhere to buy. Parents sent us to that shop to go and buy sugar and tea for them—those boys didn't allow us! They beat us! So there was war between us and them. We had to go and fight every weekend, Saturday and Sunday, fight those boys, trying to force them to allow us to go to the shop. They said it was their shop because it was in their location.[20]

To be ready for these regular fights, boys needed the proper weapons. As Mdaka summarized the situation, "There was always war between us and the Umgwali boys, so we had to have sticks, and we had to get sticks from the forest. You can't go to the river to get sticks."[21] Evidence from the archival record suggests that many boys in the region felt the same way. In 1895, for instance, one African man at the Anglican mission at St. Cuthbert's in the Tsolo District explained to the magistrate "that it is a case known [sic] boys are always anxious for sticks from the forest."[22] Yet, as many male informants made abundantly clear, not any old stick would serve the purpose. When boys harvested sticks in KwaMatiwane forests they were extremely selective, searching in specific forest areas and choosing particular tree species which, in the words of Dabulamanzi Gcanga, "were easy to cut, but not easy to break when you fight."[23] Informants and contemporary ethnographic accounts confirm that species such as umnqayi (blackwood, Gymnosporia peduncularis), ugqonci (Trichocladus ellipticus), umzane (white ironwood, Vepris lanceolata), and umnonono (hard pear, Strychnos henningsii) were all especially sought out in KwaMatiwane forests and commonly used by boys for fighting sticks in the 1910s, 1920s, and 1930s.[24] Having access to these species was important, one man explained, because "those were the trees that were tough and people loved them."[25]

Because of the reserved status of most species prized for stick-making, however, and their location in government-controlled demarcated forests, gaining access to such resources could be problematic. Several elders from different parts of the region recalled that when they were boys they would travel in groups to the forests secretly, at night, to avoid being detected and punished by state forest guards. When I asked Mdaka if he ever had any problems with the authorities for taking sticks from Engcobo area forests as a boy, he explained that he was one of the fortunate ones.

> RM: Boys would go and cut trees for their sticks. And there's always a guard looking after that forest. And if the guard caught you, then you were in trouble, you're in trouble.
>
> JT: What would they do with boys when they were in trouble?
>
> RM: Thrash them. Take them to the charge office, and there they were beaten, on their bums.
>
> JT: Did you ever have this happen to you?
>
> RM: No, no, it never happened to me.
>
> JT: So how were you successful in avoiding getting into trouble?
>
> RM: I wouldn't go daytime, I would go to the forest at night and cut wood for my sticks.[26]

Many others in the region similarly found the chance of gathering good fighting sticks well worth the risk of punishment.[27] To procure the particular types of sticks they wanted and to avoid being noticed by guards, it wasn't uncommon for local youths to devise elaborate schemes. Dabulamanzi Gcanga, for example, recalled his own solution to the problem as a boy in the Manzana area in the early 1920s.

> We used to travel here, to Umxikiko, after sunset, and we used to make our own lanterns, yes, to give light. Take a tin, put a candle inside, and when we get into the forest we light and look for sticks. And we cut them and used to have a bundle of sticks this side [motioning toward his legs], you tie them here, and another bundle tie them here . . . on your legs. And you walk [with the sticks bound to your legs], so that the police must not know that you are carrying sticks. Travel from there to here—by dawn we are back home.[28]

Even with such inventive tactics, the coast was not necessarily clear when boys reached home. Many boys would temporarily bury the "poached" sticks in their cattle kraals to keep them hidden from authorities and their parents, who might disapprove of their activities. Burying the sticks also kept them from drying out too quickly, letting them stay "soft so that they can perform their work," as Gcanga told me. "When we think they are right, we take them out."[29]

Gcanga's recollections suggest how male youths in some locales often employed elaborate strategies to maintain access to those forest resources they perceived to be most essential to their social and cultural development as men. While some foresters intentionally targeted the activities of these young males during initiation seasons, the latter's interests regarding and pursuit of certain types of wood as social resources extended well beyond the range of colonial knowledge and control. These examples reveal how very specific such interests could be in negotiations over resource access, as certain age- and gender-defined groups in particular locales quite selectively sought particular forest species for very specific social and cultural ends.

SECURING RESOURCES FOR HEALING AND PROTECTION

The historical centrality of forest resources in many people's social and cultural lives becomes even more evident as one probes more closely popular perspectives on the material and spiritual powers of different forest species. Particularly when attempting to keep affliction and misfortune at bay in the early colonial period, many KwaMatiwane residents turned to certain forest trees and plants to provide healing and protection for themselves, their family members, and their property. Access to certain species was especially crucial for differently placed individuals to ritually mark and ensure their passage and that of their kin through various stages of the life cycle. While the colonial government classified and regulated access to certain trees as reserved "timber," people thus continued to pursue their own particular understandings of many of these very same species as vital resources for physical, social, and spiritual well-being.

Informants and contemporary sources consistently emphasized the historical significance of many forest species as *amayeza* (natural medicines, sing. *iyeza*), which KwaMatiwane men and women regularly harvested in their daily pursuit of physical well-being.[30] All parts of different forest species, from leaves to roots, were commonly cut, picked, or dug by people living in the mountain forest areas as well as in the adjacent low-lying plateaus. Tsikitsiki Nodwayi, for instance, described the common practice of digging for tree roots during his youth in the forested hills of the Umtata District: "[A]mayeza were there, and the doctors were not there. We used to carry *ulugxa* [digging sticks or rods] in sacks and climb up the mountain, and go and pick them there. Then when you arrived back here, you healed sick people."[31] Another informant recalled that many people would travel from quite a distance to take advantage of the medicinal resources of the Gqogqora forests in the Tsolo District: "People from Umtata, and especially the *amaxhwele* (herbalists), were picking up amayeza from there, from the natural forests. They were peeling the bark of trees."[32]

A problem Africans in the region faced throughout the late nineteenth and early twentieth centuries was that many of the species they regularly preferred as curative amayeza, both for themselves and their domestic animals, were routinely placed on officials' reserved lists. One restricted species, umnonono, in addition to being harvested by young men for fighting sticks, as noted above, was also a common cure for digestive problems in humans and livestock. "You peel off the bark," Sampson Dyayiya recalled, "grind it on a stone, add water and then give it to the cow, when it has got gall problems."[33] Another informant added, "[people] were picking up medicine from there, from the natural forests. They were peeling the bark of trees, like umnonono, . . . which is very bitter. You use it if you're suffering from stomach-ache, they grind it and they give it to you to drink. It heals your stomach."[34] Another reserved tree, *umthathi*, the species highly prized and restricted as a building material by the government, was also an important veterinary resource for local stockowners. As Dyayiya explained, "When your cow is hurt, you let it drink umthathi. . . . It heals a cow's eyes, when it has an eye sore."[35] In certain cases, colonial authorities acknowledged the local importance of such medicinal remedies, even as they worked to restrict access to these species.[36] In his initial 1888 report on attractive "timber" prospects in the region's mountain forests, for instance, Conservator Henkel singled out the local significance of *umkhwenkwe* (cheesewood, *Pittosporum viridiflorum*): "The Gulandoda Forest is celebrated amongst the Kafirs for the bark of the Kwenkwe tree, which grows here to perfection. The bark of the tree (not too old) is stripped off and dried, and when reduced to a powder is used as a sure cure for dysentery."[37] Together with such other species as *iqwili* (*Alepidea amatymbica*) and *umhlonyane* (wild wormwood, *Artemisia caffra*), umkhwenkwe continued to be exploited as a treatment for various illnesses, including coughs, sore throats, colds, and stomach ailments.[38]

Despite officials' occasional recognition of the local medicinal value of such species, over the next few decades African men and women faced possible arrest and prosecution if caught "damaging" these resources. Such cases are seldom described in great detail in colonial records, but a few well-documented instances in the Tsolo District in 1899 describe how different individuals were caught digging out medicinal roots in government forest areas—in one case a man's iron digging rod was seized by the local officer.[39] More vivid descriptions of these encounters come from informants' personal memories.

Sampson and Notozamile Dyayiya (SD and ND) recalled how continued restrictions in the 1920s and 1930s on harvesting forest amayeza forced them and many others to employ creative ways to access desired tree roots.

> SD: We used to dig them, and we were stealing, because we were not allowed to dig. . . . We used to wake up early in the morning, or even during the day you would go in the forest, you'd identify the medicine, and then late, when the guard is asleep, you go and pick it.

> ND: Sometimes when they accompanied the cattle to the mountains, they passed through the forest, and then they'd dig the medicine and leave it there. And then in the evening they'd go and pick it up. . . . Then you'll hide it at home, though the guard won't do anything at your home, because he only guards the forest.[40]

Other informants mentioned amayeza that were particularly significant in their lives as they recalled the problems of local forest access. Eunice Matomela, for instance, explained that "the medicines for coughs, *isigcimamlilo*, and so on, all those are inside the forest. People were caught when they entered the forest."[41] For Olga Tyekela, this restricted access to the basic cures for everyday illnesses was the most memorable of impositions: "We couldn't cut the green trees. Even the umkhwenkwe, when we peeled the bark you'd get caught when doing that. . . . You go to court at Tsolo town. You face your charge, they give you a fine, you're put into prison if you don't have it. . . . That's why I was afraid."[42]

While officials sometimes targeted the harvesting of trees for amayeza in the region, they were much less aware of or concerned with the multiple other ways local people desired and utilized forest species as resources of protection and healing. Plants and trees were components of complex local understandings of natural, social, and spiritual power, as well as of repertoires of practices that attempted to harness these forces for various purposes. Africans across the region, as in other parts of the Transkei and southern Africa, utilized a multitude of floral species as resources for attempting to ensure the prosperity and health of themselves, their homesteads, their crops, and their livestock.[43] For many people in KwaMatiwane, particular species, often renowned for their physical capacities to ease pain and cure ailments, also represented important means of harnessing empowering forces or deflecting debilitating ones.

As men and women incorporated trees and plants into their ritual life, inherited taboos and prescriptions helped shape how members of different clans and families approached and exploited certain species. Some elders recalled that as children they were often instructed to follow specific prohibitions against using specific plants and trees in order to guard themselves and their families from debilitating forces. Eunice Matomela distinctly remembered taboos concerning the *umsenge* tree: "We were told not to burn the umsenge tree as firewood. We must not bring it home. . . . It brings misfortune."[44] Monwabisi Ndzungu further noted that certain tree species were "reserved" by local people for special occasions: "Yes, there were such trees. . . . But you were not supposed to cut that tree, except when you were doing a certain ritual."[45] As Joan Broster noted in the 1960s and 1970s in her research on Thembu communities in the Engcobo District, many people in the region historically relied on a specially "reserved" plant or tree, *iyeza lasekhaya* (medicine of the home) or *ubulawu* (medicine associated with a specific clan), in order to perform numerous rituals connecting living family and clan members with the ancestral spirits.[46] *Ukubethelela*, or the ritual "doctoring" of a home by a specialist to protect it from evil spirits, often required such a special iyeza lasekhaya. Tozama Gqweta recalled such ceremonies from his youth in the Baziya area of the Umtata District: "[W]hen the *igqirha* [diviner-healer] came he would *chaza* people [make incisions in the skin and rub amayeza into them], you see. There's a way of prohibiting anything that would endanger our life. What is this called now? Prevention from getting certain diseases, a precautionary measure of that nature. . . . 'Immunization.'"[47] In the early 1940s, Robert Godfrey also recorded such ukubethelela practices in the region, whereby an igqirha would use various "medicines" to protect a homestead from evil, "with the result that no wizard would venture near, knowing his presence would be detected."[48] Emmie Fiko recalled how knowledge of the required iyeza lasekhaya in such ceremonies was often the prerogative of patriarchs: "We used to hear about those things [amayeza], that some are being dug from the forest by people who knew them, like amaxhwele—do you know what I mean by amaxhwele? They dug them and then they came to our homes to make the house strong. . . . [C]hildren would be instructed to go and wash with that iyeza. We were washed with an itchy iyeza. We did not know the iyeza, it was only our fathers who knew. After we have been washed, they would chaza us, and we were told that they [the incisions] were to make us strong."[49]

One especially important tree for many such ceremonies of ritual protection was umthathi. According to one informant from the Gqogqora area:

> There is another thing people were picking [from the forest], like
> umthathi, when there was an occasion where you would brew beer

and slaughter a cattle. You go and cut the leaves of that tree. When you dished out the meat, it would be put on those leaves. There would be a reason why this is done that way—maybe the ancestors are complaining that you don't do anything for them in that particular home. It would be then that you go and pick umthathi and use it as your dishes. You use the umthathi leaves, you don't use any dishes.[50]

Olga Tyekela (OT), who grew up in the same area in the 1910s and 1920s, recalled in more detail one of the particular occasions for which umthathi was specifically required:

OT: When there was a ritual to be done at home, the home would select a person to go and cut a certain tree from the forest.

JT: Which tree do you remember?

OT: Umthathi. . . . Only the leaves, so when we were dishing out the meat we'd use the leaves. . . . We used umthathi to put down and place the roasted meat on top of it.

JT: What ritual was it for?

OT: The ritual was when we'd call the herbalists [amaxhwele] to come and make the household strong. There were herbalists in the old days. Doctors [oogqirha[51]].

JT: Why did you use umthathi? Was there something special about umthathi?

OT: It's the tree that the old herbalists used to believe would make everything fine.[52]

Other oral and written accounts of the area testify to similar historical uses of umthathi leaves and branches, for holding the meat during ritual sacrificial appeals to ancestral guardians and when "doctoring" the home and the family for protection and prosperity.[53] W. Blohm, the Moravian missionary at Baziya, described in 1933 how umthathi was the wood commonly used by abaThembu in the area for sacrifices and feasts performed when ill people in a household sought aid from their ancestors. As people roasted the sacrificial meat over umthathi branches and then ate it, the "hardness" of this species, he wrote, was "supposed to work magically on the patients and strengthen the weak bodies."[54]

Umthathi and other forest species were also particularly employed to protect individuals during their passage from one life stage to another. For many pregnant women and mothers of newborns, for instance, certain trees and plants were especially crucial to ensure successful births and the healthy development of young infants.[55] Alice Gcanga noted the practice of wearing a necklace of *umthombothi* (sandalwood, *Acalypha glabrata*) wood, popular among what she viewed as more "traditional" neighbors during her youth in the western Engcobo District: "When a person has just given birth, she would wear umthombothi. . . . I didn't wear it myself. I used to see it from my sisters with red blankets."[56] Fumanekile and Ntombizanele Sithelo (FS and NS) remembered another special forest plant from their early years in the Qelana area, used by expectant mothers as a channel of protective forces.

NS: There's a medicine that is given to somebody who is giving birth for the first time.

FS: *Isilawu.* There is isilawu in the forest, there fallen in the forest. When a young woman gives birth for the first time we pick isilawu.

NS: . . . She is given the *isiko* [custom] which belongs to this home, because this is the first child. You will see from this isilawu whether the fetus is alive or not. . . . When the pregnant woman sleeps, if she has had a bad dream, she must wake up and go to that isilawu and tell that dream to it.[57]

One type of ritual, known as *ukufutha*, is especially well documented for the early twentieth century, and draws its name from the use of smoke generated by burning particular tree branches to drive away evil spirits and enshroud newborns in a veil of protective forces.[58] In the early 1930s, Blohm described in intriguing detail some of the ways ukufutha was practiced by many Thembu communities in the Baziya area of the Umtata District. A Thembu woman would build a fire from the harvested green leaves of particular tree species—umthathi, umhlonyane, umkhwenkwe, and umzane—in order to create a steady stream of smoke through which she would wave her baby back and forth (see figure 4.2). As noted earlier, some of these species were also used for their medicinal properties. One informant recalled that during her childhood in the Manzana area "they used to smoke us with *umhlonyane* . . . so that you don't get the flu."[59] As Blohm's account reveals, the powers of these trees took on even greater significance in *ukufutha* practices. A song recorded by the missionary most directly suggests some of the local meanings of these species. During an *ukufutha* ceremony a mother would sing in call and response as other women in the community formed a circle around her and the baby:

Woshi, woshi, woshi	Whoosh, whoosh, whoosh
Malipum ihashe kumntwana	Let the impurity come out of the child
Sik umhlonyane nomkwenkwe	Cut umhlonyane and umkhwenkwe
Woshi, woshi, woshi	Whoosh, whoosh, whoosh
Malipum ihashe kumntwana	Let the impurity come out of the child
Sik umtati nomzane	Cut umthathi and umzane
He mama, he mama	Hey mama, hey mama
Woshi, woshi, woshi	Whoosh, whoosh, whoosh
Intolongo yivenkile	Prison is a shop
Woshi, woshi, woshi	Whoosh, whoosh, whoosh
Ubokanyel into yaziyo	Deny that which you know
Woshi, woshi, woshi	Whoosh, whoosh, whoosh
Into yomntu yeyako	What belongs to others is yours
Woshi, woshi, woshi	Whoosh, whoosh, whoosh.[60]

Although the particular lyrics of this song more than likely varied from person to person and from place to place as they were modified over time,[61] certain elements of the song and the ritual are illustrative here. The "bath" of smoke with which the mother "washes" her baby is intended to purify and protect the child as it begins its life course. The physical and spiritual "impurity" within the child's body is ritually brought out and its malignant power debilitated. The lines advising the child of life's truths, certainly open to a variety of interpretations, at the very least attempt to instill a lesson of self-preservation in the newborn. In his own summary of the ritual's significance, Blohm viewed its meaning in rather limited terms: "A child who is not exposed to such a bath of smoke will remain a weakling and a coward." However, this practice can be seen as a much broader process of empowering and protecting the baby's life for the future, a process in which the selected tree species were central catalysts. Oral and written sources suggest that these and other particular forest species were commonly harvested and used for ukufutha and other similar practices in the region in the early twentieth century.[62]

Some similar uses of protective smoke can be seen in contemporary ceremonies marking the entry of boys into manhood. As Eunice Matomela (EM) recalled:

EM: There's a tree called *uluthongothi* [gardenia, *Hyperacanthus amoenus*] There's a tree with that name.

Figure 4.2. An apparently posed example of *ukufutha*, c. 1930s. *Photograph courtesy of Cape Town Archives Repository, Jeffreys Collection*

JT: What do you use it for?

EM: We use it when we are having ceremonies for boys' initiation.
      Yes, it's said we must use it as firewood.

JT: When do you make fire from this tree?

EM: Any time, just to get the smoke out of it. . . . It's burnt here at
      home, just before the ceremony, and it can be burned even in
      the initiation hut.[63]

In a 1934 article, Moravian missionary W. Blohm similarly noted the use of umthathi as a means of empowering local Thembu abakhwetha during their ritual seclusion period. After an animal was slaughtered, the meat would be roasted for the initiates on a fire made from the sticks of exceptionally hard trees, usually umthathi. As Blohm explained his understanding of local beliefs, "'Hard' woods must be used for the roasting fire, because the hardness will work magically on the boys, who will eat the meat roasted on it and thereby grow hard and strong."[64] Other sources reveal that woods besides umthathi were specifically utilized for the same purposes in neighboring communities.[65]

There were many other ritual uses of specific trees in male initiation. Sticks, for example, were not just important as weapons but also played significant roles in marking male social status and identity at important stages in an individual's life cycle. Through the acquisition of sticks from particular tree species men ritually attained the material symbols of deeper socioeconomic participation and prerogative.[66] Among many Xhosa-speaking groups in the Eastern Cape and Transkei, for example, sticks of umnqayi have often ritually marked the end of male initiates' seclusion and empowered them as they advanced to a new stage of social maturity and responsibility.[67] In many Kwa-Matiwane locales, umnqayi sticks had similar purposes and connotations. As an elder Mpondomise man in the Manzana community explained to me about his youth, "Umnqayi is used for when our people are coming to manhood. When they change from boyhood to manhood, they must carry that black stick known as umnqayi, a lucky tree."[68] Charlie Banti, a sixty-year-old man living at Tabase, recalled that in his youth umnqayi was still used by Thembu communities in the Baziya and Tabase areas for various rituals involving the passage from one male life stage to another: "There's a black stick like this one [pointing to a walking stick he's holding], called umnqayi. That stick is used when there is a circumcision initiate. You need to go and cut it for him and he must carry it. That stick was also used when you'd go and ask for somebody's daughter to be somebody's wife, you used to carry that stick."[69]

Specific trees were also important in marking one final stage in the life cycle: death. As Mlungisi Ngombane (MN) explained, one particularly significant tree which many people in the Qelana area historically used to connect with their deceased relatives was *umphafa* (buffalo thorn, *Zizyphus mucronata*): "[W]hen you have dreamt of something, if you have dreamt of your grandfather whom you don't know, you take the umphafa and wash with it, and then you sleep. When you wake up in the morning you will know who you were dreaming of. . . . The ancestor will tell you who he is." Given umphafa's significance, the elder continued, it was especially important in rituals involving gravesites, performed by rural patriarchs to maintain links between the living and the dead.

> MN: When there's something that's done for the ancestors, you'd cut a branch from this tree and put it on top of the grave which belongs to the ancestors, and then on the day of performing the ritual you go and fetch that branch. And it will be hung on the kraal fence. . . . It is like when I've got a grave in my garden and maybe because of the poor state of the soil I leave this place for another one, leaving those graves behind, and go somewhere else and then leave graves behind there. Or that person was struck by lightning.

TS[70]: And then where was that person buried?

MN: A person who has been struck by lightning gets buried in that very same place. . . . The dead person is not brought to the home, so this branch of umphafa will be put on the grave of that person, and by so doing we are fetching that person's spirit back home. This branch is a "bus" or "taxi," we can term it that way [he laughs]. Then the person will be removed from there back home and we work like that. It's a "taxi."[71]

In ritual practices covering birth to death, whether a mother enveloping a newborn with protective smoke or a patriarch "taxiing" the spirits of deceased relatives back "home," these examples show just some of the many ways people sought the powers of forest trees and plants to ensure health and prosperity for themselves and their kin, especially at key stages in their life cycles. Differently situated individuals imbued the forest species of their landscapes with levels of meaning particular to their social positions and ritual needs. As men and women negotiated forest access in the early colonial period, they thus asserted diverse and complex interests in securing certain resources for their specific "quests" for therapy, healing, and protection.[72]

MEDIATING SOCIAL RELATIONS AND THE CHARMS OF EVERYDAY LIFE

People also relied on the powers of certain forest species in their everyday social interactions, and often for less benevolent purposes. To influence and protect themselves from other members of society, people employed and contended with magic and sorcery, and particular trees and plants were common, valuable mediums for such practices in the late nineteenth and early twentieth centuries. Access to forest resources was therefore tied to individuals' negotiation of their daily social tensions, whether involving disputes with relatives or neighbors, lovers' quarrels, or even run-ins with colonial authorities.

One valuable historical source for interpreting such dynamics is the record of colonial court proceedings. Cases of witchcraft accusations, regularly held in colonial courtrooms in Engcobo, Tsolo, and Umtata during the late nineteenth and early twentieth centuries, often involved the use of tree roots, bark, leaves, and twigs and other "charms" which were allegedly employed to cause pain, disease, and even death.[73] In testifying in court, African witnesses often specified the exact tree species that had been used by an accused *umthakathi* (witch or wizard) to exert harmful influences on others. In one typical case in the Engcobo District in 1890, an adulterous relationship between an African man named Jim and Nosesi, the wife of his neighbor Jonas, led to accusations that Jim and his family members used witchcraft to make Nosesi ill: "[T]hey had used two charms which he called Sindiandiya [*isindiyandiya*, bastard sneezewood, *Bersama tysoniana*] and Gazini [*umgzina*, *Curtisia dentata*], which they had placed at Jonas['s] kraal, that their intention was to kill Jonas but the charms had affected his wife instead of him."[74] In another instance in the Umtata District several years later, Nosesi Matikolo accused a man named Bom of harming her after she had refused his advances. After lying in bed "ill" for three days, dreaming of the charm Bom had allegedly used, she went to his kraal to confront him: "'[S]et me free Bom you have bewitched me you have been trying to get me to be your sweetheart.'" When Bom asked her by what means he had done this, "she said by means of a piece of a wild beast's skin stuck away in the roof of my hut, and a bark of Black ironwood, stuck away in the river somewhere."[75] These examples suggest how people commonly associated specific forest species with certain powers of sorcery. As men and women expressed and resolved their social relationships through idioms of witchcraft, they often exploited and identified certain trees as potent forces for seeking retribution and exacting revenge.

Individuals also activated the powers of different species to affect people in much less destructive ways, to influence their emotions and thoughts or counter their ability to cause harm. In cases of unrequited love or sexual interest, for instance, vengeance was not the only option: "love potions," often derived from various trees and plants, could be employed to try to influence another person's affection.[76] A witness in a Tsolo District courtroom in 1886 described some popular beliefs about certain plants at the time: "They say that if you spit the medicines and mention a girl's name she will dream of you though she does not see you."[77] In another case in the late 1890s in the Umtata District, a woman had consulted an igqirha to help release her from the magic spell of a pursuing man. Yelling out "let me alone Mhlobo I won't have intercourse with you," she proceeded to name for the igqirha different local tree species, such as umhlonyane and *uvelabahleke*, which she claimed had been used to entrance her.[78] As Robert Godfrey recorded, this latter species was also "used as a charm by girls, who smear themselves with it that they may be loved by everybody."[79]

People might also try to influence the behavior of others through charms when they were trying to conceal their actions or protect themselves from others' interventions. One such charm frequently used in some KwaMatiwane communities and in other areas of the Transkei was *umabophe* (*Plumbago capensis*). In the early 1930s, J. H. Soga noted that the charm had at one time been utilized by amagqirha in certain locales to ensure success of warriors going off to battle, by rendering their enemies harmless; over time, its use had been gradually incorporated into everyday popular experiences.[80] Some four decades later, C. M. Lamla took up this theme again in his research on ama-Qwathi communities in the Engcobo District. Lamla recorded umabophe being used by cattle thieves when trying to escape discovery by angry kraal-owners, by boys wanting to overpower their adversaries in stick-fights, and by defendants in legal cases striving to secure acquittals.[81]

This last situation is particularly evident in the sources for the region in the late nineteenth and early twentieth centuries. Defendants, plaintiffs, and witnesses in colonial and chiefly courts across the Transkei often prepared for their cases by washing in a particular "medicine" or procuring a piece of root or bark to chew in the courtroom. Different tree and plant charms served different ends, from confusing the mind of a prosecutor to silencing an incriminating witness to effecting a chief's or magistrate's favorable opinion. Magistrates themselves occasionally commented upon the regular use of such legal charms. Looking back on his career as a magistrate at various levels of the Transkeian administration in the early twentieth century, Frank Brownlee noted the following: "From time to time I found that witnesses had medicine in their mouths. Their embarrassment and confusion when they were ordered to remove it was amusing. The charm might be in the form of a button, a piece of dried twig, a bit of bark. These medicines were no doubt used for the purpose of convincing me of the truthfulness of the evidence being given or to protect the witness from being found out in some lie supposed to be necessary to the success of the case."[82]

Robert Godfrey recorded the use of one tree species in particular for such purposes—isindiyandiya, "whose root is chewed as a charm to throw the magistrate's mind into utter confusion when hearing a case; the party whose aim it is to create such confusion expels the influence of the chewed juice with a gentle smacking of his lips in the direction of the person trying the case."[83] One informant from the Tsolo District attested to the potential effects of using such tree roots:

> HS: There's another tree called *izidumo* [water tree, *Ilex mitis*], and isindiyandiya, we used to get them from the forest. . . .
>
> JT: What were you using these trees for?
>
> HS: Isindiyandiya and izidumo were used when somebody had an offense, like a trial.
>
> JT: When he has a case?

HS: Yes, he would wash with those trees before he goes to the case.
      . . . It makes good luck for that person, and when the magistrate
      looks at him he likes him. [He laughs.]

JT: . . . Did it work for you?

HS: I've never been fined. I've never been fined.[84]

 Whether or not most people facing charges in a courtroom could claim to be so lucky, there were still plenty of individuals regularly trying out such charms in the early colonial period. Men and women across KwaMatiwane dealt with a variety of sticky social situations and relationships in their everyday lives, and they often relied upon the powers of particular forest trees and plants to navigate and resolve these tensions.

⮑

These insights into people's historical resource use suggest the various ways forest species played important roles as individuals weathered their shifting social and cultural lives in colonial KwaMatiwane. People turned to forest resources to negotiate the dynamics of their wider everyday worlds, whether to demonstrate their male prowess to peers, to enlist the healing forces of ancestral spirits, or to deflect the influence of local colonial authorities. Even as the government attempted to restrict forest access through species reservation, and despite the inconvenience and risk presented by colonial forest patrols, men and women continued to pursue those trees and plants that they felt could best meet their unique social and cultural needs. As the above examples suggest, such interests and pursuits were quite specific, often reflecting individuals' particular status and life circumstances and directed very selectively at certain choice species.

 These diverse dimensions of environmental meaning do not easily fit into standard narratives of the Transkei's past or the orientation of much environmental history writing in South Africa and beyond. Yet to probe how people experienced and responded to changes in resource access, it is crucial to understand how differently situated actors viewed the diverse social and cultural stakes involved in securing forest species. For, when people in KwaMatiwane went to the forest or were restricted from doing so in these years, they did not leave their deeper social and cultural worlds behind them.

5 ⤳ The Python and the Crying Tree

Commentaries on the Nature of Colonial and Environmental Power

EXPLORING IN DETAIL THE negotiated cultural meaning of particular landscapes in the early colonial Transkei is often difficult. Oral sources provide crucial insights into local perspectives on the cultural significance of various forest resources and the history of state resource interventions, yet their representations of local residents' worldviews a century ago are understandably colored by the distance of time and more recent life experiences. Colonial sources, of course, also contain pitfalls of silence and bias, and colonial observers in the turn-of-the-century Transkei seldom commented at length, or knowledgeably, about Africans' cultural perspectives on their particular environments. Most likely people in different locales strove to minimize official and missionary intrusions into their cultural practices and ritual life by keeping certain environmental interests secret from outsiders.[1] However, there are exceptional cases when colonial accounts and oral sources converge, and even collide, to provide instructive clues for interpreting how people understood specific resources and environmental sites and the meaning of colonial interventions into these culturally potent landscapes. This chapter draws from the narratives of certain elders from the Gqogqora area of the Tsolo District in order to read beyond the limits of colonial knowledge, as embedded in archival sources, and to see how colonialism itself was the subject of Africans' commentaries on their transforming landscapes in the emerging colonial order.

One of the most renowned forest resources in KwaMatiwane, in both popular memory and written records, is one particular forest—one particular tree, in fact—in the mountain forests of the Gqogqora area. When colonial foresters began surveying the area in the late nineteenth century, they immediately encountered popular stories among local Mfengu residents of violent forests, disappearing people, and "unaccountable" wild animals. Officials quickly dismissed these narratives, but they were also keen to represent them as examples of Africans' irrationality and superstition and thus as evidence of the necessity of colonial environmental takeover. Yet, contrary to their representation in colonial discourse, such stories cannot be reduced to inconsequential rumor or mere assessments of local flora and fauna. As multiple sources reveal, such accounts in fact expressed the significance of specific forest resources and animal symbols in the practices of ritual divination and healing which were most closely associated with the well-being and protection of local communities. These stories not only served to restrict popular access to forest areas and species "reserved" for ritual specialists, but further reflected Africans' complex perspectives on the meaning of local environmental control in the region and the impact of colonial efforts to undermine it. Understanding the deeper significance of such environments can thus help us see how local people critically commented on the changes they experienced in their social, spiritual, and biophysical landscapes in the early colonial period.

COLONIAL FORESTRY AND THE "ENCHANTMENT" OF LOCAL LANDSCAPES

Colonial observers first encountered and negotiated the ritualized landscapes and the symbolic stories surrounding them in the Gqogqora area in the 1880s. As noted in chapter 1, colonial environmental control did not really expand in the Tsolo District until after the political tumult of the early 1880s stabilized. CMK Walter Stanford's new system of forest regulation in East Griqualand in the mid-1880s, the establishment of the Transkeian Conservancy in 1888, and the basing of a newly appointed conservator, Henkel, in Umtata, increasingly turned official attention toward the rich forests of the neighboring Tsolo District. In fact, in his earliest official inspections of local forest conditions, Henkel singled out the Tsolo mountain forests and their supply of such species as yellowwoods as "the most valuable and extensive" in the territory, initiating a campaign to strictly "preserve" these timber resources for local and regional European exploitation.[2]

At the same time that Henkel and other officials expanded their surveying of local forests for commercially desirable species in these years, they also commented upon their encounters with Africans' quite different perspectives on the particular forest sites being "protected." H. C. Schunke, for instance, compiling a geographical survey of the Transkei in 1885, noted that people in the adjacent mountainous region of the Engcobo District claimed that a local forest was "enchanted" and that "evil spirits . . . were supposed to dwell in the forest in a large pool, and no native dared to approach this consecrated place."[3] During Henkel's initial rounds as conservator a few years later, he described and dismissed rather similar stories told by Mfengu residents of the Gqogqora area of the Tsolo District:

> Many bush buck, blue buck, wild pigs, and tigers are found in the Qoqora Forest, a large and valuable forest, containing the finest timber of all the Transkeian Forests. Black Stinkwood and all reserved trees are found here to perfection. This forest has been protected from the fact that natives around here firmly believe that poisonous vleis exist in some part of it, stating that once a Kaffir together with his dogs went in pursuit of a bushbuck ram into this forest but never returned, nor was anything ever heard of him afterwards. The vleis I have not as yet been able to find. The people about here credit me with killing the python who had his lair in this forest, which, however, I have not yet seen, nor the large buffalo bull said by the natives to feed outside the forest. I have searched in the various forests but have found not even a trace or spoor of a buffalo, and believe they are all extinct as far as the Transkeian forests are concerned.[4]

As we will soon see, Henkel's account put an extremely simple gloss on local residents' understandings of their natural environs and the impact of colonial domination in their lives. For Henkel and other colonial visitors, such "rumors" about the landscape might fortuitously protect the forest from popular use, but ultimately could be categorized as superstition and cast aside if not confirmed by "scientific" observation. Over the next few years, Henkel put such thoughts into action. In 1893, the Gqogqora forests that Henkel described were some of the first to be demarcated and beaconed off as state reserves in the Transkei, remaining under strict state control ever since.[5]

Despite his eagerness to undermine popular access to these forests, Henkel was particularly smitten by the local beliefs he encountered and continued to write about them in his later years. In the early 1910s, while penning his unpublished and incomplete memoirs, Henkel produced an intriguing manuscript entitled "The Enchanted Vlei." Styling himself "the Oude Bosch-ranger," the retired conservator romantically recalled local stories about the Gqogqora forests, situating them within a larger tale of colonial conquest and valor. "One of the finest streams and forest of like name is the Qoqora to which has been attached for ages a great deal of romance. The Natives of the adjoining locations believe that a poisonous vlei or swamp exists somewhere in the Qoqora forest, that natives who have been hunting buck and wild pigs from time to time have never returned to their kraals, but have been swallowed up with the game they had killed and their dogs, by the enchanted vlei." The author then describes how "alive" the "magnificent" forest was with all sorts of game available for the sportsman, as well as numerous types of "valuable" timber trees. Out on a day's hunting expedition with some of his friends, while waiting for the hired African "boys" accompanying them as beaters, carriers, and cooks to prepare the evening meal, Henkel takes a momentous gaze at his surroundings from a hilltop: "Tranquil and solemn is the scene below us—the stillness of ages which sits upon the wilderness rendering its silence almost oppressive. Stupendous trees of the Yellowwood, and Black Ironwood, shielded from the ruthless violence of the tempest by yon mountains."[6] Over the course of a couple of pages, Henkel has thus condensed the story of colonial conservation on the "frontier" and his own life's work into one dramatic moment and place. His manuscript becomes a symbolic performance of civilization's conquest over primitive man and nature, with colonialism taming both the irrational, "ruthless violence" exerted by Africans on the natural beauty and value of the area and the "destructive" forces of nature itself. As a sign of colonialism's victory, Africans are rendered ineffectual yet entertaining props, harmlessly attached to their "enchanting" and "romantic" beliefs in the supernatural while dutifully serving the colonial protagonists as the latter engage in noble encounters with their natural surroundings.[7]

Dominating and "conserving" local forest landscapes for colonial consumption while relegating popular perspectives on them to the realm of "enchanting" superstition thus worked hand in hand. Henkel presented Africans' stories about extinct wild animals roaming the landscape and forests swallowing up people and game not only as entertaining footnotes to his surveys and memoirs but also as evidence of the environmental irrationality which colonial conservationism needed to overcome.[8] In the process, Henkel failed to account for or even see the fact that such stories actually conveyed much deeper commentaries about both the ritual significance of local forests and the meaning of colonial interventions in these landscapes.

BEYOND THE ENCHANTED FOREST:
RITUAL LIFE AND THE HISTORICAL MEANING OF A TREE

By the time of colonial forestry's expansion and Henkel's exploits in the Gqogqora location, Zizi Mfengu dominated the population and thus were the likely source of the forest stories he encountered. The local presence of these groups resulted from waves of migration during the social tumult of the early nineteenth century and then again in the 1860s and 1870s, as well as from the colonial resettlement policies following the 1880–81 rebellion (chapter 1). Descendants of these residents, who grew up in the Gqogqora area mountains but were subsequently relocated during the apartheid era,[9] still recall and tell stories about the special forest mentioned in Henkel's accounts, stories which they themselves heard in their youth in the 1920s and 1930s.

Informants confirm that the forest in question was known as Nocu. More specifically, many stories and practices in the late nineteenth and early twentieth centuries focused on a particular tree in one swampy area of the Nocu Forest. As with other natural places of special significance, residents circulated numerous warnings and rumors in their communities to keep people away from the site, including the stories reaching Henkel's ears about hunters disappearing when trespassing there. Mlungisi Ngombane, a seventy-eight-year-old man who was born and raised in this part of the Tsolo District, explained to me that the swamp had a special *umdlebe* tree which he and others avoided coming near:

> This tree is in the water, and it has got leaves. If there's something coming close to it, it gets pulled to the tree like to a magnet. It attracted goats, cattle, and people. And it threw them into the swamp, and they would come out just bones. Everything, yeah, it would throw it all into the swamp. . . . That thing was very confusing, so much so that we left that place not knowing the truth [referring to when he moved from the location in the late 1950s]. Nobody knows what was happening. Because we used to lose a horse, an ox, a goat, a sheep, a person, everything would be sucked in. There were so many bones around, and other ones belonging to wild animals.

When I asked whether he knew anyone who personally experienced problems in that forest, he answered: "No, I didn't know any by name, except the stories that people were disappearing. Because that thing happened like this, it's just like this: as you are going along . . . and entering the forest and pass beyond the point you are not supposed to, it will throw you inside the swamp. Then I won't know you anymore. I won't know you. So, we didn't go there to Nocu, because we knew we would be sucked in there. That's why I wouldn't know them anymore."[10]

Others I interviewed provided very similar accounts. As they suggest, there were multiple reasons for discouraging people from wandering toward the swamp. On one level, the dark and isolated swamp posed the very real threat that someone, particularly a child, might get lost, fall in the deep mud and water, or be attacked by wild animals and never be found.[11] Henkel himself once fell prey to this treacherous terrain. On the way to his memoir-inspiring hunting expedition, in the area of the Nocu Forest, "we crossed the Umhlahlani stream and a swamp in which the writer's horse nearly came to grief and we had trouble to extract the animal as it was nearly immersed in the mud of the swamp."[12] Such environmental threats to animals and humans alike were reflected in frightening stories of the tree in the swamp swallowing up everything in its path and leaving only the bones.

There were also physical dangers associated with the known properties of the umdlebe tree itself. Among neighboring Nguni groups to the north, there is a particularly rich record of very similar stories and popular perspectives concerning the species *Synadenium cupulare*, known as umdlebe in isiZulu, and in English, ominously, as the Dead Man's Tree. The latex of this euphorbia can be highly irritating to the skin and eyes, causing severe inflammation and blistering if not handled with care, and the plant is potentially fatally poisonous if ingested.[13] Perhaps the most notorious case of umdlebe's potency in the region involved the demise of members of Shaka's army, as he sparred with his rival Dingane: "Dingane came back to 'kill' King Shaka whilst the latter's regiments were away on a military expedition at kwaSoshangane, where they were killed by the poisonous synadenium tree-umdlebe hence 'ezinye ziyofa umdlebe,' and thus King Dingane is referred to as 'inkomo eyabuya yodwa kwaSoshangane,' (cow that came back alone from Soshangane land)."[14] Umdlebe's noxious physical properties were also tied to the tree's reputation as a dangerous source of magical and spiritual power. The Anglican missionary Henry Callaway recorded the following oral testimony on umdlebe in his late nineteenth-century work on Zulu religious practices: "[M]en contend with this tree; it is a powerful opponent; a man cannot pluck it before he has fought with it. It is also said that beneath it there are many bones of animals which die there; and birds if they pitch on it, die. . . . This, then, is the tree which kills people, which if cast into the midst of a village, that village perishes; a great fever arises; and a man dies with all his bones racked [*sic*] with pain."[15] As in the case of the Nocu Forest, these and more contemporary stories from the region emphasized the tree swallowing up all living creatures which approached it, leaving behind only piles of human and animal bones.[16]

Interestingly, considering these strikingly similar stories, residents in the Gqogqora area referred to a quite different species when they used the term umdlebe. From available evidence, it seems that umdlebe described a tree common in these mountain forests, most likely either *Chionanthus foveolata* (bastard ironwood or common pock ironwood) or *Chionanthus peglerae* (bastard black ironwood or giant pock ironwood), while *Synadenium cupulare* was not even present locally.[17] Although neither of these ironwood species shares the same toxic properties with the euphorbia umdlebe described above, such trees were still known in Gqogqora and neighboring communities to be tremendous sources of magical power, employed to drive away enemies or, in other instances, to bewitch and harm them.[18] More significantly, the umdlebe tree in the Nocu Forest shared the renown of certain select trees known as umdlebe north of the Transkei—trees which were off-limits to everyday exploitation and whose powers could only be harnessed safely by knowledgeable specialists.

As various informants explained, the umdlebe tree at Nocu had a particular name, Nome, which derived from the fact that the tree would cry out "meh," like a bellowing animal, when its bark was cut. As one elder recalled, after he and many others heard such things about Nome in their youth they were sure to avoid that area of the forest at all costs: "We were frightened because we were young children, but we were told that if that tree cries out you will die."[19] By emphasizing the dangerous powers of this tree, such stories both reinforced taboos against harvesting umdlebe in the Nocu Forest and suggested the special rituals and ritual knowledge required to access these forces without harm. This latter point was stressed by Fumanekile Sithelo. In the Nocu Forest, as in other forests in the Gqogqora area, if one wanted to access certain resources and survive, it was necessary to appease various spirits dwelling in the landscape. Sithelo gave one such instance involving spirits appearing in the guise of different animals: "This topic we are talking about, people have seen. For example, a brother to the chief used to go and fetch water from the forest early in the morning, at three o'clock. When we came from the river one day he saw a dog coming in front of him—it was an elephant that can convert itself into anything. When he saw this he gave respect to it, and then it let him go. If not, it would have caused danger to him."[20] In many cases, "giving respect" to ensure one's survival meant performing a ritual, such as making a special offering of maize or sacrificing an animal.[21] In the case of Nome, Sithelo continued, similar sacrifices were required: "[W]e knew as we had been brought up in that area that there's a tree called umdlebe which cries out. When you come closer to it you'd see bones from goats that died there. When you cut the tree blood comes out of it. . . . If you knew these types of trees, you'd take a goat and go there. And then when you get there, you'd stab the goat and it would cry out, yet you don't kill it. Right at the time that it's crying, you go to the tree and take the bark."[22] As Mlungisi Ngombane explained, the people who "knew these types of trees" were trained specialists, amagqirha, who alone possessed the requisite skill and fortitude to harness Nome's unique powers free of harm: "[W]hen you cut this bark, blood comes out like in a person. . . . Immediately when the blood comes out and when you cut the tree, the tree will cry out 'meh-eh-eh-eh.' . . . As we understand this, since the tree would otherwise kill the 'witchdoctor' [his English word], it takes the goat instead."[23]

Certain elements in these stories surrounding Nome further convey this particular tree's special connection with amagqirha. For instance, specific symbols call attention to the unique relationship between the igqirha and the spirits of the ancestors, the *amathongo*. The bellowing of both the goat and the tree during the sacrificial offering reflects a common feature of similar rituals in the region's history, in which the cry of the dying animal signifies a call to the clan's amathongo, to alert them to the sacrifice made in their honor and to request their protective and healing presence.[24] As other scholars have described, during many ritual sacrifices performed by amagqirha there are close symbolic connections between the slaughtered animal and the community, such as the last breath of the dying beast and its suggestion of the conduit between the living and the dead.[25] In the stories about Nome, the sparing of the igqirha's life and the tree's ingesting and bleeding of the animal's blood all add to the symbolism of the tangible connections between the amathongo and their living descendants, made possible through the unique ritual practices and environmental expertise of the igqirha.

Additional comparative evidence of historical associations between umdlebe and diviners is documented in Callaway's research on Zulu divination practices.[26] One translated narrative offers in rich detail a story of two Zulu "doctors," Usopetu and Upeteni, who wage a contest to prove who is more skilled at their craft. Usopetu taunts his colleague, "'[Y]ou are utterly unable to pick umdhlebe, though you are a doctor.' The other said, 'I can pluck it at once, as soon as I reach it.'" They then proceed to the tree's location at the lower Umlazi River to begin their contest. Taunted again by Usopetu, Upeteni rushes into action: he selects his strongest medicines, chews on them, and then repeatedly tries to stab the tree with his assegai, yet to no avail. He then quickly becomes cold and weak, calling out to Usopetu, "'O, I am conquered. Help me; I am now ill.'" Usopetu eagerly takes his turn, shouting, "'Yes! yes! You are about to be satisfied today that I am a doctor; you are [just] my boy.'" He uses his medicines to cure Upeteni and then successfully stabs the tree and harvests some of its branches, which he shares with his defeated colleague. Upeteni can only humbly reply, "'Usopetu, you are a doctor. You have conquered me this day.'" In this dramatic narrative, umdlebe is not merely associated with diviners, but specifically with only those individuals skilled enough to master the tree's powers—all lesser practitioners become feeble in its presence.

In the specific case of the umdlebe tree in the Nocu Forest, informants stress that Nome was actually "reserved" for ritual specialists at the turn of the century.[27] Such popular environmental "protection" measures were not unknown in the region. R. T. S. Mdaka, who grew up in the neighboring Engcobo District, recalled that in the early 1900s, "Some of the forests, some of the trees, were used by our doctors for curing people, and they had to be guarded."[28] More specifically, the stories describing Nome's historical significance point to the tree's reserved role in the initiation and training processes of local amagqirha, known as *ukuthwasa*.

Very similar stories have been passed down concerning other sites "reserved" for ukuthwasa practices in neighboring mountain forests. For example, Liziwe Ndzungu (LN), an elderly woman who moved to the Qelana area of the Tsolo District in the late 1930s, recalled, along with her son Monwabisi (MN), stories they heard about the nearby eLengwe Forest during the 1940s and 1950s.

> LN: We were told to be careful not to go to a certain forest.
>
> TS[29]: Why?
>
> LN: Because there were wild animals.
>
> TS: What type of wild animals?
>
> LN: Animals that would frighten us.
>
> MN: Tell them that leopards and lions were there, they were there in the eLengwe forest. . . . And a certain vlei which was in the eLengwe. And we were not supposed to just go around in that forest. . . . It was a pool where amagqirha became qualified.[30]

Joan Broster, during her research in the northern Engcobo District from the late 1940s to the early 1960s, recorded a very similar pattern of stories and taboos regarding a special site associated with ukuthwasa. In much of her writing over the years, she emphasized the spiritual importance of the Gulandoda Mountain in the training of amagqirha among local amaQwathi communities. In her 1967 *Red Blanket Valley*, she described how within the Gulandoda lay the "sacred forest of the witchdoctors."

> From time immemorial the amaQaba have kept this forest sacred.
> It is the forest of the spirits where their priests pass their final novi-
> tiate (*ukuthwasa*). For the rest of the tribe it is taboo, dreaded and
> shunned and shrouded in weird legend. No one but a witchdoctor
> or his acolyte ventures in. The outskirts of the forest are piled many
> feet high in dead brush and undergrowth, but not a piece is taken
> for firewood. Rarely is the forest disturbed. Apart from the witch-
> doctors it is visited by forest guards who keep to their well-defined
> paths.

She later explains how one woman undergoing the final stages of ukuthwasa ventured into a particular place deep in the Gulandoda Forest, "the sacred pool of the witchdoctors."[31] Despite Broster's highly romanticized framing of Qwathi rituals and beliefs, her account offers further evidence of a
broader historical pattern in the region. The stories surrounding the Gulandoda and eLengwe forests suggest that Nome was just one of many similar environmental sites of ritual empowerment in the region—locally "reserved" for
the training of amagqirha, hidden deep in the inner recesses of mountain
forests, and centered on "sacred" pools or vleis.[32]

In addition to these broad similarities with other ukuthwasa sites, there are
also particular practices mentioned in the stories about Nome that indicate
the place's special ritual significance. As scholars of African divination in this
and other areas of the Eastern Cape have described, dreaming is an important
experience in the novice's "calling" and training as an igqirha's apprentice. At
different stages in his or her training process, the novice will have recurring
dreams in which an ancestral spirit will communicate and reveal itself in
multiple guises. These and other images in the novice's dreams are instructions for him or her to go to particular natural places and engage in certain
practices, such as collecting specific plants or roots, ubulawu, revealed in the
dreams. As Manton Hirst has described for amaXhosa groups in the Grahamstown area, each clan is associated with a particular species of ubulawu root
which is a central and sacred component of both an igqirha's initiation and
his or her performance of sacred rites connecting homesteads with ancestral
spirits.[33] The stories surrounding Nocu and Nome, particularly the harvesting
of umdlebe, express these aspects of the ukuthwasa process and the training of
amagqirha in the powers of the natural world.[34] While colonial foresters began
claiming these and other forest areas in the district as "timber" reserves in the
1880s, many local residents viewed this particular tree as crucial to the development of an igqirha's skill in harnessing the necessary forces for their healing and protection.

ANIMALS, ANCESTORS, AND
COLONIAL ENVIRONMENTAL EXPANSION

With such alternative understandings of the Nocu Forest in mind, how then do we bring them to bear on the expansion of colonial forestry in the region? How did local residents in the late nineteenth century perceive and comment upon the colonial government's efforts to expand control over such a ritually significant site? Returning to Henkel's accounts, we find some suggestive clues. Recall that the conservator noted how local informants described a buffalo bull and a python inhabiting the forest and its environs. Henkel explained that he could verify neither story. In fact, being a keen sportsman and avid cataloguer of animal lore,[35] Henkel thoroughly scoured the forest margins to establish whether "even a trace or spoor of a buffalo" could be identified before categorizing the species as "extinct" in the area and completely dismissing local claims about the large bull residing there.[36] Yet Henkel's "rational" approach to these statements failed to recognize their full import. Rather than implying that Henkel or other officials could directly observe these animals, such stories instead both referred to the significance of ancestral spirits at this unique environmental site and commented critically on the impact of colonial intrusions into this landscape.

The recurring presence of ancestral spirits in various animal forms is often a significant dimension of an igqirha-in-training's experiences of the ukuthwasa process. Over time an ancestral spirit will appear in the form of one animal in particular—one of a set of animals traditionally associated with a novice's clan. This *isilo* (pl. *izilo*) will then dominate the novice's dreams, revealing a particular species of ubulawu and guiding him or her through the other dimensions of ukuthwasa.[37] Broster noted that the lion and the leopard were some of the most common izilo among the Thembu and Qwathi clans she researched in neighboring areas of the Engcobo District, occupying the dreams of women undergoing ukuthwasa and guiding them through rites of passage and explorations of nature's powers in the Gulandoda Forest. In the stories recalled by Liziwe and Monwabisi Ndzungu, lions and leopards were also said to dwell in the eLengwe Forest. The name "eLengwe" itself means "at the place of the leopard," which might have derived from the historical presence of leopards in the area but also from the historical associations of this specific isilo with this particular forest.[38] Two of the other izilo commonly dwelling in forest areas and sacred pools in the Eastern Cape have been the animals Henkel was told inhabited the Nocu Forest—the water buffalo and the Cape python, each known and respected for its unique qualities and powers.[39]

In the late 1880s, as Henkel encountered and dismissed the stories about these animals as farfetched or "irrational," local residents were identifying the presence and importance of such ancestral izilo at this forest site and directly implicating the conservator himself in changes affecting this particular landscape. Henkel's reference to the python living in the forest, and his being credited with its death, are particularly revealing in this regard. Given the fact that Henkel was a newcomer to the Transkei and the Tsolo District when he wrote his first account of the forest, we cannot be sure that his use of the term "python" accurately translated the local name for the mentioned snake. Several types of large snake existed at the turn of the century throughout this part of the Tsolo District and surrounding areas, and beliefs in the powers of several species have a long history.[40] Such snakes fall into two distinct categories: those used as "familiars" by witches and sorcerers for nefarious ends and those serving as izilo for the protection and well-being of groups and individuals. Henkel's explanation that he was told the "python" had its "lair" in the forest directs us to the latter group, as "familiars" are generally associated with the home, do not permanently dwell in forest and other resource areas, and are thus not connected with nature's tremendous powers.[41]

Perhaps the most documented snake in the region is the *majola* (mole snake), also known as *inkwakhwa*, which has historical connections with the royal clan of the amaMpondomise. Because of its embodiment of royal ancestral spirits, the majola has historically been respected and protected from harm when appearing at people's homes.[42] Another snake, *ugqoloma* (Cape python), has also been an important isilo among different groups in the Eastern Cape and has historical connections with royalty among the Gcaleka Xhosa.[43] Frans Prins has recently documented Mpondomise divination training practices in the Tsolo District involving encounters with the mythical water snake *ichanti*, which lives at the bottom of pools or rivers.[44] Such beliefs and rituals associated with ichanti were also noted in the region by Robert Godfrey in the early twentieth century.[45] As Manton Hirst has described, among some amaXhosa the appearance of ichanti has symbolized the ancestors' message of death to the living, serving as a guard to the ancestral abode lying beneath particular sacred rivers and pools, and only responding to the unique skills and interventions of amagqirha.[46] These and other snakes—izilo embodying ancestral spirits of particular clans and appearing to guide individuals through ukuthwasa—have historically represented the power of amagqirha to ward off evil and ensure security, prosperity, and health in many African communities.[47] By contrast, harm to such snakes has historically been associated with disorder, scarcity, and illness. This connection is described most vividly in A. C. Jordan's 1940 novel *Wrath of the Ancestors*, set among Mpondomise communities in the Tsolo District, where the author himself grew up. When one character dismisses local beliefs in majola's connection with the ancestors and destroys the snake, death, misery, and discord wrack the social order.[48] Henkel's being identified with killing the snake isilo at Nocu was loaded with such negative connotations.

So why would local residents impute this act to Henkel, especially when he himself claimed never to have seen or touched any snake in the forest? Part of the answer lies in the fact that Europeans were indeed slaying snakes at this time in the region, without any apparent attention to their particular meaning in local cultures.[49] Various government agents and others in the colonial community actively destroyed different snakes, particularly pythons, viewing them as sources of physical and moral danger. Henkel himself noted one such example soon after his first report on the Nocu Forest: "Several large pythons were seen by Sawyers and Kafirs in the mountain forests, and another was shot on the banks of the Umzimkulu River by Mr. Marshall of the Customs Department during the year under report."[50] A few years earlier, the leading Moravian missionary at Baziya, just across the Umtata River from the Gqogqora location, unexpectedly came upon a snake and then killed it during a historic foundation-laying ceremony for the station's new church. This "great conflict with the old serpent" then became the symbolic centerpiece of Brother Weitz's sermon about the necessary destruction of "wicked" forces in the spread of the gospel.[51] A year later, in 1887, the Tsolo magistrate came upon some very large snakes while out hunting for game in the Bulembu Forest, which he immediately shot in order to destroy "this trace of ugly monsters."[52] Such events brought about tangible links in the minds and words of many local residents between indiscriminate snake-killing and the expanding colonial and European presence in their midst. Moreover, there is some evidence of popular discourse associating Europeans not merely with the killing of such snakes but also with the appropriation of their powers for nefarious purposes. Several decades later, Robert Godfrey, a Presbyterian missionary who lived and conducted much of his research in the Tsolo District, offered the following dictionary definition of the word ugqoloma: "Python . . . , now extinct locally, but well-known through its figuring in the sayings: *ilitye likagqoloma* and *izitena zikagqoloma* (the python's stone, and the python's bricks),—imaginary charms having powerful effect and believed to be possessed and used successfully by white people."[53] Such associations may indeed have been traces of prior historical ideas about the European appropriation and exercise of power in the region.

There were also much more specific and profound reasons for Henkel to be implicated particularly in the killing of a snake isilo at Nocu. Local residents made such statements at a time when colonial domination of their physical environment was beginning to have more direct impact upon their lives than ever before. People had already experienced the broader expansion of colonial rule over the previous decade and a half, particularly after the 1880–81 rebellion. From 1886 on, people were increasingly familiar and vocal about growing government restrictions on their access to forest resources and its effect on their livelihoods (chapter 2). What was new in 1889, when the story of the python reached Henkel's ears, was the expanded effort by the colonial state to manage forest access across the newly minted Transkeian Conservancy and to intervene more immediately in forest control in the Gqogqora location itself. Local residents referred to the snake killing at Nocu during a moment of particularly hostile popular reaction to colonial forestry across the colony, as communities publicly protested the impending implementation of the comprehensive Forest Act of 1888 in their own locales (chapter 2). In this tense climate, Henkel, in his new role as conservator, began touring the mountain forests of the Gqogqora location for their potential reservation, the first representative of the colonial state to do so comprehensively. His introduction as the new chief environmental authority in the area, along with his surveying, note-jotting, and sketch mapping of local forests, all surely inspired interest and concern among local residents.[54] The popular stories about the python and the forest encountered by the conservator commented upon this broadening, yet immediately perceptible, expansion of colonial power into local landscapes.

These stories implicated Henkel, the most immediate symbol of the colonial state's environmental power, in both the domination of the ritually significant Nocu Forest and the negative consequences of this action for local residents. Henkel's activities in the area and the government's impending appropriation of local forest control together caused the demise of ancestral protection by way of this vital ritual site. Colonial power had usurped and destroyed the power of the snake isilo at Nocu to serve as a conduit between the realms of the living and the ancestral spirits and to ensure social and environmental order.[55] More specifically, this tragedy occurred at the special home of Nome, where the isilo was the guiding force enabling amagqirha to access the umdlebe tree, an essential source of their spiritual power to protect the health and welfare of the community. While Henkel would look back on this event as a simple tale of a romantically "enchanted vlei"—a footnote to the story of colonial conservation's expansion—his arrival and work in the area had been identified as a rupture in local communities' spiritual and material control over their landscapes and lives, as well as an omen of further interventions to come.

The stories surrounding the Nocu Forest and Nome offer a rare, detailed glimpse into the complex meaning of both ritually loaded landscapes and local people's commentaries on changing access to them in the early colonial Transkei. Colonial observers showcased these tales of animals and trees as amusing yet inconsequential environmental myths in order to demonstrate Africans' irrational backwardness and the need for civilized colonial dominance in the region. But as one breaks through the limits of colonial knowledge and representation, richer insights emerge into the significance of Nocu and these stories for people in the Gqogqora area. As people conveyed the message that colonial power was severing their vital environmental ties to guardian ancestral spirits, they utilized their changing physical and spiritual landscape as a dynamic symbol of their diminishing ability to control their own destinies and ensure their own well-being.[56] Moreover, the fact that informants recalled Nocu and other sites still being used as important places for ukuthwasa and other rituals in the 1920s and 1930s, well after colonialism was firmly established, and the fact that such recollections have continued decades after these individuals were removed from the Gqogqora area, together reflect the persistent resonance of these landscapes in people's lives despite their appropriation and representation by colonial officials.

The stories of Nocu thus reveal both the deep cultural meanings associated with very distinct components of specific historical landscapes—one special tree, in this case—and the connection between such meanings and people's responses to broader changes in power relations in the colonial period. Though distinct in many respects, these narratives suggest some of the ways Africans in other particular locales of the Transkei might have also interpreted and negotiated the meaning of colonial environmental expansion in their wider social and cultural lives.

Conclusion

THE VARIOUS HISTORICAL DEVELOPMENTS and processes described in this book—from people contending with political-ecological restructuring and gendered changes in economies and entitlements to the meaning and use of resources in diverse social and cultural practices—reveal many linked negotiations around environmental access, yet are also not easily resolved into a uniform, "seamless" narrative. Much as Bruce Braun has described the multiplicity of forest politics in the Pacific Northwest, forest access negotiations in the Transkei did not "coalesce to form a stable entity, nor . . . respond to a single logic."[1] Collectively, however, they recast past state and scholarly narratives of the region's history by illuminating some of the multifaceted ways in which environmental restructuring was embedded in deeper changes in power relations and wider everyday experiences in the formative colonial period. As differently situated actors in the Transkei shaped and responded to shifts in forest access in the late nineteenth and early twentieth centuries, they infused these negotiations with the diverse concerns of their particular life circumstances and positions in an evolving colonial order.

My focus on KwaMatiwane has revealed how negotiations over resource access were woven into the fabric of political and socioeconomic transformations from the earliest days of colonial administration in the Transkei. Officials incorporated environmental restrictions into their schemes for politically stabilizing colonial control in the region following the tumultuous rebellion of the early 1880s. As environmental authority was restructured in subsequent years, it was intricately implicated in the dynamic collusions and collisions of local rural politics. Magistrates and foresters contended with chiefs and headmen who negotiated their ambiguous positions in the colonial system and their influence over local forest access in unpredictable ways, whether for their own or their constituents' benefit. Within African communities, negotiations over forest access and use were drawn into ongoing socioeconomic tensions and personal disputes and often centered on local holders of environmental authority—headmen and, increasingly from the mid-1890s onward, their competitors, forest officers and guards. Following the creation of "headmen's forests" in the early 1910s, headmen continued to find space to shape the nature of their influence in matters of local resource access in individualized and often contentious ways.

The restructuring of environmental entitlements also developed within deeper colonial transformations and was similarly enveloped and shaped by the diverse responses of differently situated actors. African communities immediately responded critically to the erosion of their forest rights in the late 1880s and early 1890s, and in certain ways such protest helped alter the course of particular colonial policies. Africans' responses were also able to reverberate more profoundly in these early colonial years, given the administrative cracks created by officials' own in-fighting. Magistrates and forest officials intensely sparred over their respective power and authority in the developing colonial bureaucracy, particularly in the late 1890s amid various rural crises, enabling popular protest and strategies of evasion to have a greater debilitating impact on the enforcement of new forest restrictions. In the 1890s and early 1900s, variously situated rural men and women also negotiated the specific ways in which changes in resource entitlements affected their particular livelihoods, particularly as officials engineered popular "subsistence" rights in ways conducive to the gendered transformation of the Transkei into a labor reserve. Taxpaying men and women with unique teza rights each asserted more expansive interpretations of their environmental entitlements in daily practice, linking local changes in forest access to their wider experiences of colonial socioeconomic change.

The changes in forest rights of these early years also had lasting impacts on popular experiences, as people in different locales negotiated growing problems of wood scarcity amid escalating poverty and labor migrancy in the early twentieth century. Contributing to such predicaments was the more systematic development of colonial environmental restructuring from about 1903 onward, as officials reconciled their major differences, the Forest Department achieved more consolidated control over Transkeian forests, and the government became increasingly committed to constraining popular forest rights. In various parts of the Transkei, people questioned the legitimacy of government attempts to "wean" them from indigenous forest use and induce them to buy exotic fuel and building supplies from the state, even as many Africans increasingly turned to government tree plantations as a major wood source in the 1910s and 1920s. In daily practice, differently positioned men and women brought their particular social, economic, and ecological situations to bear on their responses to such changes. Women, increasingly targeted by official environmental policies from the late 1920s onward, manipulated local environments and reorganized their labor in certain ways to contend with *teza* restrictions as they negotiated larger pressures on their working lives in the 1920s and 1930s. The space for Africans to shape colonial policies might have shrunk or even disappeared by this point in time, but people still ameliorated the various colliding strains on their livelihoods in ways they could most effectively control.

Access to forests in the colonial Transkei was also about much more than just livelihoods. While successive governments demarcated natural spaces, reserved certain tree species as "timber," and otherwise imposed certain economic values onto these resources, people in various local settings actively pursued forest resources to meet the needs of their complex social and cultural lives. In ways particular in time, space, and the circumstances of different individuals' lives, men and women evaluated and exploited forest species very selectively in their healing practices, in ritual processes defining personhood and marking the life cycle, and in the everyday mediation of social relations. Such interests and contexts greatly informed how different people understood and responded to changes in local species access. The detailed example of the Nocu Forest and Nome suggests how the deeper cultural meanings of specific resources and resource sites could play a profound role in local responses to both changes in forest access and the expansion of colonial environmental power more broadly.

These multidirectional aspects of resource use and negotiation complicate any simple or singular readings of the meaning of environments over time or the meaning of changing environmental relations in the colonial period. They thus undermine subsequent official narrations of the Transkei's linked social and environmental history. In the late 1930s, for instance, as the period largely covered in this book transitioned to an era of more intensive and aggressive state interventions into rural livelihoods, South African forest officials began reflecting back on the preceding decades of colonial rule and environmental management in the Transkei. Arguing for the expansion of state control over rural landscapes and livelihoods, particularly through the recently passed Native Trust legislation, officials deployed images of the "crisis" of deforestation and ecological decline in African locations.[2] In the Forest Department's annual report for the 1936–37 fiscal year, now standard criticisms of Africans' "wasteful" forest use and "irresponsible" ways of managing resources were dramatically reinforced with a two-page visual representation of the "inevitable" historical outcome of such practices. Entitled "Stages of Forest Destruction," the series of photographs walked the reader through the story of how forest decline in the Transkei had unfolded in certain areas because Africans had been permitted to use these resources solely under "native ownership" and outside of state control. All of the images of progressive forest "destruction" were photographs of location forests under headmen's supervision; in stunning contrast was the "typically" robust forest which "Departmental control" and expertise had systematically assured.[3] The government's prescient action in assuming the management of the majority of forest areas in the early colonial period, for the "betterment" of the land and the people themselves, was now scientifically observed and documented. Officials particularly relied on landscape histories of the Transkei, some notably from KwaMatiwane, to typify more generally the inherent problems of African resource control across South Africa and to demonstrate the historical necessity and ability of state technocrats to remedy environmental crises in the reserves through scientifically sound management practices.[4] Such images would continue to dominate state narrations of forest control in the Eastern Cape and Transkei for decades to come.[5]

But the introduction and expansion of colonial control over forest environments in the Transkei was far from a simple, unproblematic story of conservationists waging a noble fight to protect local environments and Africans from their own destructive habits. Such narratives conveniently gloss over and "occlude" the involvement of forest policies and practices in deeper colonial transformations and social interactions from their initial days in the Transkei.[6] Moreover, they fail to account for the complex negotiations and contending interests, even within the colonial state, that surrounded and gave meaning to environmental restructuring over the late nineteenth and early twentieth centuries.

Exploring these historical dynamics seems particularly important to critically interpreting resource dilemmas and conflicts in South Africa today. Many recent environmental disputes causing such tension within and among African communities and state and development agencies—whether struggles over and contested claims to resources in forest reserves, plantations, and the protected parks of the Wild Coast and beyond,[7] the patrolling of African hunters in Kwa-Zulu Natal,[8] negotiations over the nature and availability of fuel and building materials to meet local needs,[9] or broader debates about the relationship between and the meaning of conservation, rural development, and social and environmental justice[10]—all echo themes and problems from the early years of colonial change in the Transkei. To be sure, the contexts of these various problems and the state's approach to resolving them, through such mechanisms as land restitution or community-oriented forestry schemes, have diverged tremendously from the founding era of colonialism. Yet particular socioeconomic and environmental relations that continue to shape contemporary struggles in many rural areas of South Africa have their origins in the formative development of colonial rule. Furthermore, all of these situations, whether unfolding today or a century ago, share a common feature: the embeddedness of environmental issues within deeper power dynamics, socioeconomic changes, and cultural negotiations. If current efforts to reintegrate the former homelands and disentangle the colonial legacies of social and environmental exploitation are to be meaningful, they thus must be informed by a deeper understanding of the long-term evolution of resource access in particular contexts, its location in the development of wider colonial relations, and the complex negotiations over power and meaning shaping these histories.

Appreciating such colonial transformations is particularly important in critically assessing current development policy formulations for the Eastern Cape and the historical assumptions and narratives upon which they rely. Particular versions of the history of "customary" environmental management, for instance, have been employed in ongoing discussions of one of the region's thorniest issues: the role so-called traditional authorities should play in local processes of land reform, development, and resource management in a post-apartheid society. Since 1994, successive South African government strategies of political devolution and democratization have often resulted in intense antagonism between traditional leaders and newly elected government councillors in many locales in the region. Multiple factors have been at work. The ANC governments have attempted to contain yet not alienate the alternative power bases traditional leaders represent, resulting in rather loose and unclear definitions of their scope and function in the new political dispensation. Traditional leaders, whose positions in the apartheid state apparatus were often fiercely contested in rural communities, now see democratization as a new threat to their status, and thus often compete for legitimacy and authority with local government councillors. Councillors, however, generally lack sufficient commitment, resources, and capacity from the state to counter traditional leaders' authority or deliver promised services to their rural constituents.[11] The resulting state of confusion has had significant impacts on resource use across the region. Echoing the complaints of officials during the initial expansion of colonial power in the Transkei over a century ago, state authorities, development and conservation agents, and academics have all lamented the "vacuum" of political control over resource access in many rural areas—with no clear or enforced accountability for popular resource extraction, many local forests and communal resources have been exploited more aggressively.[12]

Such diagnoses of the region's urgent environmental problems and pre-scriptions for their political solution have sometimes relied upon particular depictions of the nature and evolution of "customary" resource management. In developing its present policy paper on "sustainable forest management," for example, the Department of Water Affairs and Forestry (DWAF) cited "the eroding of traditional structures of authority and beliefs" as a major cause of current "breakdowns in the controls on natural resource utilization" in the former homelands.[13] The final white paper repeated the idea of a recently threatened "traditional control of harvesting of natural products."[14] Others have similarly implied that South Africa's post-1994 democratization process has undermined preexisting "customary" and more "sustainable" community-level resource management systems centered on local traditional leaders.[15]

There are potentially serious problems with such formulations, particu-larly if they are utilized to argue for some sort of idealized "resuscitation" of "customary" environmental authority in future development strategies, much as magistrates and foresters proposed in the late 1800s.[16] Despite over a cen-tury of complex colonial changes to environmental authority in the Transkei, and despite the ready acknowledgement by these same authors of the prob-lematic and locally contested position of chiefs during the apartheid era, such depictions suggest the existence of relatively unchanged "traditions" of re-source control and accountability until the ending of apartheid. As chapters in this book have instead demonstrated, the meaning of "traditional" environ-mental leadership and relations within African communities has long been extremely dynamic and deeply embedded in the complex evolution of wider political-economic transformations, well before the rise and fall of apartheid. Since the late nineteenth century, the nature of traditional leaders' roles in local resource management has been the historical product of shifting power negotiations among and between themselves, their rural constituents, and the state. It is thus essential to understand contemporary changes in community-level resource access and management from this deeper historical perspective.

The negotiations illustrated in this book can further infuse current dis-cussions about environment and development planning in the former Trans-kei and beyond with an historical appreciation of the significant social and cultural dimensions of resource access. Over the past decade, researchers in the Eastern Cape have understandably focused on redressing past state poli-cies by suggesting ways of enhancing rural Africans' livelihoods and socio-economic power through more equitable and democratic processes of local resource management. Analyses emphasizing livelihoods, political economy, and questions of production and consumption have, however, overshadowed a deeper understanding of important perspectives on the social and cultural meaning of resources themselves.[17] This is not to say that such issues are entirely absent in the literature: particular populations' medicinal uses of certain forest species, for instance, are often included in assessments of local communities' livelihood needs and practices.[18] Yet many other dimensions of popular interests in resources—such as their meaning in various healing, ritual, and social practices—are often marginalized in such formulations, despite their potential salience in individuals' perspectives on local dynamics of resource access. How, for instance, might a mother's wanting the proper plants to ritually protect her newborn, a young man's interests in certain types of fighting sticks, or a family's desire to visit ancestral gravesites in a state-controlled forest or privatized plantation intersect with their other resource goals and practices? And how do individuals' current and historical experiences of constrained access to such social and cultural resources then influence their wider perspectives on, and participation in, evolving environmental management strategies and development processes in local areas?[19]

Such questions are intimately connected to formulating a more expansive and responsive concept of environmental justice in this postapartheid era. The scholarship on environmental justice in South Africa continues to recognize the historical environmental injustices perpetrated by the state along lines of race, class, and gender and their profound ongoing impact on the majority population's livelihoods, health, and welfare.[20] However, as recent critics in South Africa and beyond have commented, it is crucial to move beyond traditional economic and legal concepts of justice—whether framed in terms of markets, livelihoods, basic needs, or rights—to incorporate the multiplicity of social and cultural valuations of environments in particular historical contexts. Appreciating the multiple meanings different actors have attached to both resources and the social relations surrounding them can serve a more democratic process of addressing past and present environmental grievances and alternative visions of equity and legitimacy. It thus forces an even deeper questioning of truth and valuation regimes, rooted in the colonial past, that continue to influence our perspectives today.[21]

Notes

INTRODUCTION

1. Sustainable Livelihoods in Southern Africa Team, *Decentralisations in Practice in Southern Africa* (Cape Town: Programme for Land and Agrarian Studies, University of the Western Cape, 2003); Lungisile Ntsebeza, "Decentralisation and Natural Resource Management in Rural South Africa: Problems and Prospects" (paper presented at the ISACP biannual conference, Victoria Falls, Zimbabwe, 17–21 June 2002); Ntsebeza, *Land Tenure Reform, Traditional Authorities and Rural Local Government in Post-Apartheid South Africa: Case Studies from the Eastern Cape* (Cape Town: Programme for Land and Agrarian Studies, University of the Western Cape, 1999).

2. Robin Palmer, Herman Timmermans, and Derick Fay, eds., *From Conflict to Negotiation: Nature-Based Development on South Africa's Wild Coast* (Pretoria: Human Sciences Research Council; Grahamstown: Institute of Social and Economic Research, Rhodes University, 2002), particularly chaps. 5 and 6.

3. Ashley Westaway and Gary Minkley, "Rural Restitution: The Production of Historical Truth Outside the Academy," Working Paper 45 (Fort Hare Institute of Social and Economic Research, August 2003); Sustainable Livelihoods in Southern Africa Team, *Rights Talk and Rights Practice: Challenges for Southern Africa* (Cape Town: Programme for Land and Agrarian Studies, University of the Western Cape, 2003); Zolile Ntshona, *Valuing the Commons: Rural Livelihoods and Communal Rangeland Resource in the Maluti District, Eastern Cape* (Cape Town: Programme for Land and Agrarian Studies, University of the Western Cape, November 2002); Palmer, Timmermans, and Fay, eds., *From Conflict to Negotiation*; Charlie Shackleton, Sheona Shackleton, and Ben Cousins, "The Role of Land-Based Strategies in Rural Livelihoods: The Contribution of Arable Production, Animal Husbandry and Natural Resource Harvesting in Communal Areas in South Africa," *Development Southern Africa* 18, no. 5 (December 2001): 581–604; Thembela Kepe, *Waking Up from the Dream: The Pitfalls of "Fast-Track" Development on the Wild Coast* (Cape Town: Programme for Land and Agrarian Studies, University of the Western Cape, 2001); Kepe, *Environmental Entitlements in Mkambati: Livelihoods, Social Institutions and Environmental Change on the*

Wild Coast of the Eastern Cape (Cape Town: Programme for Land and Agrarian Studies, University of the Western Cape, 1997); Land and Agriculture Policy Centre, *Overview of the Transkei Sub-Region of the Eastern Cape Province* (Johannesburg: Land and Agriculture Policy Centre, 1995); Crickmay, Venter and Associates, "Transkei Forestry Strategic Development Plan: The Future of Forestry in Transkei," unpublished paper, prepared for the Transkei Department of Agriculture and Forestry, Pietermaritzburg, 1993.

4. Thomas R. Sim, *The Forests and Forest Flora of the Cape Colony of Good Hope* (Aberdeen: Taylor and Henderson, 1907); N. L. King, "Historical Sketch of the Development of Forestry in South Africa," *Journal of the South African Forestry Association* 1 (October 1938): 4–16; King, "The Exploitation of the Indigenous Forests of South Africa," *Journal of the South African Forestry Association* 6 (April 1941): 26–48; W. K. Darrow, "Forestry in the Eastern Cape Border Region" (Bulletin 51, South African Department of Forestry, 1973); H. A. Luckhoff, "The Story of Forestry and its People," in *Our Green Heritage*, ed. W. F. E. Immelman, C. L. Wicht, and D. P. Ackerman (Cape Town: Tafelberg, 1973), chap. 4; Douglas Hey, "The History and Status of Nature Conservation in South Africa," in *A History of Scientific Endeavour in South Africa*, ed. A. C. Brown (Cape Town: Royal Society of South Africa, 1977), 132–63; F. and Jutta von Breitenbach, "Notes on the Natural Forests of Transkei," *Journal of Dendrology* 3, nos. 1 and 2 (1983): 17–53; J. D. M. Keet, *Historical Review of the Development of Forestry in South Africa* (Pretoria: Department of Environmental Affairs, 1984); Bruce McKenzie, "Ecological Considerations of Some Past and Present Land Use Practices in Transkei" (Ph.D. thesis, University of Cape Town, 1984); A. K. Shone, *Forestry in Transkei* (Umtata: Transkei Department of Agriculture and Forestry, 1985); K. H. Cooper and W. Swart, *Transkei Forest Survey* (Durban: Wildlife Society of Southern Africa, 1991).

5. International Institute for Environment and Development (London) and CSIR-Environmentek (Pretoria), *Instruments for Sustainable Private Sector Forestry* (CD-ROM) (London and Pretoria, 2003); Palmer, Timmermans, and Fay, eds., *From Conflict to Negotiation*; Sustainable Livelihoods in Southern Africa, *Natural Resources and Sustainable Livelihoods in South Africa: Policy and Institutional Framework* (Cape Town: Programme for Land and Agrarian Studies, University of the Western Cape, September 2001); Thembela Kepe, *Waking Up from the Dream*; C. Ham and J. M. Theron, "Community Forestry and Woodlot Development in South Africa: The Past, Present and Future," *Southern African Forestry Journal* 184 (March 1999): 71–79; Kepe, *Environmental Entitlements in Mkambati*; Institute of Social and Economic Research, Rhodes University, *Indigenous Knowledge, Conservation Reform, Natural Resource Management and Rural Development in the Dwesa and Cwebe Nature Reserves and Neighboring Village Settlements*, Interim Report for the Human Sciences Research Council (Grahamstown: 1997); Republic of South Africa, Department of Water Affairs and Forestry, *Sustainable Forest Development in South Africa: The Policy of the Government of*

National Unity: White Paper (Pretoria: Department of Water Affairs and Forestry, 1996); Anton Eberhard and Clive van Horen, *Poverty and Power: Energy and the South African State* (London: Pluto Press, 1995); Land and Agriculture Policy Centre, *Overview of the Transkei*; Crickmay, Venter and Associates, "Transkei Forestry Strategic Development Plan."

6. For example, Peter Delius and Stefan Schirmer, "Soil Conservation in a Racially Ordered Society: South Africa, 1930–1970," *Journal of Southern African Studies* 26, no. 4 (2000): 655–74; Peter Delius, *A Lion Amongst the Cattle: Reconstruction and Resistance in the Northern Transvaal* (Portsmouth, N.H.: Heinemann, 1996); Chris de Wet, *Moving Together, Drifting Apart: Betterment Planning and Villagisation in a South African Homeland* (Johannesburg: Witwatersrand University Press, 1995); Fred Hendricks, "The Pillars of Apartheid: Land Tenure, Rural Planning and the Chieftaincy" (Ph.D. thesis, Uppsala University, 1990); Hendricks, "Loose Planning and Rapid Resettlement: The Politics of Conservation and Control in Transkei, South Africa, 1950–1970," *Journal of Southern African Studies* 15, no. 2 (1989): 306–25; P. A. McAllister, "Resistance to 'Betterment' in the Transkei: A Case Study from Willowvale District," *Journal of Southern African Studies* 15, no. 2 (1989): 346–68; Terrence Moll, *No Blade of Grass: Rural Production and State Intervention in Transkei, 1925–1960*, Cambridge African Occasional Papers 6 (Cambridge: African Studies Centre, 1988); William Beinart, "Soil Erosion, Conservationism and Ideas about Development: A Southern African Exploration, 1900–1960," *Journal of Southern African Studies* 11, no. 1 (1984): 52–83; Beinart, "Introduction: The Politics of Colonial Conservation," *Journal of Southern African Studies* 11, no. 1 (1984): 143–62; Govan Mbeki, *South Africa: The Peasants' Revolt* (Harmondsworth: Penguin Press, 1964).

7. William Beinart has recently commented on and responded to this trend, delving into longer-term environmental conflicts—over locusts and forests—to help explain the 1960 Pondoland revolt. "Environmental Origins of the Pondoland Revolt," in *South Africa's Environmental History*, ed. Stephen Dovers, Ruth Edgecombe, and Bill Guest (Cape Town: David Philip, 2002), 76–89.

8. Contributing to this trend are two articles by Chris Tapscott—"The Rise of 'Development' Thinking in Transkei," *Journal of Contemporary African Studies* 11, no. 2 (1992): 55–75, and "Changing Discourses of Development in South Africa," in *Power of Development*, ed. Jonathan Crush (New York: Routledge, 1995), 176–91—which fail to connect the development of state "development" thought and practice in the region to preapartheid history. For a critical interpretation of the temporal framing of environmental history narratives, see William Cronon, "A Place for Stories: Nature, History and Narrative," *Journal of American History* 78 (March 1992): 1347–76.

9. For overviews of radical scholarship in this period, see Gary Minkley and Ciraj Rassool, "Orality, Memory, and Social History in South Africa," in *Negotiating the Past: The Making of Memory in South Africa*, ed. Sarah Nuttall and Carli Coetzee (Cape Town: Oxford University Press, 1998); Belinda Bozzoli and Peter

Delius, "Radical History and South African Society," in *History from South Africa: Alternative Visions and Practices*, ed. Joshua Brown, Patrick Manning, Karin Shapiro, and Jon Wiener (Philadelphia: Temple University Press, 1991), 3–25; Ken Smith, *The Changing Past: Trends in South African Historical Writing* (Athens: Ohio University Press, 1988), chap. 5; Christopher Saunders, *The Making of the South African Past: Major Historians on Race and Class* (Cape Town: David Philip, 1988).

10. Colin Bundy, *The Rise and Fall of the South African Peasantry* (London: Heinemann, 1979); William Beinart, *The Political Economy of Pondoland, 1860–1930* (Cambridge: Cambridge University Press, 1982); Beinart, "Settler Accumulation in East Griqualand," in *Putting a Plough to the Ground: Accumulation and Dispossession in Rural South Africa, 1850–1930*, ed. William Beinart, Peter Delius, and Stanley Trapido (Johannesburg: Ravan Press, 1986), 259–310; William Beinart and Colin Bundy, *Hidden Struggles in Rural South Africa: Politics and Popular Movements in the Transkei and Eastern Cape, 1890–1930* (London: James Currey, 1987). More recent histories of the colonial Transkei have instead generally treated the environment as backdrop: Sean Redding, "Legal Minors and Social Children: Rural African Women and Taxation in the Transkei, South Africa," *African Studies Review* 36, no. 3 (1993): 49–74, and "South African Women and Migration in Umtata, Transkei, 1880–1935," in *Courtyards, Markets, City Streets: Urban Women in Africa*, ed. Kathleen Sheldon (Boulder: Westview Press, 1996), 31–46; Clifton Crais, *The Politics of Evil: Magic, State Power, and the Political Imagination in South Africa* (Cambridge: Cambridge University Press, 2002).

11. William Beinart, *The Rise of Conservation in South Africa: Settlers, Livestock, and the Environment 1770–1950* (Oxford: Oxford University Press, 2003); Beinart, "Vets, Viruses and Environmentalism at the Cape," in *Ecology and Empire*, ed. Tom Griffiths and Libby Robin (Edinburgh: Keele University Press, 1997), 87–101; Beinart, "The Night of the Jackal: Sheep, Pastures and Predators in the Cape," *Past and Present* 158 (February 1998): 172–206; Beinart, "Soil Erosion, Animals and Pasture over the Longer Term: Environmental Destruction in Southern Africa," in *The Lie of the Land*, ed. Melissa Leach and Robin Mearns (London: International African Institute, 1996), 54–72; Lance van Sittert, "'Our Irrepressible Fellow Colonist': The Biological Invasion of Prickly Pear (*Opuntia ficus-indica*) in the Eastern Cape, c. 1890–c. 1910," in Dovers, Edgecombe, and Guest, *South Africa's Environmental History*, 139–59; van Sittert, "'The Seed Blows About in Every Breeze': Noxious Weed Eradication in the Cape Colony, 1860–1909," *Journal of Southern African Studies* 26, no. 4 (2000): 655–74; van Sittert, "'Keeping the Enemy at Bay': The Extermination of Wild Carnivora in the Cape Colony, 1889–1910," *Environmental History* 3, no. 3 (July 1998): 333–56; Richard Grove, "Early Themes in African Conservation: The Cape in the Nineteenth Century," in *Conservation in Africa*, ed. David M. Anderson and Richard Grove, 21–39 (Cambridge: Cambridge University Press, 1987); Grove, "Scottish Missionaries, Evangelical Discourses and the Origins of Conservation Thinking

in Southern Africa 1820–1900," *Journal of Southern African Studies* 15, no. 2 (1989): 163–87; Grove, "Scotland in South Africa: John Croumbie Brown and the Roots of Settler Environmentalism," in Griffiths and Robin, *Ecology and Empire*, 139–53; Karen Brown, "The Conservation and Utilisation of the Natural World: Silviculture in the Cape Colony, c. 1902–1910," *Environment and History* 7, no. 4 (November 2001): 427–47; Dan Gilfoyle, "The Heartwater Mystery: Veterinary and Popular Ideas about Tick-Borne Animal Diseases at the Cape, c. 1877–1910," *Kronos: The Journal of Cape History* 29 (2003): 139–60.

12. A notable exception is Farieda Khan, "Rewriting South Africa's Conservation History—The Role of the Native Farmer's Association," *Journal of Southern African Studies*, 20, no. 4 (1994): 499–516. For a pointed critique of Beinart's recent *Rise of Conservation* in this regard, see Lance van Sittert, "The Nature of Power: Cape Environmental History, the History of Ideas, and Neoliberal Historiography," *Journal of African History* 45, no. 2 (2004): 305–13.

13. Nancy Rose Hunt, *A Colonial Lexicon of Birth Ritual, Medicalization, and Mobility in the Congo* (Durham, N.C.: Duke University Press, 1999), 24, 323.

14. Jesse C. Ribot and Nancy Lee Peluso, "A Theory of Access," *Rural Sociology* 68, no. 2 (2003): 153–81. See also Jesse C. Ribot, "Theorizing Access: Forest Profits along Senegal's Charcoal Commodity Chain," *Development and Change* 29, no. 2 (April 1998): 307–41, and "From Exclusion to Participation: Turning Senegal's Forestry Policy Around?" *World Development* 23, no. 9 (1995): 1587–99.

15. Sara Berry, *No Condition Is Permanent: The Social Dynamics of Agrarian Change in Sub-Saharan Africa* (Madison: University of Wisconsin Press, 1993), 3, 12–13, 15–17; Berry, "Social Institutions and Access to Resources," *Africa* 59, no. 1 (1989): 41–55.

16. Ribot and Peluso, "Theory of Access," 158–60.

17. Hunt, *Colonial Lexicon*, 23–24.

18. For recent critiques of the focus on conflict in the environment and development literature, see J. Peter Brosius, "Analyses and Interventions: Anthropological Engagements with Environmentalism," *Current Anthropology* 40, no. 3 (1999): 277–309; Anna Tsing, "Becoming a Tribal Elder and Other Green Development Fantasies," in *Transforming the Indonesian Uplands: Marginality, Power and Production*, ed. Tania Murray Li (Amsterdam: Harwood Academic Publishers, 1999), 159–60.

19. Sara Berry describes how observers have often misunderstood and misdiagnosed such negotiations and "negotiability" in African societies, in "Resource Access and Management as Historical Processes: Conceptual and Methodological Issues," in *Access, Control and Management of Natural Resources in Sub-Saharan Africa—Methodological Considerations*, ed. Christian Lund and Henrik Secher Marcussen, Occasional Paper 13 (International Development Studies, Roskilde University, Denmark, 1994), 24.

20. Some seminal texts include Terence Ranger, "The Invention of Tradition in Colonial Africa," in *The Invention of Tradition*, ed. Eric Hobsbawm and

Terence Ranger (Cambridge: Cambridge University Press, 1983), 211–62; Martin Chanock, *Law, Custom and Social Order: The Colonial Experience in Malawi and Zambia* (Cambridge: Cambridge University Press, 1985); Sally Falk Moore, *Social Facts and Fabrications: "Customary" Law on Kilimanjaro, 1880–1980* (Cambridge: Cambridge University Press, 1986); Nicholas Thomas, "The Inversion of Tradition," *American Ethnologist* 19, no. 2 (May 1992): 213–32; Terence Ranger, "The Invention of Tradition Revisited: The Case of Colonial Africa," in *Legitimacy and the State in Twentieth-Century Africa*, ed. Terence Ranger and Olufemi Vaughan (London: Macmillan Press, 1993), 62–111; Peter Pels, "The Pidginization of Luguru Politics: Administrative Ethnography and the Paradoxes of Indirect Rule," *American Ethnologist* 23, no. 4 (1996): 738–61; Peter Pels, "The Anthropology of Colonialism: Culture, History and the Emergence of Western Governmentality," *Annual Review of Anthropology* 26 (1997): 176–77.

21. Emmanuel Kreike critiques this temporal focus in *Re-Creating Eden: Land Use, Environment, and Society in Southern Angola and Northern Namibia* (Portsmouth, N.H.: Heinemann, 2004), 4–5. Some recent analyses in this vein are Thomas Spear, *Mountain Farmers* (Berkeley: University of California Press, 1997); Fiona Mackenzie, *Land, Ecology and Resistance in Kenya, 1880–1952* (Edinburgh: Edinburgh University Press, 1998); Terence Ranger, *Voices from the Rocks: Nature, Culture and History in the Matopos Hills of Zimbabwe* (Oxford: James Currey, 1999); Sara Berry, *Chiefs Know Their Boundaries: Essays on Property, Power, and the Past in Asante, 1896–1996* (Portsmouth, N.H.: Heinemann, 2001); Ingrid Yngstrom, "Representations of Custom, Social Identity and Environmental Relations in Central Tanzania," in *Social History and African Environments*, ed. William Beinart and JoAnn McGregor (Athens: Ohio University Press; Oxford: James Currey, 2003), 175–95.

22. Various authors have examined the ambiguous position of African chiefs and headmen within colonial power structures. Some prominent examples in the southern African literature include Allen Isaacman, "Chiefs, Rural Differentiation and Peasant Protest: The Mozambican Forced Cotton Regime, 1938–61," *African Economic History* 14 (1985): 15–56, and *Cotton Is the Mother of Poverty: Peasants, Work, and Rural Struggle in Colonial Mozambique, 1938–1961* (Portsmouth, N.H.: Heinemann, 1996), 225–29; Beinart and Bundy, *Hidden Struggles*, 78–105, 160–62; Ivan Evans, *Bureaucracy and Race: Native Administration in South Africa* (Berkeley: University of California Press, 1997), chap. 6.

23. An insightful discussion of these themes in the Indian context is K. Sivaramakrishnan, "Colonialism and Forestry in India: Imagining the Past in Present Politics," *Comparative Studies in Society and History* 37, no. 1 (January 1995): 3–40. Bruce Berman notes that such infighting between officials of specialist and technical departments and the central administrative staff commonly plagued state bureaucracies in many colonial spheres, in *Control and Crisis in Colonial Kenya: The Dialectic of Domination* (London: James Currey, 1990), 85–88. For a useful review of negotiation and contestation within colonial bureaucracies, see

Ann Laura Stoler and Frederick Cooper, "Between Metropole and Colony: Rethinking a Research Agenda," in *Tensions of Empire: Colonial Cultures in a Bourgeois World*, ed. Frederick Cooper and Ann Laura Stoler (Berkeley: University of California Press, 1997), 1–56.

24. For some diverse colonial examples of African leaders similarly invoking "customary" authority to serve various interests, see Pauline Peters, "Struggles over Water, Struggles over Meaning: Cattle, Water and the State in Botswana," *Africa* 54, no. 3 (1984): 29–49, and *Dividing the Commons: Politics, Policy, and Culture in Botswana* (Charlottesville: University Press of Virginia, 1994), chap. 3; Richard Roberts, "The Case of Faama Mademba Sy and the Ambiguities of Legal Jurisdiction in Early Colonial French Soudan," in *Law in Colonial Africa*, ed. Kristin Mann and Richard Roberts (Portsmouth, N.H.: Heinemann, 1991), 185–98; Berry, *Chiefs Know Their Boundaries*.

25. For conceptual frameworks of environmental access and entitlements as well as their applications, see Ribot and Peluso, "Theory of Access"; Melissa Leach, Robin Mearns, and Ian Scoones, "Environmental Entitlements: Dynamics and Institutions in Community-Based Natural Resource Management," *World Development* 27, no. 2 (1999): 225–47; Thembela Kepe and Ian Scoones, "Creating Grasslands: Social Institutions and Environmental Change in Mkambati Area, South Africa," *Human Ecology* 27, no. 1 (1999): 29–53; Kepe, *Environmental Entitlements in Mkambati*; Ribot, "Theorizing Access"; Robin Mearns, "Institutions and Natural Resource Management: Access to and Control over Woodfuel in East Africa," in *People and Environment in Africa*, ed. Tony Binns (Chichester: John Wiley and Sons, 1995), 103–14; Berry, "Social Institutions and Access to Resources."

26. See, for example, Louise Fortmann, "The Tree Tenure Factor in Agroforestry with Particular Reference to Africa," and Dianne Rocheleau, "Women, Trees, and Tenure: Implications for Agroforestry," in *Whose Trees?: Proprietary Dimensions of Forestry*, ed. Louise Fortmann and John Bruce (Boulder: Westview Press, 1988), 16–33 and 254–72, respectively; Gerald Leach and Robin Mearns, *Beyond the Woodfuel Crisis: People, Land and Trees in Africa* (London: Earthscan Publications, 1988); Judith Carney and Michael Watts, "Disciplining Women?: Rice, Mechanization, and the Evolution of Mandinka Gender Relations in Senegambia," *Signs* 16, no. 4 (1991): 651–81; Melissa Leach, *Rainforest Relations: Gender and Resource Use among the Mende of Gola, Sierra Leone* (Washington, D.C.: Smithsonian Institution Press, 1994); Mearns, "Institutions and Natural Resource Management"; Barbara Thomas-Slayter and Dianne Rocheleau, eds., *Gender, Environment, and Development in Kenya: A Grassroots Perspective* (Boulder: Lynne Rienner, 1995); Donald S. Moore, "Marxism, Culture, and Political Ecology: Environmental Struggles in Zimbabwe's Eastern Highlands," in *Liberation Ecologies: Environment, Development, Social Movements*, ed. Richard Peet and Michael Watts (New York: Routledge, 1996), 125–47; Richard A. Schroeder, *Shady Practices: Agroforestry and Gender Politics in the Gambia* (Berkeley: University of

California Press, 1999); Sian Sullivan, "Gender, Ethnographic Myths and Community-Based Conservation in a Former Namibian 'Homeland,'" in *Rethinking Pastoralism in Africa: Gender, Culture and the Myth of the Patriarchal Pastoralist,* ed. Dorothy Hodgson (Oxford: James Currey, 2000), 142–64.

27. For example, Terence Ranger, "Women and Environment in African Religion: The Case of Zimbabwe," in Beinart and McGregor, *Social History and African Environments,* 72–86; Heidi Gengenbach, "'I'll Bury You in the Border!': Women's Land Struggles in Post-War Facazisse (Magude District), Mozambique," *Journal of Southern African Studies* 24, no. 1 (1998): 7–36; David Lee Schoenbrun, *A Green Place, A Good Place: Agrarian Change, Gender, and Social Identity in the Great Lakes Region to the 15th Century* (Portsmouth, N.H.: Heinemann, 1998); Melissa Leach and Cathy Green, "Gender and Environmental History: From Representation of Women and Nature to Gender Analysis of Ecology and Politics," *Environment and History* 3, no. 3 (1997): 343–70; Henrietta Moore and Megan Vaughan, *Cutting Down Trees: Gender, Nutrition, and Agricultural Change in the Northern Province of Zambia, 1890–1990* (Portsmouth, N.H.: Heinemann, 1994); Carney and Watts, "Disciplining Women?"; Elias Mandala, *Work and Control in a Peasant Economy: A History of the Lower Tchiri Valley in Malawi, 1859–1960* (Madison: University of Wisconsin Press, 1990).

28. A few notable exceptions are Anne Mager, "'The People Get Fenced': Gender, Rehabilitation and African Nationalism in the Ciskei and Border Region, 1945–1955," *Journal of Southern African Studies* 18, no. 4 (1992): 761–82; Pamela Maack, "'We Don't Want Terraces!': Protest and Identity under Uluguru Land Usage Scheme," and Jamie Monson, "Canoe-Building under Colonialism: Forestry and Food Policies in the Inner Kilombero Valley 1920–40," in *Custodians of the Land: Ecology and Culture in the History of Tanzania,* ed. Gregory Maddox, James Giblin, and Isaria Kimambo (Athens: Ohio University Press, 1996), 152–74 and 200–12, respectively; F. Mackenzie, *Land, Ecology and Resistance in Kenya, 1880–1952.*

29. See, for example, Clive van Horen and Anton Eberhard, "Energy, Environment and the Rural Poor in South Africa," *Development Southern Africa* 12, no. 2 (April 1995): 197–211; Eberhard and van Horen, *Poverty and Power;* Barry Munslow et al., *The Fuelwood Trap: A Study of the SADCC Region* (London: Earthscan Publications, 1988).

30. Examples of such neglect include Beinart, *Rise of Conservation,* and Dovers, Edgecombe, and Guest, *South Africa's Environmental History,* which latter even explicitly concedes: "Gender is a field which is not dealt with here" (7). By contrast, Nancy Jacobs incorporates a discussion of gender for certain historical periods in her recent *Environment, Power, and Injustice: A South African History* (Cambridge: Cambridge University Press, 2003).

31. Roderick Neumann describes similar official struggles in *Imposing Wilderness: Struggles over Livelihood and Nature Preservation in Africa* (Berkeley: University of California Press, 1998), 97–121. Very recently, scholars involved in

contemporary land reform and development processes along the southern Wild Coast in the Transkei have begun reflecting on such historical themes. See Palmer, Timmermans, and Fay, *From Conflict to Negotiation*, chap. 4; Derick Fay, "Paternalist and Commercial Visions of Forest Conservation in South Africa: Intergovernmental Disputes and African Agency in the Coastal Transkei, ca. 1890–1936" (paper presented at "Producing and Consuming Narratives," American Society for Environmental History annual meeting, Denver, Colorado, 20–24 March 2002).

32. For analysis of the contested meaning of colonial interventions regarding other types of forest resources—such as wildlife, livestock pasturage, and soils for cultivation—see Jacob Tropp, "Dogs, Poison and the Meaning of Colonial Intervention in the Transkei, South Africa," *Journal of African History* 43, no. 3 (2002): 451–72; Tropp, "The Contested Nature of Colonial Landscapes: Historical Perspectives on Livestock and Environments in the Transkei," *Kronos: The Journal of Cape History* 30 (November 2004): 118–37; Tropp, "Roots and Rights in the Transkei: Colonialism, Natural Resources, and Social Change, 1880–1940" (Ph.D. thesis, Department of History, University of Minnesota, June 2002), esp. chap. 4.

33. On Africans debating and negotiating the meaning of colonial "customary" law in daily practice, see Thomas McClendon, *Genders and Generations Apart in South Africa: Labor Tenants and Customary Law in Segregation-Era Natal, 1920s to 1940s* (Portsmouth, N.H.: Heinemann, 2002). See also Neumann, *Imposing Wilderness*, 40–44, for some similar forest conflicts in Tanzania.

34. Frederick Cooper discusses how struggles within colonizing groups often created openings and opportunities for popular contestation, which in turn could force "European adaptation (and resistance) to the initiatives of the colonized." "Conflict and Connection: Rethinking Colonial African History," *American Historical Review* 99, no. 5 (December 1994): 1530–32.

35. Some important works in these areas include Paul Richards, *Indigenous Agricultural Revolution: Ecology and Food Production in West Africa* (London: Unwin, 1985), and "Ecological Change and the Politics of African Land Use," *African Studies Review* 26, no. 2 (June 1983): 1–72; Robert Harms, *Games against Nature: An Eco-Cultural History of the Nunu of Equatorial Africa* (New Haven, Conn.: Yale University Press, 1987); David Cohen and E. S. Atieno Odhiambo, *Siaya: The Historical Anthropology of an African Landscape* (Athens: Ohio University Press, 1989); K. Wilson, "Trees in Fields in Southern Zimbabwe," *Journal of Southern African Studies* 15, no. 2 (1989): 369–83; Kojo Sebastian Amanor, *The New Frontier: Farmer Responses to Land Degradation: A West African Study* (Atlantic Highlands, N.J.: Zed Books, 1994); James Giblin, *The Politics of Environmental Control in Northeastern Tanzania, 1840–1940* (Philadelphia: University of Pennsylvania, 1992); H. Moore and Vaughan, *Cutting Down Trees*; James McCann, *People of the Plow: An Agricultural History of Ethiopia, 1800–1990* (Madison: University of Wisconsin Press, 1995); Michele Wagner, "Environment, Community and History: 'Nature in the Mind' in Nineteenth and Early Twentieth-Century Buha, Tanzania," in *Custodians of the Land: Ecology and Culture in the History*

of Tanzania, ed. Gregory Maddox, James Giblin, and Isaria Kimambo (Athens: Ohio University Press, 1996), 175–99; James Fairhead and Melissa Leach, *Misreading the African Landscape: Society and Ecology in a Forest-Savanna Mosaic* (Cambridge: Cambridge University Press, 1996); Ute Luig and Achim von Oppen, eds., "The Making of African Landscapes," special issue, *Paideuma: Mitteilungen zur Kulturkunde* 43 (1997); Spear, *Mountain Farmers*; Ranger, *Voices from the Rocks*; Tamara Giles-Vernick, *Cutting the Vines of the Past: Environmental Histories of the Central African Rain Forest* (Charlottesville: University Press of Virginia, 2002); Sandra Greene, *Sacred Sites and the Colonial Encounter: A History of Meaning and Memory in Ghana* (Bloomington: Indiana University Press, 2002).

36. On the problems of reifying "indigenous knowledge" and the historically contingent and situational nature of local environmental practices and belief systems, see Kojo Sebastian Amanor, "Managing the Fallow: Weeding Technology and Environmental Knowledge in the Krobo District of Ghana," in "Indigenous Agricultural Systems and Development," special issue, *Agriculture and Human Values* 8, no. 1 (1991): 5–13; Jane Jacobs, "Earth Honoring: Western Desires and Indigenous Knowledges," in *Writing Women and Space: Colonial and Postcolonial Geographies*, ed. Alison Blunt and Gillian Rose (New York: Guilford Press, 1994), 169–96; B. B. Mukamuri, "Local Environmental Conservation Strategies: Karanga Religion, Politics and Environmental Control," *Environment and History* 1 (1995): 297–311; K. Wilson, "'Water Used to Be Scattered in the Landscape': Local Understandings of Soil Erosion and Land Use Planning in Southern Zimbabwe," *Environment and History* 1, no. 3 (1995): 281–96; J. Peter Brosius, "Endangered Forest, Endangered People: Environmentalist Representations of Indigenous Knowledge," *Human Ecology* 25, no. 1 (1997): 47–69; Akhil Gupta, *Postcolonial Developments: Agriculture in the Making of Modern India* (Durham, N.C.: Duke University Press, 1998); Roy Ellen, Peter Parkes, and Alan Bicker, eds., *Indigenous Environmental Knowledge and Its Transformations: Critical Anthropological Perspectives* (Amsterdam: Harwood Academic, 2000); Bruce Willems-Braun, "Buried Epistemologies: The Politics of Nature in (Post)colonial British Columbia," *Annals of the Association of American Geographers* 81, no. 1 (1997): 3–31; Tamara Giles-Vernick, "We Wander Like Birds: Migration, Indigeneity, and the Fabrication of Frontiers in the Sangha River Basin of Equatorial Africa," *Environmental History* 4, no. 2 (April 1999): 168–97.

37. Melissa Leach, for instance, explores some of these dimensions extensively in *Rainforest Relations*.

38. Historians have generally bypassed such opportunities in several recent publications, including N. Jacobs, *Environment, Power, and Injustice*; Beinart, *Rise of Conservation*; Dovers, Edgecombe, and Guest, *South Africa's Environmental History*; and *Kronos: The Journal of Cape History* 29 (2003), special issue on environmental history. See Lance van Sittert's recent overview of this literature in "Nature of Power." By contrast, scholars involved in contemporary land reform and development have begun to touch on such themes: Daniel Lieberman,

"Ethnobotanical Assessment of the Dwesa and Cwebe Nature Reserves," in Institute of Social and Economic Research, *Indigenous Knowledge, Conservation Reform* . . . , 40–84; Kepe, *Environmental Entitlements in Mkambati*. On the environment as an underexploited avenue for exploring cultural history in African studies more generally, see William Beinart, "African History and Environmental History," *African Affairs* 99, no. 395 (April 2000): 298.

39. Giles-Vernick, *Cutting the Vines of the Past*.

40. The following description synthesizes multiple sources: Elsa Pooley, *The Complete Field Guide to the Trees of Natal, Zululand and Transkei* (Pietermaritzburg: Natal Flora Trust Publications, 1993); Cooper and Swart, *Transkei Forest Survey*; Sizwe Cawe, "Coastal Forest Survey: A Classification of the Coastal Forests of Transkei and an Assessment of Their Timber Potential" (unpublished paper, Umtata, University of Transkei, 1990), and "A Quantitative and Qualitative Survey of the Inland Forests of Transkei" (M.S. thesis, University of Transkei, 1986); J. M. Feeley, "The Early Farmers of Transkei, Southern Africa before A.D. 1870," *Cambridge Monographs in African Archaeology* 24, Bar International Series 378 (1987); W. D. Hammond-Tooke, *Command or Consensus: The Development of Transkeian Local Government* (Cape Town: David Philip, 1975), chap. 1; H. C. Schunke, "The Transkeian Territories: Their Physical Geography and Ethnology," *Transactions of the South African Philosophical Society* 8, part 1 (1893): 1–15.

41. These fall into the category of mist belt mixed *Podocarpus* forests.

42. Mangrove communities and swamp forests exist in the northernmost areas of the coastal forests, for instance. Deep gorges also interrupt some major forest blocks in this region as rivers wind their way to the sea.

43. The following paragraphs, describing the colonial incorporation of African societies, draw from Crais, *Politics of Evil*; Les Switzer, *Power and Resistance in an African Society: The Ciskei Xhosa and the Making of South Africa* (Madison: University of Wisconsin Press, 1993); Beinart and Bundy, *Hidden Struggles*; Christopher Saunders, *The Annexation of the Transkeian Territories*, Archives Year Book for South African History 39 (Pretoria: Government Printer, 1978); Bundy, *Rise and Fall*; Robert Ross, *Adam Kok's Griquas: A Study in the Development of Stratification in South Africa* (Cambridge: Cambridge University Press, 1976); Hammond-Tooke, *Command or Consensus*.

44. I use the term "Mfengu" here with caution, as the origins of the groups traditionally designated as "Fingo" in the colonial literature, or as "Mfengu" in the majority of scholarly interpretations, have recently come under intense scrutiny. See Alan Webster, "Unmasking the Fingo: The War of 1835 Revisited," in *The Mfecane Aftermath: Reconstructive Debates in Southern African History*, ed. Carolyn Hamilton (Johannesburg: Witwatersrand University Press, 1995), 241–76; Timothy Stapleton, "Oral Evidence in a Pseudo-Ethnicity: The Fingo Debate," *History in Africa* 22 (1995): 358–69, and "The Expansion of a Pseudo-Ethnicity in the Eastern Cape: Reconsidering the Fingo 'Exodus' of 1865," *International Journal of African Historical Studies* 29, no. 2 (1996): 233–50.

45. See Stapleton, "Expansion of a Pseudo-Ethnicity," for a critical analysis of this Mfengu "exodus" from the Cape.

46. Saunders, *Annexation*, 168–73.

47. In 1890, the chief magistracies of Thembuland and Transkei were combined.

48. Pondoland's early administration went through unique changes. Upon annexation, Eastern Pondoland was administered by the chief magistrate of East Griqualand, while Western Pondoland came under the Thembuland chief magistracy; after 1896, the latter managed the whole of Pondoland.

49. Grove, "Scottish Missionaries"; Richard Grove, "Colonial Conservation, Ecological Hegemony and Popular Resistance: Towards a Global Synthesis," in *Imperialism and the Natural World*, ed. John Mackenzie (Manchester: Manchester University Press, 1990), 37; Grove, "Scotland in South Africa." See also Grove's broader work on imperial expansion and environmentalism: "Conserving Eden: The (European) East India Companies and their Environmental Policies on St. Helena, Mauritius and in Western India, 1660 to 1854," *Comparative Studies in Society and History* 35, no. 2 (April 1993): 318–51; *Green Imperialism: Colonial Expansion, Tropical Island Edens and the Origins of Environmentalism, 1600–1860* (Cambridge: Cambridge University Press, 1995).

50. Grove, "Scottish Missionaries"; King, "Historical Sketch," 5–7; Sim, *Forests and Forest Flora*, 85–86.

51. The Forest Department in the Transkei functioned as a subdepartment in other administrative branches until 1906, when it first became an independent entity. Until 1891, department staff reported to the superintendent of woods and forests, based in Cape Town, who in turn reported to the commissioner of crown lands and public works (CLPW). When the superintendent's position was abolished in 1891, the head of each conservancy under Cape control, including the Transkeian conservator, then reported directly to the commissioner. In 1892, the Forest Department became a subdivision of the Department of Lands, Mines, and Agriculture (LMA), which then was reorganized as the Department of Agriculture in 1893. This organizational structure continued until the creation of an independent Forest Department in the mid-1900s. Sim, *Forests and Forest Flora*, 83–84, 86; King, "Historical Sketch," 4–16; Keet, *Historical Review*, 57. For the sake of clarity, I will be referring to all annual reports of the Forest Department in the late nineteenth and early twentieth centuries simply as *Annual Report* followed by the year concerned.

52. Beinart, "Soil Erosion, Conservationism"; Moll, *No Blade of Grass*; Hendricks, "Loose Planning and Rapid Resettlement"; Native Trust and Land Act, No. 18 of 1936; General Trust Regulations, General Notice (GN) 494 of 1937; Cape Town Archives Repository (CTA), Archives of the Chief Magistrate of the Transkeian Territories (CMT) 3/1333, 24/24/1, Circular No. H.2200/I, "Forests on Land Transferred to the South African Native Trust," Director of Forestry, Pretoria, to all officers of the Division of Forestry, 16 January 1937; Conservator of Forests, Transkeian Conservancy (FCT) to Director of Forestry, Pretoria, 6 August

1937; *Annual Report 1936–37*, 10–11; W. J. G. Mears, "A Study in Native Administration: The Transkeian Territories, 1894–1943" (D.Litt., University of South Africa, 1947), 193–94, 212.

53. Ranger, *Voices from the Rocks*, 1–5.

54. I use the term "coloured" here to describe those individuals, identified as "Hottentots" in the colonial literature, who migrated to KwaMatiwane from the Eastern Cape in the late nineteenth century, principally as woodcutters and sawyers.

55. Kate Crehan, *The Fractured Community: Landscapes of Power and Gender in Rural Zambia* (Berkeley: University of California Press, 1997), 8–10, 227.

56. Ibid., 9.

57. The following description is based on Cawe, "A Quantitative and Qualitative Survey," A8–9, A13–14; Colin T. Johnson and Sizwe Cawe, "Analysis of the Tree Taxa in Transkei," *South African Journal of Botany* 53, no. 5 (1987): 388–94; Sizwe Cawe and Bruce McKenzie, "The Afromontane Forests of Transkei, Southern Africa. II: A Floristic Classification," *South African Journal of Botany* 55, no. 1 (1989): 31–39; Mlungisi Ndima, "A History of the Qwathi People from Earliest Times to 1910" (M.A. thesis, Rhodes University, 1988), 10–11.

58. Isaacman, *Cotton Is the Mother of Poverty*, 14; Adam Ashforth, *The Politics of Official Discourse in Twentieth-Century South Africa* (Oxford: Clarendon Press, 1990).

59. Godfrey regularly contributed articles on natural history in the Tsolo District and elsewhere to missionary and natural history periodicals. In the preface to the dictionary, he thanks a few key individuals at the Somerville Mission Station in the Tsolo District for their constant assistance in compiling the manuscript. South African Library, Manuscripts collection, MSB 783, unpublished manuscript, 3rd edition of "A Xhosa-English Dictionary," file 1, "Preface," April 9, 1946, Pirie Mission, King Williams Town.

60. Albert Kropf, *A Kaffir-English Dictionary* (Stutterheim: Lovedale Mission Press, 1899); James McLaren, *Concise Kaffir-English Dictionary* (London: Longmans, Green, and Co., 1915).

61. While Godfrey's attention to detail is helpful, this is not to say that such dictionary sources are unproblematic. See Derek Peterson, "Translating the Word: Dialogism and Debate in Two Gikuyu Dictionaries," *Journal of Religious History* 23, no. 1 (February 1999): 31–50.

62. Jan Vansina, "Epilogue: Fieldwork in History," in *In Pursuit of History: Fieldwork in Africa*, ed. Carol Keyes Adenaike and Jan Vansina (Portsmouth, N.H.: Heinemann, 1996), 135; Paul la Hausse, "Oral History and South African Historians," in Brown, Manning, Shapiro, and Wiener, *History from South Africa*, 342–50.

63. Vansina, "Epilogue," 135–38.

64. See Luise White, "Tsetse Visions: Narratives of Blood and Bugs in Colonial Northern Rhodesia, 1931–39," *Journal of African History* 36, no. 2 (1995): 219–45.

65. For some recent perspectives on the problems of treating oral sources uncritically, see Elizabeth Tonkin, *Narrating Our Pasts: The Social Construction of Oral History* (Cambridge: Cambridge University Press, 1992), 113–15; Luise White, "'They Could Make Their Victims Dull': Genders and Genres, Fantasies and Cures in Colonial Southern Uganda," *American Historical Review* 100, no. 5 (December 1995): 1379–1402"; Minkley and Rasool, "Orality, Memory, and Social History in South Africa," 89–99.

66. Vansina, "Epilogue," 138–39.

67. Tonkin, *Narrating Our Pasts*, 113–15; L. White, "They Could Make Their Victims Dull," 1382; Luise White, *Speaking with Vampires: Rumor and History in Colonial Africa* (Berkeley: University of California Press, 2000), 91; Marjorie Mbilinyi, "'I'd Have Been a Man': Politics and the Labor Process in Producing Personal Narratives," in *Interpreting Women's Lives: Feminist Theory and Personal Narratives*, ed. Personal Narratives Group (Bloomington: Indiana University Press, 1989), 204–27.

68. Interview with Mambanjwa Kholwane, Preston Farms, Umtata District, 24 February 1998.

69. Given the local political situation, I have chosen not to name the individuals involved here.

70. For a variety of examples of how such African fieldwork experiences and contexts help shape the production of historical narratives, see Carol Keyes Adenaike and Jan Vansina, eds., *In Pursuit of History: Fieldwork in Africa* (Portsmouth, N.H.: Heinemann, 1996).

71. I explore this particular history in greater detail in "Displaced People, Replaced Narratives: Forest Conflicts and Historical Perspectives in the Tsolo District, Transkei," *Journal of Southern African Studies* 29, no. 1 (2003): 207–33.

72. Isabel Hofmeyr, *"We Spend Our Years as a Tale That Is Told": Oral Historical Narrative in a South African Chiefdom* (Portsmouth, N.H.: Heinemann, 1993), 160.

73. D. S. Moore, "Marxism, Culture, and Political Ecology," 135–36; and Moore, "Subaltern Struggles and the Politics of Place: Remapping Resistance in Zimbabwe's Eastern Highlands," *Cultural Anthropology* 13, no. 3 (August 1998): 368–70.

74. In some places in their new locations in the Umtata District, people who were relocated are still able to see the hills of their former settlements in the distance. At one point during my conversation with Jemimah Dubo at her home in Payne, she guided me to the doorway and pointed toward the horizon to the northwest: "I still see it even from here," she explained. "Yes. . . . The last mountain with fog, those last mountains. . . . We are from that one with the mist. That is Qelana there." Interview with Jemimah Dubo, Payne, Umtata District, 24 February 1998.

75. On how oral narratives produce moral arguments regarding changes in the social and natural order, through contrasting representations of life "then" and

"now," see Tonkin, *Narrating Our Pasts*, 36–37; Tamara Giles-Vernick, "Na lege ti guiriri (On the Road of History): Mapping Out the Past and Present in M'Bres Region, Central African Republic," *Ethnohistory* 43, no. 2 (1996): 245–75; JoAnn McGregor, "Conservation, Control and Ecological Change: The Politics and Ecology of Colonial Conservation in Shurugwi, Zimbabwe," *Environment and History* 1, no. 3 (1995): 257–79; Corinne Kratz, "We've Always Done It Like This . . . Except for a Few Details: 'Tradition' and 'Innovation' in Okiek Ceremonies," *Comparative Studies in Society and History* 35, no. 1 (1993): 40–42. For similar processes in the recollection of resettlement experiences in South Africa, see de Wet, *Moving Together*; Andrew D. Spiegel, "Struggling with Tradition in South Africa: The Multivocality of Images of the Past," in *Social Construction of the Past: Representation as Power*, ed. George Bond and Angela Gilliam (London: Routledge, 1994), 194; Hofmeyr, "*We Spend Our Years as a Tale That Is Told*," 98–101.

76. Hofmeyr, "*We Spend Our Years as a Tale That Is Told*," 101.

77. Luise White, "True Stories: Narrative, Event, History, and Blood in the Lake Victoria Basin," in *African Words, African Voices: Critical Practices in Oral History*, ed. Luise White, Stephan Miescher, and David Cohen (Bloomington: Indiana University Press, 2001), 281–99.

78. Tonkin, *Narrating Our Pasts*, 1; Michel-Rolph Trouillot, *Silencing the Past: Power and the Production of History* (Boston: Beacon Press, 1995), 26–27.

CHAPTER 1

1. Thembuland was not formally annexed until 1885. Elsie J. C. Wagenaar, "A History of the Thembu and Their Relationship with the Cape, 1850–1900" (Ph.D. thesis, Rhodes University, 1988), 209.

2. Saunders, *Annexation*, 26; Act 38 of 1877; Hammond-Tooke, *Command or Consensus*, 21–2.

3. For instance, see Cape of Good Hope, Department of Native Affairs, *Blue Book on Native Affairs (BBNA) 1878*, 84–85, Alexander R. Welsh, Magistrate with Umditshwa; Bundy, *Rise and Fall*, 99–101; Beinart and Bundy, *Hidden Struggles*, 106–12.

4. Ndima, "History of the Qwathi People," 1, 8, 41–43, 49, 52–54; W. D. Hammond-Tooke, *The Tribes of Umtata District*, Ethnological Publications 35 (Pretoria: Union of South Africa, Department of Native Affairs, 1956), 11.

5. Wagenaar, "History of the Thembu," 171–72; Ndima, "History of the Qwathi People," 56–60, 65, 80–82, 85, 87–88, 96–101, 104–6.

6. Ndima, "History of the Qwathi People," 41–42, 56–59; Bundy, *Rise and Fall*, 57, 69; Hammond-Tooke, *Command or Consensus*, 13; Beinart and Bundy, *Hidden Struggles*, 112; Stapleton, "Expansion of a Pseudo-Ethnicity." Some Bhele, Zizi, and Hlubi groups had settled among Qwathi, Thembu, and Mpondomise communities decades before, amid the local upheavals of the 1820s.

7. *BBNA 1878*, 84–85, Alexander R. Welsh, Magistrate with Umditshwa; Wagenaar, "History of the Thembu," 178; Bundy, *Rise and Fall*, 99–101; Beinart, *Political Economy of Pondoland*, 139–43; Beinart and Bundy, *Hidden Struggles*, 106–12.

8. For a fuller account of the rebellion and its extent, see Crais, *Politics of Evil*, 35–70, 94–95; Sean Redding, "Sorcery and Sovereignty: Taxation, Witchcraft, and Political Symbols in the 1880 Transkeian Rebellion," *Journal of Southern African Studies* 22, no. 2 (1996): 249–69; Christopher Saunders, "The Transkeian Rebellion of 1880–81: A Case-Study of Transkeian Resistance to White Control," *South African Historical Journal* 8 (1976): 32–39; Saunders, *Annexation*, 91–109.

9. See, for example, official histories of the resettlements that followed the 1857 Cattle-Killing episode. *BBNA 1879*, 115, RM M. B. Shaw, report on No. 1 Kentani division, Western Gcalekaland; *BBNA 1884*, 154–55, Chief Magistrate East Griqualand (CMK) Charles Brownlee; CTA, Archives of the Secretary for Agriculture (AGR) 144, 595, part 1, FCT to Asst. Commissioner CLPW, 1 February 1892; CTA AGR 739, F1930, RM Willowvale Liefeldt to CMT, 5 February 1897; Switzer, *Power and Resistance*, 72–73.

10. Cape of Good Hope, Ministerial Department of Crown Lands and Public Works, G.2–1884, *Report of the Griqualand East Land Commission*; *BBNA 1884*, 154–58, CMK Charles Brownlee; *BBNA 1885*, 159–60, George M. Theal; W. E. Stanford, *The Reminiscences of Sir Walter Stanford*, ed. J. W. Macquarrie (Cape Town: van Riebeeck Society, 1958), 1:176–86; Ndima, "History of the Qwathi People," 157–58; Saunders, *Annexation*, 104–5.

11. For this and related forest practices during wartime, see CTA AGR 766, F1930, Henkel to CMT, 25 October 1895; CTA CMT 3/87, RM Engcobo to CMT, 5 December 1895; Stanford, *Reminiscences of Sir Walter Stanford*, 1:185–86; Ndima, "History of the Qwathi People," 42–43, 135–40, 192–94; Rev. John Robert Lewis Kingon, "Some Place-Names of Tsolo," *Reports of the South African Association for the Advancement of Science* 4 (1916): 606–7; interview with Anderson Joyi, Mputi, Umtata District, 23 December 1997. See also Jeffrey Peires, *The House of Phalo: A History of the Xhosa People in the Days of Their Independence* (Berkeley: University of California Press, 1981), 25, 60, 66; Archie Mafeje, "Religion, Class and Ideology in South Africa," in *Religion and Social Change in Southern Africa*, ed. Michael Whisson and Martin West (Cape Town: David Philip, 1975), 180.

12. Cape of Good Hope, A.25-1881, *Reports from Chief Magistrates and Resident Magistrates in Basutoland, Transkei, etc.*, 11, RM Engcobo W. E. Stanford to CM Tembuland; *BBNA 1882*, 33, RM Umtata A. H. Stanford; 51–52, "Report on Native Affairs, District of Engcobo," RM W. E. Stanford; Ndima, "History of the Qwathi People," 135–40, 143.

13. *BBNA 1882*, 33, RM Umtata A. H. Stanford; 51–52, "Report on Native Affairs, District of Engcobo," RM W. E. Stanford.

14. For example, see CTA CMT 1/36, statement of Gangelizwe before RM Umtata A. H. Stanford, 9 May 1881.

15. See also Beinart, *Political Economy of Pondoland,* 44.

16. Ndima, "History of the Qwathi People," 135–40; *BBNA 1882,* 33, RM Umtata A. H. Stanford; 51–52, "Report on Native Affairs, District of Engcobo," RM W. E. Stanford.

17. For an interesting account of a similar, yet failed relocation scheme after the colonial suppression of rebellion in Southern Rhodesia, see Ranger, *Voices from the Rocks,* 69–72.

18. CTA Papers of the Resident Magistrate of the Engcobo District (1/ECO) 5/1/3/1, RM Engcobo W. E. Stanford to CM Tembuland, letters of 27 April, 8 June, and 11 July 1881; *BBNA 1882,* 51–52, "Report on Native Affairs, District of Engcobo," RM W. E. Stanford; *Report and Proceedings of the Tembuland Commission,* "Appendix P," 12–13, RM W. E. Stanford to CM Tembuland, 18 September 1882; Ndima, "History of the Qwathi People," 150–52; Wagenaar, "History of the Thembu," 177, 197; Beinart and Bundy, *Hidden Struggles,* 50.

19. *BBNA 1882,* 33, RM Umtata A. H. Stanford; *Report of the Griqualand East Land Commission; BBNA 1884,* 154–58, CMK Charles Brownlee; *BBNA 1885,* 159–60, George M. Theal; Wagenaar, "History of the Thembu," 195–96.

20. *Report and Proceedings of the Tembuland Commission; Report of the Griqualand East Land Commission;* Wagenaar, "History of the Thembu," chap. 5; Ndima, "History of the Qwathi People," 150–55; Saunders, *Annexation,* 104–5; CTA Papers of the Resident Magistrate of the Tsolo District (1/TSO) 3/1/7/1, RM Maclear to RM Tsolo, 31 January 1885; CTA 1/TSO 3/1/7/2, Henry Willard, Maclear, to RM Tsolo, 26 August 1888. See also Tropp, "Roots and Rights," 50–63 and 236–42.

21. For more on ethnic conflicts in colonial Africa emerging from different groups' shifting access to opportunities and resources, see Bruce Berman, "Ethnicity, Patronage and the African State: The Politics of Uncivil Nationalism," *African Affairs* 97, no. 388 (July 1998): 324–25; John Lonsdale, "The Moral Economy of Mau Mau: Wealth, Poverty and Civic Virtue in Kikuyu Political Thought," in *Unhappy Valley: Conflict in Kenya and Africa,* ed. Bruce Berman and John Lonsdale (Athens: Ohio University Press, 1992), chap. 12; Leroy Vail, ed., *The Creation of Tribalism in Southern Africa* (London: James Currey, 1989); Sara Berry, *No Condition Is Permanent: The Social Dynamics of Agrarian Change in Sub-Saharan Africa* (Madison: University of Wisconsin Press, 1993), 32–39.

22. On the headmanship as a locus of increasing local political struggle in the colonial Transkei, see William Beinart, "Chieftaincy and the Concept of Articulation: South Africa ca. 1900–1950," *Canadian Journal of African Studies* 19, no. 1 (1985): 95; Evans, *Bureaucracy and Race,* 208; Beinart and Bundy, *Hidden Struggles,* 160–62.

23. Bundy, *Rise and Fall,* 86–87; Cape of Good Hope, A.25-'81, *Reports from Chief Magistrates and Resident Magistrates in Basutoland, Transkei, etc.,* 4–5, "Supplementary Report to Annual Report for Blue-Book for the Year 1880, Dated Umtata, 1st January, 1881," CM Elliot, and 5, RM Umtata A. H. Stanford; *BBNA*

1882, 33, RM Umtata A. H. Stanford; 51–52, "Report on Native Affairs, District of Engcobo," RM W. E. Stanford; 71, CMK Charles Brownlee; 88, RM Tsolo A. R. Welsh.

24. CTA 1/ECO 5/1/3/1, RM W. E. Stanford to CM Tembuland, 8 June 1881; Leslie Farrant, office of RM Engcobo, to CM, 15 June 1881, quoting a report from Headman Mandela.

25. CTA Papers of the Resident Magistrate of the Umtata District (1/UTA) 2/1/1/4, case 270, Qumpu v. Mshikile, 24 October 1882; BBNA 1883, 173, RM Emjanyana James F. Boyes; CTA 1/UTA 4/1/8/1/3, T. Weiz to RM Umtata, letters of 7 November 1883 and 14 December 1885; CTA CMT 1/143, meeting between CM Elliot, Dalindyebo, and other Thembu leaders, 11 January 1886; CTA 1/TSO 3/1/7/1, RM Umtata J. R. Merriman to CM Tembuland, 23 August 1887; Wagenaar, "History of the Thembu," 207.

26. This paragraph is based on the following studies: Monica Hunter, Reaction to Conquest: Effects of Contacts with Europeans on the Pondo of South Africa (London: Oxford University Press, 1936), 427; Hammond-Tooke, Command or Consensus, 54, 105, 113–18, 124; Beinart, Political Economy of Pondoland, 36–39, 115–22; Evans, Bureaucracy and Race, 207–13.

27. Beinart, Political Economy of Pondoland, 117.

28. Hammond-Tooke, Command or Consensus, 50–53; Evans, Bureaucracy and Race, 207–13.

29. William Beinart, "Conflict in Qumbu: Rural Consciousness, Ethnicity and Violence in the Colonial Transkei," 111–12, and William Beinart and Colin Bundy, "Introduction: 'Away in the Locations,'" 8–10, in Beinart and Bundy, Hidden Struggles; see also Evans, Bureaucracy and Race, 210–11.

30. Ndima, "History of the Qwathi People," 177–78.

31. CTA CMT 1/143, Minutes of Meetings, "Meeting regarding complaint by Mavuma, assistant to Samuel Mazwe, headman at Tabase," 17 November 1885; Statement of Dalindyebo, 20 November 1885; Meeting of CM and Dalindyebo's messengers, 16 December 1885; CTA 1/UTA 4/1/8/1/4, P. Nkala, headman at Kambi, to Rev. R. Jenkin, Wesleyan Mission Station, 13 March 1889; Hammond-Tooke, Tribes of Umtata District, 51–52.

32. CTA 1/UTA 1/1/1/24, case 645, R. v. Tyili, 4 December 1899. For similar problems in other locales, see BBNA 1884, 180, RM Tsolo A. R. Welsh; CTA 1/ECO 4/1/5, Samuel Mazwe, Tabase, to CM Tembuland, 16 November 1885; CTA CMT 1/37, RM Umtata A. H. Stanford to CM, 26 February 1891; CTA CMT 3/170, RM Umtata to CMT, 21 May 1897; W. D. Cingo, Ibali laba Tembu (Palmerton, Pondoland: Mission Printing Press, 1927), 59–62; and recollections of this fractious period in CTA CMT 3/589, 44/24, RM Tsolo to CMT, 7 April 1908; CMT to SNA, 12 May 1908; CTA 1/TSO 3/1/7/7, copy of letter from Key and Key, Solicitors, Indwe, to RM Tsolo, 10 March 1910; CTA CMT 3/589, 44/24, CMT A. H. Stanford to SNA, 19 May 1910; Umtata Archives Repository (UAR), CMT 190, Tsolo district, 3/22/3/35A, Ngcele location, RM Tsolo to CMT, 13 June 1923.

33. CTA 1/UTA 4/1/8/1/3, Paul Nkala, Kambi, to RM Umtata, 14 November 1885; R. Jenkin, Wesleyan Parsonage, Umtata, to RM Umtata, 27 May 1887; CTA 1/UTA 4/1/8/1/4, Paul Nkala, Kambi, to Rev. Jenkin, 13 March 1889; R. Jenkin to RM Umtata, 19 March 1889.

34. For examples, see CTA CMT 3/89, T. H. Kelly, law agent, Engcobo, to RM Engcobo, "General Complaints against Headman Vetu," 3 February 1897, attaching "Memo on complaint against Headman Vetu"; CTA CMT 3/91, RM Engcobo Charles E. Warner to CMT, 25 April 1902; CTA CMT 3/589, 44/24, RM Tsolo to CMT, 7 April 1908; CMT to Secretary for Native Affairs (SNA), 12 May 1908; CTA 1/TSO 3/1/7/7, copy of letter from Key and Key, Solicitors, Indwe, to RM Tsolo, 10 March 1910; CTA CMT 3/589, 44/24, CMT to SNA, 19 May 1910; CTA CMT 3/589, 44/24, H. Gush & Co., Umtata, to CMT, 9 May 1919; UAR CMT 190, Tsolo district, 3/22/3/35A, Ngcele location, RM Tsolo to CMT, 13 June 1923.

35. For representative examples, see CTA 1/UTA 1/1/1/18, case 25, R. v. Ncapukiso Ncombo, 20 January 1896; CTA CMT 3/90, RM Engcobo to CMT, 14 January 1899, attaching statement of Mfenenkomo, Mdoda, Nyhebese and others, 30 December 1898; Fenenkomo, Mdoda and Nyebese, at Nxamagele, to CMT, 29 July 1899; CTA 1/UTA 1/1/1/22, case 524, R. v. Benjamin Sinukela and 8 others, 11 November 1898; CTA 1/UTA 1/1/1/27, case 579, R. v. Memani Mtohebo, 27 September 1900; CTA Archives of the Secretary for Native Affairs (NA) 630, B2071, Carey Miller, Solicitor, All Saint's, Engcobo district, to Attorney General, 28 December 1906.

36. *BBNA* 1882, 86, 90, RM Tsolo A. R. Welsh; CTA 1/TSO 2/1/9, "Matters concerning the Fingo Headman at Umjika Thomas Ngudhle," "Notes of Inquiry into *complaint* made by *William Maqenyana*," before RM Tsolo J. O'Connor, 7 October 1886; CTA 1/TSO 9/1, Asst. RM Tsolo to CMK, 6 November 1888; CTA 1/TSO 3/1/7/2, Headman Magwaxaza to RM Tsolo, 10 April 1889; John Ayliff and Joseph Whiteside, *History of the Abambo, Generally Known as Fingos* (Butterworth, Transkei: 1912), 91; CTA 1/TSO 5/1/114, N1/1/5(20), Chiefs and Headmen, Location No. 20, Kambi, Octavius Ngudle, Lower Mjika, to RM Tsolo, 4 March 1957.

37. *BBNA* 1879, 52–53, Annexure D, "St. John's Territory"; *BBNA* 1880, 98, RM Tsolo Alexander Welsh; *Report of the Griqualand East Land Commission*; CTA 1/TSO 3/1/7/2, Thomas Ngudhle, Kambi, to RM Tsolo, 20 January 1885; CTA 1/UTA 4/1/8/1/3, Headman Paul Nkala, Kambi, to RM Umtata, 14 November 1885; CTA 1/TSO 9/1, Asst. RM Tsolo to CMK, 6 November 1888.

38. CTA 1/TSO 9/2, Minutes of meeting between RM and Thomas Ngudhle's people, 29 May 1884; CTA 1/TSO 3/1/7/2, Thomas Ngudhle to RM Tsolo, 20 January 1885; CTA 1/UTA 4/1/8/1/3, Headman Paul Nkala, Kambi, to RM Umtata, 14 November 1885; CTA 1/TSO 1/1/3, case 50, R. v. Bobotyana, 15 October 1886; CTA CMK 1/139, Thomas Ngudhle to RM Tsolo, 12 March 1888, attachment to RM Tsolo to CMK, 15 March 1888. See also Beinart, "Conflict in Qumbu," for a similar dynamic.

39. CTA 1/TSO 9/2, Minutes of meeting between RM Tsolo and Thomas Ngudhle's people, 29 May 1884; CTA 1/TSO 3/1/7/2, Thomas Ngudhle, Umjika, to RM Tsolo A. R. Welsh, 13 September 1884; CTA 1/TSO 9/2, Meeting between CMK and Tsolo people, 2 July 1886; CTA 1/TSO 1/1/3, case 14, R. v. Matanzima and 2 others, 2 March 1886; CTA 1/TSO 2/1/3, Thomas Ngudle v. Matanzima Xayimpi, 10 September 1886; CTA 1/TSO 2/1/9, "Matters concerning the Fingo Headman at Umjika Thomas Ngudhle," "Notes of Inquiry into complaint made by William Maqenyana," 7 October 1886, and "Garden Case. In Location under Thomas Ngudhla," 7–21 October 1888; CTA CMK 1/139, Thomas Ngudhle to RM Tsolo, 12 March 1888, attachment to RM Tsolo to CMK, 15 March 1888.

40. CTA 1/TSO 3/1/7/2, Thomas Ngudhle, Kambi, to RM Tsolo, 20 January 1885; Thomas Ngudhle, Kambi, to RM Tsolo, 18 June 1886.

41. CTA CMK 1/139, RM Tsolo J. T. O'Connor to CMK, "Final report regarding Subdivision of the Pondomise Location: Tsolo District," 30 December 1888; CTA 1/TSO 3/1/7/1, FCT Henkel to RM Tsolo, 1 May 1889; FCT Henkel to Asst. Commissioner of CLPW, 22 May 1889.

42. See, for example, the forest politics surrounding Headman Longden Sotyato, in CTA CMT 3/92, 3/93, and 3/94, as well as in UAR CMT 71, Engcobo district, 3/4/3/1, Part 1, Clarkebury, Location no. 1.

43. Ranger, "Invention of Tradition Revisited"; Carolyn Hamilton, *Terrific Majesty: The Powers of Shaka Zulu and the Limits of Historical Invention* (Cambridge, Mass.: Harvard University Press, 1998); Pels, "Pidginization of Luguru Politics"; Kate Crehan, "'Tribes' and the People Who Read Books: Managing History in Colonial Zambia," *Journal of Southern African Studies* 23, no. 2 (1997): 203–18.

44. Crais, *Politics of Evil*, 88–92; Beinart, *Political Economy of Pondoland*, 36–39, 115–22; Beinart and Bundy, *Hidden Struggles*, 151; Evans, *Bureaucracy and Race*, 22–23, 35–37. See also Mitzi Goheen, "Chiefs, Sub-Chiefs and Local Control: Negotiations over Land, Struggles over Meaning," *Africa* 62, no. 3 (1992): 390; Berry, *Chiefs Know Their Boundaries*, chap. 1.

45. CTA CMT 1/143, "Minutes of a Meeting between the Chief Magistrate of Tembuland, and 'Dalindyebo' . . ." and others, 10 March 1885; University of Cape Town, Manuscripts and Archives Library (UCT), BC293 Sir W. E. M. Stanford Papers (Stanford Papers), Diary, A10, 19 March 1885; Wagenaar, "History of the Thembu," 224–25.

46. *Report and Proceedings, with Appendices, of the Government Commission on Native Laws and Customs*, 20; Beinart, *Political Economy of Pondoland*, 40; Hammond-Tooke, *Command or Consensus*, 78; Martin Chanock, *The Making of South African Legal Culture 1902–1936: Fear, Favour and Prejudice* (Cambridge: Cambridge University Press, 2001), 252–57; Mahmood Mamdani, *Citizen and Subject: Contemporary Africa and the Legacy of Late Colonialism* (Princeton, N.J.: Princeton University Press, 1996), 115–17.

47. CTA CMT 1/144, Statement of Pagati, before RM Umtata, 12 June 1889; CTA Papers of the Resident Magistrate of the Mqanduli District (1/MQL) 1/1/1/4, Criminal Cases, Dalindyebo v. Ngxishe, 31 July 1889.

48. CTA 1/MQL 1/1/1/4, Criminal Cases, Dalindyebo v. Ngxishe, 31 July 1889; CTA CMT 1/144, Statement of Pagati, before RM Umtata, 12 June 1889; Statement of Vakale, messenger for Chief Dalindyebo, before CMT, 2 August 1889; Statement of Dalindyebo before CMT, 7 August 1889; Meeting between CMT, Dalindyebo, and about 100 followers, 13 November 1889.

49. CTA 1/MQL 1/1/1/4, Criminal Cases, Dalindyebo v. Ngxishe, 31 July 1889; CTA CMT 1/144, Statement of Dalindyebo before CMT, 7 August 1889; Meeting between CMT, Dalindyebo, and about 100 followers, 13 November 1889.

50. CTA CMT 1/144, "Investigation into dispute between Holomise & Sandile, re Forest Laws," 17 May 1889.

51. Ibid.; Statement of Holomisa of the Mqanduli district before CMT, 12 August 1889. Such heated negotiations among chiefs, magistrates, headmen, and commoners over the redrawing of boundaries of resource control were replicated in other areas of the Transkei following annexation. See CTA Papers of the Resident Magistrate of Mount Fletcher District (1/MTF) 2/1/1/6, Civil Cases, case 34, Andries Nkwali v. Ramafole, 12–20 June 1895; Beinart, *Political Economy of Pondoland*, 36–39, 115–22; Beinart and Bundy, *Hidden Struggles*, 151.

52. John Henderson Soga, *The Ama-Xosa: Life and Customs* (Lovedale: Lovedale Press, 1932), 32–33; Hunter, *Reaction to Conquest*, 378–82, 384–89, 417–19; Hammond-Tooke, *Command or Consensus*, 37, 50, 59–61, 63–65, 71; Peires, *House of Phalo*, 27–42; Beinart, *Political Economy of Pondoland*, 39–41; Redding, "Sorcery and Sovereignty," 254.

53. CTA 1/UTA 2/2/1/1/3, Civil Cases Record Book, case 66, Dalindyebo v. Mbombo and 12 others, 26 August 1889.

54. CTA CMT 3/87, RM Engcobo to CM, 4 December 1894; CTA 1/ECO 4/1/8, Asst. CM John H Scott to RM Engcobo, 17 December 1894.

55. Hamilton, *Terrific Majesty*, 3–4, 26–27; Pels, "Pidginization of Luguru Politics."

56. Chanock, *Law, Custom, and Social Order*; S. F. Moore, *Social Facts and Fabrications*; Mann and Roberts, *Law in Colonial Africa*.

57. BBNA 1876, 101–2, Chas. J. Levey, Tembu Agent, Cofimvaba, to SNA; BBNA 1877, "Annual Report," Tembuland Proper, CM W. Wright, Emjanyana; *Annual Report 1881*, 38; *Annual Report 1882*, 56, RM St. Mark's A. H. Stanford to CM Tembuland, 10 April 1883; *Report and Proceedings of the Tembuland Commission*, "Appendix P," 11, RM Southeyville Chas. J. Levey to Chairman of the Tembuland Commission; BBNA 1884, 135, RM St. Mark's A. H. Stanford; BBNA 1885, 109, George M. Theal, "Tembuland"; Ndima, "History of the Qwathi People," 85, 104–6, 183; Christopher Saunders, "Transkeian Rebellion," 5–6; Redding, "Sorcery and Sovereignty," 254–55, 260–61; Tropp, "Roots and Rights," 41–46.

58. "Journey of Br. Th. Weitz from Shiloh to Baziya and Entumasi, during the months of September, October, and November, 1872," *Periodical Accounts Relating to the Missions of the Church of the United Brethren, Established among the Heathen,* September 1874, 179; *BBNA 1874,* 45, James Ayliff, Office of the Agent with Kreli, Transkei; *BBNA 1876,* 101–2, Chas. J. Levey, Tembu Agent, Cofimvaba, to SNA; *Annual Report 1881,* 38; Cape of Good Hope, G.66-'83, *Report and Proceedings of the Tembuland Commission* (1883), "Appendix P," 11, RM Southeyville Chas. J. Levey to Chairman of the Tembuland Commission; *Annual Report 1882,* 56, RM St. Mark's A. H. B. Stanford to CM Tembuland, 10 April 1883; *BBNA 1884,* 135, RM St. Mark's A. H. Stanford; *BBNA 1885,* 109, George M. Theal, "Tembuland"; *Annual Report 1884,* Eastern Conservancy, 56–57; H. C. Schunke, "Kaffraria und die Östlichen Grenzdistrikte der Kapkolonie," *Doktor A. Petermanns Mitteilungen aus Justus Perthes' Geographischer Anstalt* (Gotha: Justus Perthes) 31 (1885): 169; *Annual Report 1889,* 94; Sim, *Forests and Forest Flora,* 7; Grove, "Scottish Missionaries."

59. Sally Falk Moore describes a similar phenomenon of chiefs in Tanganyika exploiting colonial institutions to maintain access to fees and fines. *Social Facts and Fabrications,* 176–90.

60. Stanford's actions were recalled in *BBNA 1885,* 109, George M. Theal, "Tembuland"; CTA CMT 2/13, FCT to CMT, 18 November 1890; CTA FCT 2/1/1/1, FCT to Secretary LMA (SLMA), 27 October 1892.

61. CTA FCT 2/1/1/1, FCT to Secretary LMA (SLMA), 27 October 1892.

62. CTA FCT 2/1/1/1, FCT to Secretary LMA (SLMA), 27 October 1892; CTA 1/ECO 1/1/1/11, case 122, R. v. Twala, 11 June 1890; CTA 1/UTA 1/1/1/12, case 338, R. v. Magxa and 3 others, 29 November 1892; Cape of Good Hope, G.62-'93, *Report on the Extent, Value, and Administration of the Forests of the Transkei and Griqualand East* (1893), 8; CTA FCT 1/1/3/1, H. L. Caplen, Mtinthloni, to FCT, 12 March 1894; H. L. Caplen to Henkel, 6 April 1894; CTA FCT 2/1/1/1/, FCT to SLMA, 27 October 1892; CTA AGR 223, F1930, FCT to CMK, 16 April 1895; Proclamation 388 of 1896, section 14; CTA 1/UTA 1/1/1/27, case 579, R. v. Memani Mtohebo, 27 September 1900; Proclamation 170 of 1901, section 7; CTA NA 510, A262, RM Willowvale to CMT, 10 October 1901; CTA 1/ECO, 1/2/9, Criminal record book, 30, Willowvale, October 1902; Proclamation 135 of 1903, section 8; Proclamation 288 of 1908, section 17; CTA 1/TSO A3/3, RM Tsolo to Secretary of the Transkeian Territories General Council, 16 February 1909.

63. *Annual Report 1891,* 72; Hammond-Tooke, *Command or Consensus,* 80–81, 123–24; Beinart, *Political Economy of Pondoland,* 37–39.

64. See, for example, CTA 1/TSO 9/2, Meeting between RM Hook and headmen of Tsolo district, 17 May 1886; CTA 1/TSO 1/1/4, case 48, R. v. Mabungwana and Sinapu, 30 June 1887; CTA 1/TSO 3/1/7/2, Jonah Gece, Ncambele, to RM Tsolo, 14 May 1889; CTA 1/TSO 1/1/6, case 33, R. v. Poswa and 19 others, 7 June 1889; CTA CMT 1/31, RM Engcobo A. H. Stanford to CM, 28 November 1889, attaching statement by Headman Tafa in RM's court, 4 November 1889; CTA

CMT 1/144, Meeting between CM and Dalindyebo and other Thembu chiefs, 12 December 1889; CTA CMT 3/520, "Minutes of Public Meeting held at Engcobo on the 14th May 1892." For contemporary examples in other parts of the Eastern Cape and Transkei, see Beinart and Bundy, *Hidden Struggles*, 143–44; Switzer, *Power and Resistance*, 94–96.

65. *Annual Report 1888*, 62; CTA 1/UTA, 4/1/8/1/4, FCT Henkel to RM Umtata, 11 October 1889.

66. CTA 1/UTA, 4/1/8/1/4, George Stokes, acting for J. Snodgrass, Bolompo, to RM Umtata, 30 August 1889; Henkel to RM Umtata, 9 September 1889; CTA 1/UTA 2/1/1/10, case 16, Philip Charles, acting for Nqwiliso, v. Edward A. Smith, 11 October 1889.

67. CTA 1/TSO 1/1/5, case 19, R. v. Mgamle and 7 others, 6 April 1888; *Annual Report 1888*, 61; CTA 1/TSO 3/1/7/1, FCT to CMK, 23 April 1889.

68. CTA CMK 1/139, RM Tsolo J. T. O'Connor to CMK, "Final report regarding Subdivision of the Pondomise Location: Tsolo District," 30 December 1888; CTA 1/TSO 3/1/7/1, FCT Henkel to RM Tsolo, 1 May 1889; FCT Henkel to Asst. Commissioner of CLPW, 22 May 1889.

69. Proclamations 209 and 308.

70. *Annual Report 1891*, 72, 143–44; CTA FCT 2/1/1/1, FCT to SLMA, 13 July 1893; CTA AGR 223, F1930, FCT to CMK, 16 April 1895; *Annual Report 1896*, 165, 169; *Annual Report 1897*, 145.

71. *Annual Report 1889*, 94; *Annual Report 1891*, 69; *Annual Report 1892*, 89.

72. CTA FCT 2/1/1/3, FCT Heywood to Under-secretary for Agriculture (USA), 20 November 1899.

73. CTA 1/TSO 1/1/12, case of 10 September 1896; CTA 1/TSO 1/1/13, R. v. Thomas Adams, 2 April 1897; CTA 1/TSO 1/1/15, case 259, R. v. Nozinto, 26 September 1899, and case 283, R. v. Thomas Adams, 24 October 1899; CTA 1/TSO 1/1/17, case 351, R. v. Ncotsho, 15 November 1900; CTA CMT 3/161, RM Tsolo to CMT, 9 September 1902; CTA CMT 3/40, FCT to CMT, 10 March 1904; NAR FOR 19, A24, FCT Heywood to CCF, letters of 11 February and 5 March 1907.

74. CTA 1/UTA 1/1/1/17, R. v. Jongimpi, 6 June 1895; CTA CMT 3/40, FCT Henkel to CMT, 5 September 1896, enclosing report from Forester Greenaway, Kambi forest, to Henkel, 21 August 1896.

75. CTA 1/TSO 1/1/13, case of 4 August 1897; R. v. Joel Nkwiniza, 22 April 1897; CTA AGR 703, District Forest Officer Umtata (FDU) to FCT, 20 March 1897; *Annual Report 1897*, 135; CTA AGR 750, F2839, FCT to CMT, 25 June 1897.

76. CTA 1/TSO 1/1/13, cases of 8 April and 8 June 1897; CTA AGR 750, F2839, FCT Henkel to CMT, 25 June 1897. Henkel also noted here that many other assaults on forest officers and guards in the area were never officially reported.

77. CTA CMT 3/40, FCT to CMT, 24 October 1894; CTA CMT 3/192, W. J. Clarke, Engcobo, to CMT, 19 November 1894; CTA CMT 3/87, RM Engcobo to CMT, 1 March 1895; CTA 1/ECO 5/1/3/2, RM Engcobo to CMT, 30 July 1895; CTA CMT 3/170, RM Umtata to CMT, 30 August 1897; UCT, Stanford Papers,

B203.3, A. H. Stanford, Umtata, to Walter Stanford, 2 September 1897; CTA CMT 3/89, RM Engcobo to CMT, 22 November 1897; RM Engcobo to CMT, 23 December 1897; RM Engcobo, Annual report for Blue Book, 4 January 1898.

78. CTA AGR 223, F1930, vol. 1, RM Tsolo and FCT to CMK, 13 September 1893; CTA AGR 766, F1930, CMK to USNA, 19 June 1894; CTA CMT 3/86, RM Engcobo to CMT, 18 June 1894; *Annual Report 1894*, 130–33; CTA CMT 3/520, regular file, "Meeting held at Umtata 19th April 1894 between the Hon. the Premier (C. J. Rhodes) and Dalindyebo and Tembus of Umtata, Mqanduli and Engcobo Districts"; CTA FCT 2/1/1/2, FCT to USA, 1 March 1895; CTA CMT 3/520, "Meeting with Dalindyebo 21st October 1895"; CTA AGR 223, F1930, vol. 4, CMT to USNA, 5 March 1895, enclosing "Extract from Minutes of a meeting with the Tembu Chief Dalindyebo and Major H. G. Elliot C.M.G. Chief Magistrate, at Umtata on the 4th March, 1895"; CTA CMT 3/40, FCT to CMT, 25 October 1895; CTA CMT 3/87, RM Engcobo to CMT, 5 December 1895; CTA AGR 766, F1930, RM Engcobo to FCT, 23 September 1895; *Annual Report 1895*, 171, 173; CTA FCT 2/1/1/2, FCT to USA, 1 May 1896; CTA CMT 3/520, Minutes of Meetings, Engcobo, Thomas Poswayo, spokesman of Qwati, 21 August 1896; *Annual Report 1896*, 155.

79. CTA CMT 3/192, W. J. Clarke, Engcobo, to CMT, 19 November 1894; CTA CMT 3/87, RM Engcobo to CMT, 1 March 1895; "The Next Delusion," *Umtata Herald*, 23 March 1895; CTA 1/ECO 5/1/3/2, RM Engcobo to CMT, 30 July 1895; Beinart and Bundy, *Hidden Struggles*, 147. See also Tropp, "Dogs, Poison and the Meaning of Colonial Intervention."

80. CTA AGR 749, F2786, part 1, SNA to USA, 10 November 1900; CTA CMT 3/171, Robert Welsh, District Surgeon, Health Report, Umtata district, 16 January 1901; *BBNA 1901*, 35, "Report of the CM for 1900," CM Elliot; 42, RM Umtata A. H. B. Stanford; 42–43, Report from R. H. Welsh, District Surgeon; 46, RM Engcobo J. G. Leary; 61, RM Tsolo J. S. Simpson; *BBNA 1902*, 48, RM Engcobo C. E. Warner; *BBNA 1903*, 56, RM Engcobo C. A. King; 60, RM Umtata A. H. B. Stanford; 81, RM Tsolo Edwin Gilfillan; *BBNA 1904*, xiv, Section III, Transkeian Territories; 40–41, "Report of the Chief Magistrate for the Transkeian Territories for the Year 1903," W. E. Stanford; *BBNA 1905*, 62–63, 60, RM Engcobo C. A. King; 89, RM Tsolo Arthur Gladwin; 91, RM Umtata A. H. B. Stanford.

81. *Annual Report 1899*, 136, 138; CTA NA 559, Part I, A963, FCT to RM Tsolo, 4 September 1902; RM Tsolo and FCT to CMT, 31 October 1902; *Annual Report 1906*, 18, giving the history of plantation work in previous years.

82. *Annual Report 1901*, 143; *Annual Report 1902*, 142; *Annual Report 1903*, 124.

83. *Annual Report 1901*, 143.

84. CTA AGR 760, F3553, USA to FCT, 12 March 1903, enclosing several letters on the subject: FCT to CMT, 13 January 1903; RM Tsolo to CMT, 19 January 1903; CMT to SNA, 27 January 1903; SNA to USA, 27 February 1903.

85. CTA NA 549, A844, FCT to USA, 7 August 1902; SNA, Schedule submitted to Prime Minister (PM), 25 August 1902; CTA AGR 760, F3553, SNA to USA,

24 July 1903; CTA CMT 3/40, Circular from CM W. E. Stanford to RMs, 4 August 1903; CTA AGR 760, F3553, FCT to USA, 20 February 1904; SNA to USA, 29 February 1904.

86. *Annual Report 1902*, 142; *Annual Report 1903*, 124. In his annual report for 1904, RM Tsolo Arthur Gladwin cited forest offenses as one of the two most common crimes in the district, *BBNA 1905*, 89.

87. CTA 1/TSO 1/1/17, case 263, R. v. Mofukana and another, 11 September 1900.

88. CTA 1/UTA 4/1/8/1/11, FDU to RM Umtata, 11 June 1904, attaching letter from Forester T. C. P. Adams, Bazeya, to FDU, and the latter's enclosure, "Statement of Forest Guard G. C. Roux, Bazeya, Umtata" to Forester Adams; FCT to RM Umtata, 7 September 1904. For similar cases in the area, see CTA 1/UTA 4/1/8/1/11, case 288, R. v. Raga, 26 September 1900; case 324, R. v. Uxala and Tamenani, 19 October 1900; case 335, R. v. Diamond and 5 others, 26 October 1900; case 348, R. v. Umgangani, 15 November 1900; CTA 1/TSO 1/1/19, case 149, R. v. William Ndumdum and 45 others, 30 August 1901; case 228, R. v. Utintweni and Tsibantu, 27 December 1901; CTA Archives of the District Forest Officer Umtata (FDU) 3/2, Bazeya station, 6 June 1903, R. v. Mbezo and 2 others, tried by RM Engcobo on 7/20/1903; CTA FDU 3/3, Kambi station, 16 January 1904, R. v. Boza, tried by RM Tsolo on 11 March 1904; Kambi station, 10 May 1904, R. v. Lampuloni, tried by RM Tsolo, 8 July 1904; CTA 1/UTA 4/1/8/1/11, FDU to RM Umtata, 11 June 1904, attaching letter from Forester T. C. P. Adams, Bazeya, to FDU, and the latter's enclosure, "Statement of Forest Guard G. C. Roux, Bazeya, Umtata" to Forester Adams; FCT to RM Umtata, 7 September 1904; CTA 1/TSO 1/2/3, case 172, R. v. Nyangiwe, 18 August 1905.

NB: "Bazeya" reflects period spelling of "Baziya."

89. See, for example, condemnations of Africans collaborating with government livestock inoculation amid the rinderpest epidemic: UCT, Stanford Papers, B203.3, A. H. Stanford, Umtata, to Walter Stanford, letters of 2 and 26 September 1897; CTA 1/UTA 1/1/1/19, Circuit Court, R. v. Rolinyati Mgudli and Magopeni Rolinyati, 26 October 1897. For similar complications surrounding the simultaneous "insider" and "outsider" status of African national park guards in Tanzania, see Neumann, *Imposing Wilderness*, 195–201.

90. CTA 1/TSO 1/1/16, case 105, R. v. Hlabizulu, 4 May 1900.

91. CMT 3/40, FCT Henkel to CMT, 5 September 1896, enclosing report from Forester Greenaway, Kambi forest, to Henkel, 21 August 1896; CTA 1/UTA 5/1/1/15, RM Umtata to CMT, 7 October 1896.

92. For example, CTA 1/UTA 1/1/1/11, case 149, R. v. Ishefu, 16 September 1885; CTA 1/ECO, 1/2/4, Criminal Record Book, 1889, case 271; CTA 1/ECO 1/1/1/11, case 65, R. v. Galeni, 19 March 1890; CTA 1/TSO 1/1/7, case 32, R. v. Vinjwa, 8 May 1890; CTA 1/TSO 1/1/8, case 65, R. v. Poni and Umtshwana, 9 July 1891; CTA 1/UTA 1/1/1/15, case 105, R. v. Ntaka, 13 April 1894; CTA 1/TSO 1/1/14, case 116, R. v. Sicubasetswele, 10 August 1898; CTA 1/TSO 1/2/3, case 83, R. v. Sixeku and

Zenzile, 10 May 1904; CTA 1/TSO 1/2/5, case 105, R. v. Gqugula, 14 May 1907; CTA 1/TSO 1/2/6, case 330, R. v. Cuku, 21 December 1909.

93. CTA 1/TSO 1/1/7, case 35, R. v. Luvani and 7 others, 17 June 1890; CTA 1/UTA 1/1/1/15, case 105, R. v. Ntaka, 13 April 1894; Transkeian Territories, *Proceedings and Reports of Select Committees at the Session of the Transkeian Territories General Council* (TTGC) 1908, xii–xiii, "Unlawful Killing or Wounding of Animals and Burning of Huts"; *TTGC* 1909, xlvii–xlviii, "Unlawful Maiming of Stock and Hut Burning"; CMT 3/679, Returns labeled "Social Conditions" for Blue Book 1910, cited in Sean Redding, "Government Witchcraft: Taxation, the Supernatural, and the Mpondo Revolt in the Transkei, South Africa, 1955–1963," *African Affairs* 95, no. 381 (October 1996): 559.

94. CTA 1/TSO 1/2/3, case 83, R. v. Sixeku and Zenzile, 10 May 1904; *BBNA* 1905, 53, RM Mqanduli H. H. Bunn; CTA 1/UTA 1/1/2/1/2, R. v. Matsiyo Makaula and 18 others, 26 April 1917, at Mandileni in Mount Frere district; Godfrey, "Xhosa-English Dictionary," "H," 17.

95. CTA 1/TSO 3/1/7/3, Headman Bikwe Ndlebe, Ncembu, to RM Tsolo, 17 February 1904; FCT to RM Tsolo, 17 March 1904.

96. CTA CMT 3/40, RM Tsolo to CMT, 6 July 1903. See also CTA 1/UTA C1/1/1, No. 124, Petition to Chairman Umtata DC, signed by Mdukiswa (Tabase), Ndwakuse (Esiqubadwini), Jeremiah Yengwa (Baziya Mission), and Nosali (Baziya), 4 June 1904.

97. CTA 1/TSO 1/1/16, case 105, R. v. Hlabizulu, 4 May 1900; CTA 1/TSO 1/1/17, case 374, R. v. Nkumanda and Umtenki, 18 December 1900; case 325, R. v. Sogoni and 10 others, 19 October 1900; CTA 1/TSO 2/1/29, case 185, Nkumbi v. Nompintsho, 10 December 1901. Roderick Neumann describes similar relations between Arusha National Park staff and Meru villagers' lives in *Imposing Wilderness*, 197–201.

98. By 1896, the general incidence of forest staff extortion and fraud had reached such proportions, particularly in KwaMatiwane, that Henkel pushed for increasing their salaries in order to reduce economic temptations. *Annual Report* 1889, 95; CTA FCT 2/1/1/2, FCT Henkel to USA, 10 April 1896. See also CTA NA 510, A262, J. Storr Lister, "Report by the Conservator of Forests, King William's Town on the Transkeian Forest Administration," 11 March 1898; Knut A. Carlson, *Transplanted: Being the Adventures of a Pioneer Forester in South Africa* (Pretoria: Minerva Drukpers, 1947), 100.

99. CTA 1/TSO 1/2/5, case 86, R. v. Nondlwana, 9 April 1907; see also CTA 1/TSO 1/1/13, case of 8 June 1897.

100. CTA 1/TSO 3/1/7/5, F. W. Puller, St. Cuthbert's, to RM Tsolo, 1 May 1907. Thembela Kepe notes how some women in the area surrounding the Mkambati Nature Reserve have themselves used sexual favors as "bribes" to secure access to certain protected resources from guards. Kepe, *Environmental Entitlements in Mkambati*, 44.

101. The term "bargaining power" is borrowed from Neumann, *Imposing Wilderness*, 197–201.

102. Interview with Rev. G. Vika, Etyeni, Tsolo District, 2 February 1998.

103. For some typical examples at the turn of the century, see CTA 1/ECO 1/3/1, Statement of Umtenjane at RM Engcobo's office, 9 September 1887; CTA FCT 2/1/1/1, FCT to Asst. Commissioner CLPW, 12 December 1891; CTA 1/ECO 1/1/1/13, case 280, R. v. Mene, 14 November 1892; CTA FCT 2/1/1/1, FCT to SLMA, 3 June 1893; the many letters between the USA, FCT, and the District Forest Officer Kokstad (FKS) in late 1893 to late 1895, concerning Forester Ernst Ericsson and Forest Guard Mdha, in CTA FCT 1/1/1/2, 1/1/2/1, 2/1/1/1, and 2/1/1/2, and CTA Archives of the District Forest Officer Kokstad (FKS) 1/1/1; CTA 1/TSO 1/1/11, case 108, R. v. Mabuto, 24 August 1894; CTA 1/TSO 1/1/12, R. v. Monsete, 21 January 1896; CTA 1/TSO 1/1/14, criminal case of 5 April 1898; CTA 1/TSO 1/1/14, case 110, R. v. Mcam, 21 July 1898; CTA 1/TSO 1/1/15, case 196, R. v. Maqutu and Nomankwebe, 19 July 1899; CTA 1/TSO 1/1/16, case 47, R. v. Comance and Donga Dwasta, 20 March 1900; CTA AGR 703, FCT to USA, 30 October 1902; CTA 1/UTA 1/1/1/35, case 383, R. v. Forest Guard July, 17 July 1903; CTA 1/TSO 1/2/5, case 86, R. v. Nondlwan, 9 April 1907; CMT 3/648, 91/31/1, FCT to CMT, 16 June 1916; CTA 1/TSO 5/1/8, file 31, vol. 5, "Enquiry held into charges made against Forest Guard Paulus Komani during the case of R. vs. Witchell Jangi," case 164, 3 June 1920 and 3 July 1920; CTA 1/UTA T601/1, "Shooting of Native by Watcher Links," copy of case before RM Tsolo, R. v. Ntukaye, 22 May 1922; copy of case before RM Tsolo, R. v. Charlie Links (Lengisi or Lynx), begun 12 June 1922.

104. CTA 1/TSO 1/1/17, case 349, R. v. Sam Ndayi, 15 November 1900. See also Headman Thomas Ntaba's authority disputes with local forest guards: CTA 1/TSO 1/1/17, case 295, R. v. Zaza and 4 others, 27 September 1900; R. v. Zaza and 2 others, 17 October 1900.

105. CTA 1/TSO 3/1/7/3, Headman Bikwe Ndlebe, Ncembu, to RM Tsolo, 17 February 1904; FCT to RM Tsolo, 17 March 1904.

106. For another detailed example, see CMT 3/90, RM Engcobo C. Warner to CM, 7 January 1899, enclosing letter from J. H. Chapman, Solicitor, Engcobo, to RM Warner, 21 December 1898; Ndima, "History of the Qwathi People," 179.

107. For representative examples, see CTA 1/TSO 1/1/15, case 183, R. v. Winnie and 13 others, 5 July 1889; CTA 1/UTA 1/1/1/27, case 579, R. v. Memani Mtohebo, 27 September 1900; CTA 1/TSO 1/1/17, case 295, R. v. Zaza and 4 others, 27 September 1900; CTA NA 559, Part I, A963, SNA to PM, 20 November 1902; CTA 1/UTA 4/1/8/1/14, Statement of Mlotsana Mbande at RM Umtata's office, 3 May 1907; FDU to FCT, 14 May 1907; CTA 1/TSO 3/1/7/7, Reg. Goff, Gungululu, to RM Tsolo, 19 June 1909.

108. Thus, violence against foresters and other acts of "protest" were indeed "polyvalent," combining resistance to colonial interventions with multiple additional intentions and interests. See Allen Isaacman, "Peasants and Rural Social Protest in Africa," *African Studies Review* 33, no. 2 (September 1990): 1–120; Donald Crummey, ed., *Banditry, Rebellion and Social Protest in Africa* (London: James Currey, 1986). Forest officers and guards continued to be targets of popular

frustration in subsequent, more sporadic incidents of violence: for example, CTA NA 692, B2690, J. J. Boocock, Asst. District Forest Officer (DFO), Kambi, to FCT, 12 August 1906; NAR FOR 18, A24, Forest Guard A. B. Bowen, Baziya, to FDU, 1 July 1907; CTA 1/TSO 1/2/5, case 285, R. v. Hluku, 17 October 1907; CTA 1/TSO 5/1/8, file 31, vol. 1, FDU to Asst. FCT, 10 August 1910.

109. W. D. Hammond-Tooke, "The Transkeian Council System 1895–1955: An Appraisal," *Journal of African History* 9, no. 3 (1968): 455–77; Beinart and Bundy, *Hidden Struggles*, 140–41, 149; Beinart, *Political Economy of Pondoland*, 85–86. Roger Southall has called the councils "toothless." *South Africa's Transkei: The Political Economy of an "Independent" Bantustan* (New York: Monthly Review Press, 1983), 91–95.

110. Colin Bundy, "Mr. Rhodes and the Poisoned Goods: Popular Opposition to the Glen Grey Council System, 1894–1906," in Beinart and Bundy, *Hidden Struggles*, 138–65.

111. Western Pondoland launched its own Pondoland General Council (PGC) in 1911, and in 1927 the Eastern Pondoland districts joined the system. In the early 1930s the Transkeian Territories General Council combined with the Pondoland General Council to become the United Transkeian Territories General Council (UTTGC). Hammond-Tooke, "The Transkeian Council System," 461–65.

112. Andrew J. E. Charman, "Progressive Élite in Bunga Politics: African Farmers in the Transkeian Territories, 1903–1948" (Ph.D. thesis, Cambridge University, 1999); Beinart and Bundy, *Hidden Struggles*, 11–12, 161–62, 215–16; Southall, *South Africa's Transkei*, 93–95; Beinart, *Political Economy of Pondoland*, 86–87.

113. Bundy, *Rise and Fall*, 96–99; Beinart and Bundy, *Hidden Struggles*, 161–62; Evans, *Bureaucracy and Race*, 215–16; Beinart, *Political Economy of Pondoland*, 112–22.

114. TTGC 1909, lxviii; Transkeian Territories, *Proceedings and Reports of Select Committees at the Session of the Pondoland General Council* (PGC) 1914, 5, 18, 28.

115. For insightful examples of African authorities' ambiguous position within colonial power structures, see Shula Marks' analysis of Solomon ka Dinuzulu in *The Ambiguities of Dependence in South Africa: Class, Nationalism, and the State in Twentieth-Century Natal* (Johannesburg: Ravan Press, 1986), and Richard Roberts, "The Case of Faama Mademba Sy and the Ambiguities of Legal Jurisdiction in Early Colonial French Soudan," in Mann and Roberts, *Law in Colonial Africa*, 185–98.

116. For an early example, see CTA AGR 749, F2786, part 2, Chairman of TTGC to SNA, 3 March 1904, commenting on the TTGC proceedings of 25 January 1904. For similar cases of African leaders employing such self-legitimating logic, see Peters, "Struggles over Water," and *Dividing the Commons*, chapter 3; Crehan, "'Tribes' and the People Who Read Books."

117. This concession involved 15 out of the 26 Transkeian districts, with a total of about 19,000 morgen of undemarcated forest areas (out of approximately 59,000 morgen of such forests in the Territories as a whole). CTA AGR 630, T202, A. H. Harrison, Secretary for the PM's office, to Acting USA, 5 March 1908.

118. CTA NA 753, F127, SNA to CMT, 20 December 1907; Memorandum of interview with SNA, by Col. W. Stanford and Mr. C. Struben, on behalf of the National Society for the Preservation of Objects of Historic Interest and Natural Beauty in South Africa, 27 June 1908.

119. *Annual Report 1907*, 1; CTA NA 753, F127, USA to Chief Conservator of Forests (CCF) J. S. Lister, 11 December 1907; SNA to CMT, 20 December 1907; CTA AGR 630, T202, CCF to Acting USA, 5 February 1908; A. H. Harrison, Secretary for the PM's office, to Acting USA, 5 March 1908; *Annual Report 1908*, 4; *Annual Report 1909*, 2–3.

120. CTA NA 753, F127, USA to CCF, 11 December 1907; SNA to CMT, 20 December 1907; CTA AGR 630, T202, Acting Provincial Secretary Barry McMillan to CCF, 5 February 1908, enclosing extract from letter of Arthur Fowler, Tsolo, to Mr. Malan, 29 January 1908; CTA NA 753, F127, Memorandum of interview with SNA, by Col. W. Stanford and Mr. C. Struben, on behalf of the National Society for the Preservation of Objects of Historic Interest and Natural Beauty in South Africa, 27 June 1908; SNA to private secretary to AGR, 31 July 1909.

121. *TTGC 1908*, xxvii.

122. Ibid.; Proclamation 288 of 1908, sections 3 and 17; CTA FDU 1/5, 107, RM Mqanduli to DFO Kaufmann, 23 December 1908; CTA 1/TSO A3/3, RM Tsolo to Secretary of the TTGC, 16 February 1909; *TTGC 1909*, xlix–l.

123. *TTGC 1910*, 85, RM Kentani N. O. Thompson.

124. Proclamation 91 of 1910; *TTGC 1910*, 85–86; *TTGC 1911*, 36–39, 91.

125. *TTGC 1911*, 36–39, 91; *Annual Report 1910*, 6; *Annual Report 1911*, 7; CTA CMT 3/648, 91/3, Circular no. 33, Walter Carmichael, by direction of CMT, to all RMs, 30 August 1912; *Annual Report 1913*, 7–8.

126. *TTGC 1911*, 36–38. Some leaders, however, were in favor of devising clearer laws for managing headmen's forests both to bolster headmen's authority and guard against potential abuse of power. Ibid., 37–39.

127. *TTGC 1911*, 37.

128. See Andrew Charman's detailed analysis of council politics in "Progressive Élite."

129. See an analogous situation in Peters, "Struggles over Water," 37.

130. Ibid., 37; *TTGC 1911*, 37–38.

131. *TTGC 1911*, 37–38.

132. National Archives Repository (NAR), Pretoria, South Africa, Archives of the Native Affairs Department (NTS) 6937, 50/321, CMT to USNA, 12 January 1911; Walter Carmichael, for CMT, to SNA, 1 August 1911.

133. *TTGC 1913*, 16–17, comments by councillors Nyoka and Mamba; Proclamations 192 of 1912 and 212 of 1913.

134. See, for example, the well-documented incidents in the Kentani District at this time, in CTA CMT 3/648, 91/8/1, "Forests: Kentani," DFO Butterworth to Asst. FCT, 9 December 1912.

135. CTA CMT 3/648, 91/3, DFO Butterworth to Asst. FCT, 23 April 1912; Asst. FCT to CMT, 17 July 1912; DFO Butterworth to Asst. FCT, 19 March 1913; RM Engcobo to CMT, 25 April 1913; CTA CMT 3/648, 91/8/1, DFO Butterworth to Asst. FCT, 9 December 1912; Asst. FCT to CMT, 23 December 1912; CMT to RM Engcobo, 8 May 1913; CTA 1/TSO 5/1/8, file 31, vol. 5, FDU to Public Prosecutor, Tsolo, 20 September 1916; Acting RM Tsolo to FDU, 11 October 1916.

136. CTA FDU 1/5, 138, A. G. Potter, for Asst. FCT, to FDU, 30 June 1911; CTA FCT 3/1/18, T8, List of forests recommended for demarcation in Engcobo, 6 March 1911; CTA FCT 3/1/34, T102/26, RM Umtata and FDU to Chairman TTGC, n.d. December 1911; *TTGC 1920*, 63–65, "Headmen's Stipends"; CTA FCT 3/1/50, T601, DFO Butterworth to FCT, 4 August 1922; CTA FCT 3/1/51, T700, M. 624, J. Keet, Director of Forestry, Transkei, to Inspector of Forestry, Natal and Transkei, "Management of Indigenous Forests in the Transkei," 28 December 1933; CTA FCT 3/1/24, T102/5, DFO Butterworth to RM Engcobo, 14 December 1935; CTA FCT 3/1/51, T700, N. L. King, "Report on the Indigenous Forests in the Transkei and Suggestions for Their Management," 21 August 1936; King, "Exploitation of the Indigenous Forests"; *UTTGC 1938*, 72–74, interviews with Polisile Maka, Manzana, Engcobo District, 28 January 1998 and R. T. S. Mdaka, Manzana, Engcobo District, 11 February 1998.

137. Interviews with Dabulamanzi Gcanga, Manzana, Engcobo District, 5 February 1998; Adolphus Qupa, Baziya Mission Station, Umtata District, 8 January 1998; NAR NTS 6928, 27/231, 2/20/1918, RM Tsolo to CMT; CTA FCT 3/1/50, T601, DFO Butterworth to FCT, letters of 4 August and 6 December 1922.

138. CTA 1/TSO 5/1/8, file 31, vol. 1, RM Tsolo to CMT, 13 September 1910; CMT to RM Tsolo, 23 January 1911; NAR FOR 13, A17/8, Acting USNA to CMT, 10 January 1911.

139. CTA FCT 3/1/50, T601, DFO Butterworth to FCT, 4 August 1922; Acting RM Engcobo to CMT, 30 November 1922; DFO Butterworth to FCT, 6 December 1922.

140. Beinart and Bundy, *Hidden Struggles*, 216–17; Evans, *Bureaucracy and Race*, 210–11.

141. Proclamation 192 of 1912, section 3; *TTGC 1913*, 16–17.

142. UAR CMT 75, Engcobo district, 3/4/3/27, Zadungeni location, RM Engcobo to CMT, 19 August 1915.

143. Interview with Rev. G. Vika. See also Beinart, *Political Economy of Pondoland*, 127; Evans, *Bureaucracy and Race*, 209–11; Hammond-Tooke, *Command or Consensus*, 136.

144. Redding, "Legal Minors and Social Children"; Redding, "Beer Brewing in Umtata: Women, Migrant Labor, and Social Control in a Rural Town," in *Liquor and Labor in Southern Africa*, ed. Jonathan Crush and Charles Ambler, 235–51 (Athens: Ohio University Press, 1992). On the broader gendered processes affecting reserves across South Africa, see Belinda Bozzoli, "Marxism, Feminism and South African Studies," *Journal of Southern African Studies* 9, no. 2 (1983): 139–71;

Cherryl Walker, "Gender and the Development of the Migrant Labour System c. 1850–1930: An Overview," in *Women and Gender in Southern Africa to 1945*, ed. Cherryl Walker (Cape Town: David Philip, 1990), 168–96.

145. *TTGC 1911*, 37; *TTGC 1918*, 36–37; *TTGC 1924*, 87.

146. *TTGC 1918*, 36–37.

147. Ibid., my emphasis.

148. *TTGC 1920*, 62–63, RM Tsolo; interviews with Noheke Rangana, Manzana, Engcobo District, 28 January 1998; R. T. S. Mdaka.

149. Veliswa Tshabalala, a research assistant.

150. Interview with R. T. S. Mdaka.

CHAPTER 2

1. For example, see CTA FCT 2/1/1/2, FCT to USA, 8 April 1895; FCT to USA, 10 May 1895; Carlson, *Transplanted*, 100.

2. *Annual Report 1888*, 64. For similar examples of such invocations of environmental "precedent" to justify colonial "domain claims" in British India and the Malay States, see Sivaramakrishnan, "Colonialism and Forestry in India"; Nancy Lee Peluso and Peter Vandergeest, "Genealogies of the Political Forest and Customary Rights in Indonesia, Malaysia, and Thailand," *Journal of Asian Studies* 60, no. 3 (August 2001): 777–79. For a more detailed analysis of this selective and problematic rendering of Sarhili's environmental policies, see Tropp, "Roots and Rights," 47–50.

3. Cape of Good Hope, *Report of the Select Committee on Forests Bill* (1888), 72–73; *Annual Report 1888*, 63; *Annual Report 1889*, 95; *Annual Report 1890*, 12–13, 139, original emphasis; CTA FCT 2/1/1/1, Henkel to Asst. Commissioner CLPW, 3 February 1890.

4. Richard Parry, "'In a Sense Citizens, But Not Altogether Citizens . . .': Rhodes, Race, and the Ideology of Segregation at the Cape in the Late Nineteenth Century," *Canadian Journal of African Studies* 17, no. 3 (1983): 377–91; Saul Dubow, *Racial Segregation and the Origins of Apartheid in South Africa, 1919–1936* (Oxford: Oxford University Press, 1989), 29–31.

5. CTA CMT 1/143, Miscellaneous Papers, "Minutes of a Meeting between the Chief Magistrate of Tembuland, and 'Dalindyebo,'" 10 March 1885; CTA 1/TSO 9/2, "Meeting between RM Hook and Headmen," 17 March 1886; UCT, Stanford Papers, Diaries, A11, "Meeting with Bacas and Chief Makaula of Mount Frere district," 28 June 1886; CTA CMK 1/100, RM Tsolo to CMK, 3 August 1886; *BBNA 1889*, 31, RM Willowvale W. T. Hargreaves.

6. *BBNA 1888*, 36–37, CM Elliot; *BBNA 1890*, 43, Griqualand East, "Report of the CM for 1890," W. E. Stanford; CTA FCT 2/1/1/1, Henkel to Asst. Commissioner CLPW, 11 April 1890; CTA AGR 187, 1066, FCT to Asst. Commissioner CLPW, 19 May 1890; CTA FCT 1/1/1/1, CM Tembuland H. G. Elliot, CMK W. E.

Stanford, and Acting CM Transkei W. M. Liefeldt to USNA, 20 May 1890; CTA FCT 2/1/1/1, FCT to Percy Nightingale, Secretary to Treasurer, 18 August 1891. See also Ashforth, *Politics of Official Discourse*, chap. 2.

7. Proclamation 209 extended the forest regulations of Act 28 to the territories of Transkei, Tembuland, and East Griqualand; Proclamation 308 extended these to Emigrant Thembuland, Bomvanaland, Gcalekaland, and Port St. John's.

8. In addition to the *Imvo Zabantsundu* (*Imvo*) editorials noted below, see also those of 15 May 1890, 24 July 1890, 4 September 1890, 4 January 1893, 20 September 1893, 27 June 1895, and 20 November 1906. See also Farieda Khan, "Rewriting South Africa's Conservation History."

9. *Imvo*, 4 January 1889, 3.

10. *Imvo*, 1 May 1890, 3.

11. CTA FCT 2/1/1/1, FCT to Asst. Commissioner CLPW, 29 May 1890.

12. *BBNA 1889*, 31, RM Willowvale W. T. Hargreaves.

13. CTA FCT 1/1/1/1, CM Tembuland H. G. Elliot, CMK W. E. Stanford, and Acting CM Transkei W. M. Liefeldt to USNA, 20 May 1890.

14. Ibid.; UCT, Stanford Papers, Diaries, A15, 19 May 1890; "Tembu Meeting," *Umtata Herald*, 20 May 1890; Proclamation 209 of 1890; CTA NA 454, "Visit of P. H. Faure, SNA to Territories, 1890," "Minutes of a meeting held at Umtata between the SNA, Paramount Chief Dalindyebo and his councillors, and about 400 followers," 1 October 1890.

15. CTA NA 454, "Visit of P. H. Faure, SNA to Territories, 1890," "Minutes of meeting held at Idutywa with chiefs and headmen of the Willowvale district," 29 September 1890; CTA AGR 1/19, RM Mount Frere W. G. Cumming to CMK, 22 October 1890; *BBNA 1890*, 48, RM Qumbu W. T. Brownlee; *BBNA 1891*, 46–47, RM Tsolo J. P. Cumming; CTA CMT 3/520, "Minutes of meeting at Willowvale between the Chief Magistrate, the Resident Magistrate, and people," 20 February 1892; "Minutes of a meeting held at Nqamakwe between the Chief Magistrate, the Resident Magistrate, and people," 8 April 1892; "Minutes of Public Meeting held at Engcobo on the 14th May 1892"; CTA AGR 144, 595, part 1, copy of letter, RM Mount Ayliff W. Power Leary to CMK, 27 May 1892; part 2, "Complaints by natives of Mqanduli district, 1892," FCT to SLMA, 28 November 1892, enclosing "Minutes of meeting of chiefs and people at Mqanduli," 24 November 1892; CTA AGR 144, 601, CM Elliot to USNA, 27 December 1892; *BBNA 1892*, 46, CM Elliot; *BBNA 1893*, 70, RM Engcobo A. H. Stanford; T. R. Beattie, *A Ride through the Transkei* (Kingwilliamstown: S. E. Rowles, 1891), 60; Wagenaar, "History of the Thembu," 344.

16. CTA AGR 223, F1930, vol. 4, CM Elliot to USNA, 5 March 1895, enclosing "Extract from Minutes of a meeting with the Tembu Chief Dalindyebo and Major H. G. Elliot C.M.G. Chief Magistrate, at Umtata on the 4th March, 1895"; "Tembu Meeting," *Umtata Herald*, 9 March 1895.

17. Quotations are from CTA NA 510, A262, SNA W. E. M. Stanford, memorandum, "Control of Forests in Transkeian Territories," 1 October 1897. See also UCT, Stanford Papers, D12, "Precis of questions from Chiefs and Headmen on

meeting w/C. J. Rhodes at Qumbu," 17 April 1894, comments from Qumbu, Tsolo, and Mount Fletcher district leaders; CTA CMT 3/520, regular file, "Meeting held at Umtata 19th April 1894 between the Hon. the Premier (C. J. Rhodes) & Dalindyebo & Tembus of Umtata, Mqanduli & Engcobo Districts"; CTA NA 510, A262, AGR Charles Currey, "Forest Matters—Transkei," 24 February 1899; CTA NA 706, B2897, part I, copy of telegram, Stanford to Native Affairs, Cape Town, 7 November 1905.

18. UCT, Stanford Papers, A22, diary of Rhodes' trip through the Territories, April 1894; CTA FCT 2/1/1/2, FCT to USA, 8 April 1895; *Annual Report 1895*, 169; CTA Archives of the Prime Minister's Office (PMO) 256, vol. 3, "Resolutions Petition etc presented to PM on his tour through the Native Terr during Nov and Dec 1896," n.d., petition to Prime Minister Sprigg, "Complaints and Requests for the Baca Nation and our Chief Makaula" (English translation of the Xhosa original); CTA NA 510, A262, 1 October 1897, SNA Stanford, memorandum, "Control of Forests in Transkeian Territories."

19. CTA NA 510, A262, AGR Charles Currey, "Forest Matters—Transkei," 24 February 1899; CTA AGR 750, F2839, USA to USNA, 16 December 1897; CTA NA 706, B2897, part I, copy of telegram, Stanford to Native Affairs, Cape Town, 7 November 1905; Sim, *Forests and Forest Flora*, 89.

20. UCT, Stanford Papers, A22, Diary of Rhodes's trip through the Territories, April 1894; CTA FCT 2/1/1/1, Henkel to USA, 24 August 1894; CTA FCT 2/1/1/2, Henkel to USA, 8 April 1895; Henkel to USA, 30 October 1895; *Annual Report 1895*, 169.

21. *Annual Report 1888*, 60, 64; *Annual Report 1889*, 94–96; *Annual Report 1890*, 139; CTA FCT 2/1/1/1, FCT to Asst. Commissioner CLPW, 11 April 1890; CTA AGR 144, 595, part 1, FCT to CM Transkei, 7 August 1890; Henkel to Asst. Commissioner CLPW, 1 February 1892.

22. Knut A. Carlson, "Forestry in the Transkei," *Cape Agricultural Journal*, June 11, 1896, 303–4.

23. Quotation from CTA NA 510, A262, Conservator of Forests, Eastern Conservancy (FCE) to SLMA, 29 May 1893. For other examples, see CTA FCT 2/1/1/1, FCT to USA, 7 February 1894; CTA AGR 766, F1930, FCT to SNA Rhodes, 2 May 1894; CTA FCT 2/1/1/2, FCT to USA, 30 October 1895; FCT to USA, 29 January 1897; FCT to USA, 25 February 1897; CTA AGR 750, F2839, AGR W. Hammond-Tooke to USNA, 22 September 1897; CTA NA 510 A262, J. Storr Lister, "Report by the Conservator of Forests, King William's Town on the Transkeian Forest Administration," 11 March 1898.

24. CTA NA 510, A262, FCE to SLMA, 29 May 1893; CTA AGR 766, F1930, FCT to USA, 30 May 1894. See also Brown, "Conservation and Utilisation of the Natural World."

25. CTA FCT 2/1/1/1, p. 723, No. 723/447, FCT to AGR, 30 December 1893. See also CTA NA 510, A262, FCE to SLMA, 29 May 1893; CTA AGR 207, F1563, FCT to USA, 26 May 1894; CTA CMT 3/40, "Petition of the Foresters of the

Transkeian Territories," A. Louis Raymond, Frank Willard, A. L. Caplen, John R. Rushmen, G. A. Freemantle, J. A. Keevy, and A. W. F. Greenaway, to PM and SNA Rhodes, n.d., received 19 April 1895; CTA AGR 750, F2839, FCT to CMT, 25 June 1897.

26. CTA FCT 1/1/1/1, CM Tembuland H.G. Elliot, CMK W. E. Stanford, and Acting CM Transkei W. M. Liefeldt to USNA, 20 May 1890; CTA AGR 766, F1930, CMK to USNA, 19 June 1894; USNA J. Rose Innes, memorandum, "Forests in the Territories," 2 July 1894. On magistrates' views of their "expertise" in the late-nineteenth-century Cape, see Chanock, *Making of South African Legal Culture*, 252–57.

27. CTA AGR 766, F1930, USNA Innes, memorandum, "Forests in the Territories," 2 July 1894.

28. CTA FCT 1/1/1/1, CM Tembuland H. G. Elliot, CMK W.E. Stanford, and Acting CM Transkei W. M. Liefeldt to USNA, 20 May 1890.

29. Chanock, *Making of South African Legal Culture*, 243–44, 255.

30. *Annual Report 1895*, 172; CTA FCT 2/1/1/1, FCT to USA, 10 January 1896; CTA FCT 2/1/1/2, FCT to USA, 19 May 1896; Proclamation 388 of 1896.

31. Tropp, "Dogs, Poison and the Meaning of Colonial Intervention"; Pule Phoofolo, "Epidemics and Revolutions: The Rinderpest Epidemic in Late Nineteenth-Century Southern Africa," *Past and Present* 138 (1993): 112–43; Charles van Onselen, "Reactions to Rinderpest in Southern Africa, 1896–97," *Journal of African History* 13, no. 3 (1972): 473–88.

32. CTA AGR 750, F2839, SNA W. E. Stanford, memorandum, "Control of Forests in Transkeian Territories," 1 October 1897. Native Affairs colleagues heartily endorsed his ideas: CTA AGR 750, F2839, CM Tembuland Elliot, to SNA Stanford, "Report: Control of Forests in Transkeian Territories," 2 November 1897; CTA CMT 3/40, report by CMK John Scott, "Administration of Forests in Transkeian Territories," forwarded by Asst. CMT W. G. Cumming to CMT, 13 November 1897. For a related discussion in colonial Natal, see Carolyn Hamilton's examination of "Shepstone as Shaka" in *Terrific Majesty*, chap. 3.

33. CTA AGR 750, F 2839, AGR Hammond-Tooke to USA, letters of 22 September 1897 and 16 December 1897.

34. Ibid.

35. Ibid.

36. CTA AGR 1/39, FCT Heywood to AGR, 15 March 1899, and FCT to USA, 30 April 1900; Principal Clerk for USA to USA, 10 May 1900; *BBNA 1903*, 8, "Summary of Reports in Sections I., II., and III."; *BBNA 1904*, 49–50, RM Kentani N. O. Thompson.

37. CTA AGR 1/39, FCT Heywood to AGR, 15 March 1899, and FCT to USA, 30 April 1900; Principal Clerk for USA to USA, 10 May 1900; Proclamation 135 of 1903; Keet, *Historical Review*, 57; Sim, *Forests and Forest Flora*, 90.

38. Cooper, "Conflict and Connection," 1530–32.

39. CTA FCT 2/1/1/1, FCT to Secretary to Treasurer, 18 August 1891; *Annual Report 1891*, 68.

40. *Annual Report 1892*, 90, 94; *Annual Report 1893*, 137, 141; CTA CMT 3/40, FCT to CMT, 6 April 1893; *Annual Report 1894*, 130; CTA FCT 1/1/3/1, A. Louis Raymond, Forester at Manina, Engcobo district, to FCT, 16 April 1894; *Annual Report 1895*, 168, 172; Carlson, "Forestry in the Transkei."

41. *Annual Report 1891*, 5, Eastern Conservancy, DFO Kingwilliamstown (KWT) and Peddie districts; CTA FCT 2/1/1/1, FCT to Percy Nightingale, Secretary to Treasurer, 18 August 1891; *Annual Report 1893*, 103–4, Eastern Conservancy, and 137, 141, Transkeian Conservancy; CTA FCT 2/1/1/1, FCT to SLMA, 30 June 1893; CTA Archives of the Conservator of Forests, Eastern Conservancy (FCE) 3/1/52, file 640, C. C. Henkel to FCE J. S. Lister, 27 January 1893; FCT to FCE, 4 October 1893; CTA FCT 2/1/1/1, FCT to USA, 21 October 1893; FCT to USA, 7 February 1894; CTA CMT 3/40, FCT to CMT, 8 March 1894; CTA FCT 1/1/3/1, A. Louis Raymond, Forester at Manina, to FCT, 16 April 1894; DFO Caplen to Henkel, 21 April 1894; CTA AGR 766, F1930, FCT to SNA Rhodes, 2 May 1894.

42. Jamie Monson has described a similar logic employed by conservation-minded officials in colonial Tanzania, in "Canoe-Building under Colonialism."

43. See, for example, CTA AGR 766, F1930, FCT to SNA C. G. Rhodes, 2 May 1894; CTA CMT 3/40, FCT to CMT, 6 April 1893.

44. CTA NA 510, A262, FCE to SLMA, 29 May 1893; CTA FCT 2/1/1/1, p. 723, no. 723/447, FCT to AGR, 30 December 1893; CTA CMT 3/40, FCT to CMT, 8 March 1894; CTA AGR 766, F1930, FCT to SNA C. G. Rhodes, 2 May 1894; CTA FCT 1/1/1/2, USA to FCT, 17 January 1894; W. Hammond-Tooke for USA to FCT, 16 June 1894; CTA AGR 223, F1930, vol. 4, FCT to CMK, 16 April 1895; *Annual Report 1895*, 172.

45. CTA AGR 207, F1563, FCT to USA, 26 May 1894; CTA AGR 766, F1930, CMK W. Stanford to USNA, 19 June 1894; USNA J. Rose Innes, memorandum, "Forests in the Territories," 2 July 1894; DFO Butterworth M. Krausz to FCT, 30 October 1894; CTA FCT 2/1/1/1, FCT to USA, 24 August 1894; CTA AGR 766, F1930, DFO Butterworth M. Krausz to FCT, 30 October 1894; CTA CMT 3/40, "Petition of the Foresters of the Transkeian Territories," A. Louis Raymond, Frank Willard, A. L. Caplen, John R. Rushmen, G. A. Freemantle, J. A. Keevy, and A. W. F. Greenaway to PM and SNA Rhodes, n.d., received 19 April 1895; CTA FCT 2/1/1/2, FCT to USA, 19 November 1895. DFO Knut Carlson reflected on the situation in his autobiographical *Transplanted*, 100.

46. CTA AGR 766, F1930, DFO Butterworth M. Krausz to FCT, 30 October 1894.

47. CTA NA 463, memorandum, Transkei General Council, August Meeting, 1899, 3, report by Superintendent of Plantations Caplen; CTA NA 510, A262, Acting RM Tabankulu to CMT, 5 October 1901; RM Cofimvaba to CMT, 11 October 1901; RM Flagstaff to CMT, 15 October 1901; RM Willowvale to CMT, 10 October 1901; RM Kentani to CMT, 11 October 1901; RM Umzimkulu to CMK, 25 October 1901.

48. On such common peasant responses to state intervention in other contexts, see Allen Isaacman, *Cotton Is the Mother of Poverty: Peasants, Work, and Rural*

Struggle in Colonial Mozambique, 1938–1961 (Portsmouth, N.H.: Heinemann, 1996), 207–9, describing hidden or everyday forms of resistance; Michael Adas, "From Footdragging to Flight: The Evasive History of Peasant Avoidance Strategies in South and South East Asia," *Journal of Peasant Studies* 13, no. 2 (1986): 64–86.

49. CTA FCT 1/1/3/1, Forester H. L. Caplen to FCT Henkel, 9 April 1894; Forester Caplen to FCT Henkel, 14 May 1894.

50. Interviews with Albertina Ntwalana, Lindile, Umtata District, 24 February 1998; Emmie Fiko, Egerton, Umtata District, 23 February 1998; Lindiwe Gcanga, Nobantu Nomganga, and Mrs. Ncedani, Manzana, Engcobo District, 5 February 1998; Olga Tyekela, Fairfield, Umtata District, 23 February 1998; Martha Ngadlela, Manzana, Engcobo District, 11 February 1998.

51. Interviews with Eunice Matomela, Springvale, Umtata District, 23 February 1998; Sampson and Notozamile Dyayiya, Silverton, Umtata District, 27 February 1998; W. M. Ngombane, Mputi, Umtata District, 8 January 1998; Rev. Matthew Gqweta, Baziya Mission Station, Umtata District, 12 January 1998; Samuel Qina, Tabase, Umtata District, 27 January 1998; Headman Sithelo, Kaplan, Umtata District, 25 February 1998; Polisile Maka; Noheke Rangana; Rev. G. Vika; Dabulamanzi Gcanga; R. T. S. Mdaka; CTA FCT 3/1/50, T60111/30/22, RM Reg. C. Heathcote to CM Umtata.

52. *Annual Report 1895*, 173; CTA AGR 760, F3553, FCT Heywood to USA, 23 March 1903. For representative cases, see CTA 1/TSO 1/1/19, case 149, R. v. William Ndumdum and 45 others, 30 August 1901; CTA 1/TSO 1/2/9, case 134 of 1917; CTA 1/TSO 3/1/7/7, A. D. Braums, ganger in charge, Ncolosi, to Gladwin, 24 October 1909.

53. For some illuminating examples, see CTA 1/TSO 1/1/5, case 32, R. v. Maketa and Mhlakaza, 7 June 1888; CTA 1/TSO 1/1/10, case 112, R. v. Qoqo, 23 November 1893; CTA 1/TSO 1/1/13, criminal case of 2 March 1897; CTA 1/TSO 1/1/17, case 263, R. v. Mofukana and another, 11 September 1900, and case 295, R. v. Zaza and 4 others, 27 September 1900; CTA 1/ECO 1/1/1/35, case 98, R. v. Makabeni, 22 March 1901; CTA 1/TSO 1/1/17, CTA FCT 3/1/50, T601, DFO Butterworth to FCT, 3 August 1923.

54. CTA 1/TSO 1/1/13, criminal case of 12 May 1897.

55. CTA 1/TSO 1/1/12, criminal case of 3 April 1895; criminal case of 21 January 1896.

56. CTA 1/TSO 1/1/4, case 34, R. v. Samatyumtyum, 10 May 1897.

57. CTA 1/TSO 1/1/4, case 32, R. v. Samuel Langley, Nikani, Nhleleni, J. Silwani, and John Jonas, 4 May 1897.

58. CTA 1/TSO 1/1/5, case 19, R. v. Mgamle and 7 others, 6 April 1888; case 31, R. v. Gonfela and 6 others, 7 June 1888; CTA 1/TSO 1/1/19, case 149, R. v. William Ndumdum and 45 others, 30 August 1901.

59. CTA CMT 3/520, "Minutes of Public Meeting held at Engcobo on the 14th May 1892."

60. CTA FCE 3/1/52, file 628, FCT to FCE, 16 May 1893.

61. CTA 1/TSO 1/1/3, case 34, R. v. Mxokyeli, 5 July 1886; CTA 1/TSO 1/1/4, case 34, R. v. Samatyumtyum, 10 May 1887; CTA 1/TSO 1/1/6, case 31, R. v. Matyida, 17 May 1889; CTA 1/TSO 1/1/10, case 65, R. v. Rangana, 7 July 1893; CTA 1/UTA 1/1/1/14, case 375, R. v. Thomas, 27 September 1893; case 417, R. v. Charlie, 3 November 1893; case 462, R. v. Funda, 27 December 1893; CTA 1/UTA 1/1/1/15, case 152, R. v. Kusa, 26 May 1894; CTA 1/TSO 1/1/11, case 72, R. v. Gqihitale, 13 June 1894; CTA 1/TSO 1/1/12, case of 17 June 1896; CTA FCT 2/1/1/2, FCT to USA, 25 November 1897; CTA 1/TSO 1/1/14, R. v. Ntuli and another, 2 June 1898; CTA 1/TSO 1/1/15, case 51, R. v. Kufa, 12 April 1899; CTA 1/TSO, 1/1/16, case 49, R. v. Mbangwa and Gaziana, 20 March 1900; CTA 1/TSO 2/1/29, case 170, Jack v. Mehlomane, 22 October 1901; Wagenaar, "History of the Thembu," 351. For an extensive discussion of these traders, see Tropp, "Roots and Rights," 306–15.

62. Each adult married male was obliged to pay hut tax on the household of each of his wives, regardless of whether he lived in his own homestead or in one of a senior member of his family. See Beinart, *Political Economy of Pondoland*, 97.

63. Redding, "Legal Minors and Social Children."

64. CTA NA 510, A262, RM Umzimkulu to CMK, 25 October 1901; CTA 1/UTA, 4/1/8/1/9, circular no. 12/185, FCT to RMs, 29 October 1902; CTA NA 648, B2356, FCT to CCF, 28 September 1906; CTA NA 753, F127, CCF to SNA, 13 April 1908.

65. As Sara Berry has described, Africans in many colonial spheres exploited the formal rules and institutions of colonial rule by redefining them, particularly to gain access to natural resources in the new colonial order. *No Condition is Permanent*, 34–35, 40–41.

66. Redding, "Legal Minors and Social Children"; Redding, "Sorcery and Sovereignty." Redding also invokes the symbolic role of taxation in "Government Witchcraft: Taxation, the Supernatural, and the Mpondo Revolt in the Transkei, South Africa, 1955–1963," *African Affairs* 95 (1996): 555–79.

67. *Report and Proceedings, with Appendices, of the Government Commission on Native Laws and Customs*, Part I, 508–9, 6 May 1882; Part II, Appendices to Report and Proceedings, "Notes on Preceding Abstract of Colonial Laws Affecting Natives, and Native Laws and Customs," 35, CMK C. Brownlee, 30 September 1881.

68. CTA 1/TSO 9/2, Meeting at Tsolo between RM and headmen, 26 February 1883.

69. CM Elliot in his 1884 annual report, CTA CMT 1/82, cited in Redding, "Sorcery and Sovereignty," 268.

70. *BBNA 1878*, 79, Asst. RM Matatiele C. M. Liefeldt to CMK Capt. Blyth. See also *BBNA 1878*, 84, Alexander R. Welsh, Magistrate with Umditshwa.

71. Redding, "Legal Minors and Social Children," 59–60; Redding, "South African Women," 40.

72. CTA FCT 2/1/1/2, FCT Henkel to USA, 23 January 1895, original emphasis. See also CTA CMT 3/40, Asst. CMT to CMT, forwarding report by CMK John

Scott, "Administration of Forests in Transkeian Territories," 13 November 1897; Godfrey Callaway, *Pioneers in Pondoland* (Lovedale: Lovedale Press, 1939), 111–12.

73. *Annual Report 1888*, 58–59.

74. CTA Papers of the Resident Magistrate of the Kentani District (1/KNT) 5/1/1/20, RM Kentani to CMT, 16 September 1908.

75. Ibid., RM Butterworth to CMT, 17 October 1901; Carlson, "Forestry in the Transkei," 304.

76. For example, CTA CMT 3/40, Asst. CMT to CMT, forwarding report by CMK John Scott, "Administration of Forests in Transkeian Territories," 13 November 1897.

77. *Annual Report 1890*, 139. See also CTA AGR 750, F2839, AGR to USNA, letters of 22 September and 16 December 1897.

78. CTA AGR 750, F2839, CM Tembuland Elliot to SNA, report on "Control of Forests in Transkeian Territories," 2 November 1897.

79. In very different contexts, scholars have shown how development agencies in the late twentieth century have similarly naturalized women's "special" relationship with the environment or their "customary" environmental practices in order to exploit their labor in the service of broader environmental and developmental agendas: Schroeder, *Shady Practices*, viii–xxix, 133–36; Carney and Watts, "Disciplining Women?"

80. For a typical official account of "customary" fuelwood collection, see *BBNA 1894*, 76, CMK W. E. Stanford. Sian Sullivan similarly scrutinizes development agencies' gendered assumptions regarding Damara herders' natural resource management in "Gender, Ethnographic Myths and Community-Based Conservation."

81. "From Br. R. Baur," Baziya, 6 November 1871, *Periodical Accounts Relating to the Missions of the Church of the United Brethren, Established among the Heathen (Periodical Accounts)* 28, no. 294 (1872): 193.

82. Interviews with Cyprian Mvambo, Umtata, 3 February 1998; R. T. S. Mdaka; Olga Tyekela; Jemimah Dubo; Dabulamanzi Gcanga; *TTGC 1924*, 87. See also *TTGC 1930*, 225, describing similar practices in the Nqamakwe District.

83. Griqualand East Forest Regulations, section 2 and 4, published in *Annual Report 1888*, 65–66; UCT, Stanford Papers, Diaries, A10, 1 September 1885. The regulations set the following tariffs for firewood removal: 5s. per wagon load, 2s.6d. per sledge-load.

84. The following laws also regulated access to other types of "minor forest produce," such as bark, fibers, and creepers. Proclamation 209 of 1890, section 6; Proclamation 308 of 1890; Proclamation 388 of 1896, section 3; Proclamation 135 of 1903, Schedule B, Part II, section 5(b) and Part III, section 14. These regulations further stipulated that only African residents of a given magisterial district were entitled to collect fuelwood found in that district.

85. Similar processes are evident in the invention of "Indian fishery" in British Columbia, described in Diane Newell, *Tangled Webs of History: Indians and the*

Law in Canada's Pacific Coast Fisheries (Toronto: University of Toronto Press, 1993), and in Monson, "Canoe-Building under Colonialism."

86. CTA 1/TSO 3/1/7/1, FCT to CMK, 23 April 1889; CTA 1/UTA, 4/1/8/1/4, FCT to RM Umtata, 20 May 1889; CTA CMT 1/19, FCT to CMT, 15 May 1889.

87. CTA 1/TSO 1/1/5, case 43, R. v. Sigidi and Mangqabitshana, 6 June 1888.

88. CTA CMK 1/100, RM Tsolo to CMK Stanford, 3 August 1886, and Stanford's response, 5 August 1886. Elizabeth Schmidt identifies similar problems facing poorer households in early colonial Zimbabwe in *Peasants, Traders, and Wives: Shona Women in the History of Zimbabwe, 1870–1939* (Portsmouth, N.H.: Heinemann, 1992), 79–80.

89. CTA 1/TSO 9/2, statement of headman Mehlo, 12 April 1889.

90. CTA AGR 766, F1930, FCT to USA, 16 December 1893; RM Engcobo to FCT, 23 September 1895; CTA FCT 2/1/1/2, FCT to USA, letters of 29 June, 30 June, and 25 November 1897; *Annual Report 1897*, 131; CTA CMT 3/170, RM Umtata to CMT, 30 August 1897; CTA CMT 3/89, RM Engcobo to CMT, 22 November 1897; RM Engcobo, Annual Report for Blue Book, 4 January 1898.

91. Proclamations 209 and 308 of 1890; Proclamation 388 of 1896.

92. *BBNA 1900*, 34–35, RM Umtata A. H. B. Stanford; 37, RM Engcobo C. E. Warner; *BBNA 1901*, 35, "Report of the CM for 1900," CMT Elliot; 42, RM Umtata Stanford; 46, RM Engcobo J. G. Leary; 61–62, RM Tsolo J. S. Simpson; *BBNA 1902*, 48, RM Engcobo Warner; *BBNA 1903*, 56, RM Engcobo C. A. King; RM Umtata Stanford; 81, RM Tsolo Edwin Gilfillan; *BBNA 1904*, 40–41, CMT W. E. Stanford; *BBNA 1905*, 62, RM Engcobo King; 89, RM Tsolo Arthur Gladwin; 91, RM Umtata Stanford; Redding, "South African Women," 35–37.

93. CTA NA 559, Part I, A963, RM Tsolo and FCT to CMT, 31 October 1902.

94. Proclamation 119 of 1905. The tariffs now jumped from 5s. per wagonload of green firewood and 2s.6d. per dry load, the rates that had been in effect since Proclamation 209 of 1890, to 7s.6d. and 5s., respectively.

95. *TTGC 1910*, 64–65, 7 April 1910; *TTGC 1908*, xxviii; CTA NA 753, F127, 3/3/1911, SNA to CMT, 3 March 1911; *TTGC 1911*, 89.

96. CTA 1/TSO 3/1/7/1, FCT to CMK, 4/23 April 1889; CTA 1/UTA, 4/1/8/1/4, FCT to RM Umtata, 20 May 1889.

97. CTA CMT 1/19, FCT to CMT, 15 May 1889.

98. CTA CMT 1/31, RM Engcobo, Arthur Stanford, to CMT, 28 November 1889.

99. Interview with Rev. G. Vika; *TTGC 1913*, 71, Councillor Qotongo.

100. Proclamation 209 of 1890, section 6; Proclamation 388 of 1896, section 3.

101. See, for example, CTA 1/TSO 1/1/11, case 45, R. v. Nohalf and Nofile, 24 April 1894; CTA 1/TSO 1/1/12, R. v. Nselelo, 3 April 1895; CTA AGR 766, F1930, FCT to CMT, 25 October 1895; CTA 1/TSO 1/1/15, case 283, R. v. T. C. P. Adams, 24 October 1899; case 108, R. v. Xayimpi, 19 May 1899; case 259, R. v. Nozinto, 26 September 1899; CTA 1/TSO 1/1/17, case 325, R. v. Sogoni and 10 others, 19 October 1900; CTA 1/ECO 1/1/1/35, case 147, R. v. Qunde, 3 May 1901; CTA 1/TSO 1/1/19, case 149, R. v. William Ndumdum and 45 others, 30 August 1901.

102. CTA 1/TSO 1/1/15, case 108, R. v. Xayimpi, 19 May 1899.

103. Proclamation 135 of 1903, section 9.

104. CTA FDU 1/2, 40, Baziya Plantation, "Report for October, 1903," T. C. P. Adams to FDU, 2 November 1903.

105. CTA 1/UTA C1/1/1, No. 124, Petition to Chairman Umtata DC, signed by Mdukiswa (Tabase), Ndwakuse (Esiqubadwini), Jeremiah Yengwa (Baziya Mission), and Nosali (Baziya), 4 June 1904, and response from FCT to RM and Chairman Umtata DC, 25 June 1904.

106. CTA CMT 3/40, FDU to CMT, 7 March 1903; CTA CMT 3/40, Asst. CMT to FCT, 30 October 1903.

107. Griqualand East Forest Regulations, sections 2 and 4, published in *Annual Report 1888*, 65–66; UCT, Stanford Papers, Diaries, A10, 1 September 1885; Proclamation 209 of 1890, section 6; Proclamation 308 of 1890; Proclamation 388 of 1896, section 3; Proclamation 135 of 1903, Schedule B, Part II, section 5(b), and Part III, section 14; Proclamation 288 of 1908, section 6.

108. "Report of an Exploratory Journey into Kaffraria by the Brethren R. Baur and H. Hartman, between August 27th and October 9th, 1862," *Periodical Accounts* 24, no. 258 (1861): 490–91.

109. "Journey of Br. Th. Weitz from Shiloh to Baziya and Entumasi, during the months of September, October, and November, 1872," *Periodical Accounts*, September 1874, 178.

110. "Tembu Meeting," *Umtata Herald*, 9 March 1895; see also *BBNA 1879*, 79, H. G. Elliot.

111. *Annual Report 1888*, 64.

112. CTA 1/TSO 1/1/9, case 127, R. v. Nomanti and 3 others, 27 July 1892; CTA 1/TSO 1/1/9, case 128, R. v. Fazine and 3 others, 27 July 1892; CTA CMT 3/83, 18 June 1896; CTA 1/ECO 1/1/1/35, case 174, R. v. Nosamana and Nowayiti, 20 May 1901.

113. CTA 1/TSO 3/1/7/6, R. J. M. Muggleton, Inxu, to RM Tsolo, received in RM Tsolo's office 19 October 1908; CTA 1/TSO 3/1/7/4, John Thomas Michie and William McDougall Watson to RM Tsolo, 9 June 1906.

114. See Isaacman, *Cotton is the Mother of Poverty*, 8–10, for a discussion of peasants' partial autonomy in the colonial era.

CHAPTER 3

1. Keet, *Historical Review*, 57; Sim, *Forests and Forest Flora*, 90.

2. In his later years, former DFO Knut Carlson claimed due credit for first proposing the "weaning" scheme. Knut A. Carlson, "Weaning the Natives from the Natural Forests in the Native Territories," *Journal of the South African Forestry Association* 2 (April 1939): 75–76, "Correspondence" section, and *Transplanted*, 127–28. See also Sim, *Forests and Forest Flora*, 87–88; *Annual Report 1931*, 19–20; King, "Historical Sketch," 7–8; Shone, *Forestry in Transkei*, 5.

3. *Annual Report 1890*, 145; *Annual Report 1891*, 71; *Annual Report 1892*, 91–92; *Annual Report 1893*, 142–43; *Annual Report 1894*, 132; *Annual Report 1895*, 171; *Annual Report 1896*, 156; *Annual Report 1897*, 133–34; *Annual Report 1898*, 124; Sim, *Forests and Forest Flora*, 87–88; King, "Historical Sketch"; Carlson, "Weaning the Natives."

4. For an early articulation of this argument, see Carlson, "Forestry in the Transkei," 303, which is also repeated in Sim, *Forests and Forest Flora*, 8.

5. *Annual Report 1904*, 134–36; *Annual Report 1905*, 120.

6. *Annual Report 1901*, 118–19, Eastern Conservancy; *Annual Report 1902*, 116–17, Eastern Conservancy, and 148, Transkeian Conservancy; *Annual Report 1903*, 101, Eastern Conservancy, and 125–27, Transkeian Conservancy; CTA NA 648, B2356, RM Kentani and DFO Butterworth Carlson to CMT, 3 August 1903; CTA FDU 1/2, 43, Nqadu Heights Plantation, DFO to foreman, 6 April 1903; *Annual Report 1904*, 111–12, Eastern Conservancy, and 140–41, Transkeian Conservancy; CTA FCT 3/1/19, T30, DFO Kokstad to FCT, 23 July 1904, on coastal forests in Bizana district; *Annual Report 1905*, 93, 106, Eastern Conservancy, and 127–28, Transkeian Conservancy; *Annual Report 1906*, 27.

7. In the Transkei, the plan was to affect all of the Territories except for Pondoland, which had only just come under legislation for state-controlled forests and begun the process of forest demarcation after 1903. CTA NA 706, B2897, Part I, SNA to USA, July [n.d.], 1905.

8. CTA AGR 683, 42, J. B. Hartley, Inspector, Location A, to RM Peddie, 25 April 1905; R. J. Dick, Special Magistrate KWT to Civil Commissioner KWT, 12 May 1905; Civil Commissioner KWT to SNA, 16 May 1905; SNA to USA, 3 June 1905; FCE to USA, 27 June 1905.

9. See, for instance, CTA AGR 707, F150, RM Kentani to Asst. CMT, 3 May 1905; CTA 1/KNT, 10/3, "Minutes of Mtg held at Kentani on Monday 20th November 1905 to meet Col. Stanford"; CTA NA 703, B2854, Asst. CMT to SNA, 19 July 1905; Acting RM Mqanduli to SNA, 5 August 1905; CTA NA 706, B2897, Part I, RM Tabankulu to Asst. CMK, 15 August 1906; Asst. CMK to SNA, 29 August 1906.

10. CTA NA 706, B2897, Part I, copy of telegram, Stanford to Native Affairs, Cape Town, 7 November 1905; telegram, Conservator Lister to Heywood, relaying Stanford's thoughts, 5 February 1906.

11. Edward Roux, *Time Longer than Rope: A History of the Black Man's Struggle for Freedom in South Africa*, 2nd ed. (Madison: University of Wisconsin Press, 1964), 88–100; Shula Marks, *Reluctant Rebellion: The 1906–8 Disturbances in Natal* (Oxford: Clarendon Press, 1970); Sean Redding, "A Blood-Stained Tax: Poll Tax and the Bambatha Rebellion in South Africa," *African Studies Review* 43, no. 2 (2000): 29–54; Benedict Carton, *Blood from Your Children: The Colonial Origins of Generational Conflict in South Africa* (Charlottesville: University Press of Virginia, 2000).

12. CTA NA 706, B2897, Part I, copy of telegram, RM Willowvale to CMT, 9 April 1906; RM Willowvale to SNA, 1 June 1906; SNA to Asst. CMT and Asst.

CMK, 19 June 1906; RM Willowvale to SNA, 3 July 1906; Proclamation 237 of 1906, rescinding Proclamation 43.

13. *Annual Report 1909*, 21.

14. *Annual Report 1907*, photograph page between text pages 16 and 17; Sim, *Forests and Flora*, between pages 22 and 23.

15. On rethinking the application of "carrying capacity" to Africans' livelihood practices, see H. Moore and Vaughan, *Cutting Down Trees*, chap. 2, and William Beinart's literature review in "African History and Environmental History," 278–80.

16. For a similar process in a North American context, see Newell, *Tangled Webs of History*.

17. *Annual Report 1906*, 27; *Annual Report 1907*, 16; *Annual Report 1908*, 17; *Annual Report 1909*, 20–22; *Annual Report 1910*, 20; *Annual Report 1911*, 3, 22–24.

18. *Annual Report 1907*, 16.

19. *Annual Report 1909*, 20–22, my emphasis. See also similar complaints in *Annual Report 1906*, 27; *Annual Report 1907*, 16; *Annual Report 1908*, 17; *Annual Report 1910*, 20; *Annual Report 1911*, 3, 22–24.

20. *Annual Report 1906*, 7–8, 18–19; *Annual Report 1909*, 7; *Annual Report 1910*, 4, 6; *Annual Report 1911*, 6–8; *Annual Report 1913*, 7–8, 24. By the mid-1910s, Lusikisiki District was the only one in which the free permit system still operated, pending the completion of forest demarcation. *Annual Report 1917*, 29.

21. Interview with Anderson Joyi.

22. *Annual Report 1893*, 142; CTA FCT 2/1/1/1, Henkel to USA, 28 August 1894; *Annual Report 1899*, 98, 104, 108, 111–12; *Annual Report 1901*, 159; *Annual Report 1902*, 144; *Annual Report 1904*, 154–57; *Annual Report 1905*, 129, 139, 143; *Annual Report 1908*, 15; TTGC 1912, iv; Carlson, *Transplanted*, 100–101. On similar challenges in colonial Natal, see Harald Witt, "The Emergence of Privately Grown Industrial Tree Plantations," in Dovers, Edgecombe, and Guest, *South Africa's Environmental History*, 93–94.

23. *Annual Report 1906*, 18–19, including totals from both "timber" and "wattle" plantations.

24. CTA FCT 2/1/1/1, FCT to USA, 28 August 1894; *Annual Report 1904*, 144; *Annual Report 1905*, 128; TTGC 1912, iv.

25. BBNA 1904, 49–50, RM Kentani N. O. Thompson; CTA NA 703, B2854, USA to SNA, 31 May 1904; CTA 1/KNT, 10/3, "Minutes of Mtg held at Kentani on Monday 20th November 1905 to meet Col. Stanford," statement of Sandile.

26. For example, CTA NA 703, B2854, Asst. CMT to SNA, 19 July 1905; Acting RM Mqanduli to SNA, 5 August 1905; CTA 1/UTA, 4/1/8/1/14, statement of Headman Mlotsana Mbande, before RM Umtata, 3 May 1907.

27. CTA NA 706, B2897, Part I, RM Tabankulu to Asst. CMK, 29 August 1906; Asst. CMK to SNA, 15 August 1906.

28. Southall, *South Africa's Transkei*, 74; Bundy, *Rise and Fall*, 127.

29. Beinart, *Political Economy of Pondoland*, 167–68; Southall, *South Africa's Transkei*, 73–64; Bundy, *Rise and Fall*, 127.

30. Southall, *South Africa's Transkei*, 72.

31. *TTGC 1911*, 36–39, particularly the comments of Langa Sokapase, Nqamakwe District, and 65; CTA CMT 3/648, 91/8/1, RM Kentani to CMT, 25 November 1912; *TTGC 1913*, 16–17, 70–71; CTA FCT 3/1/7, F110/10, RM Lusikisiki to CMT, 24 September 1914; *PGC 1914*, 27–28; CTA CMT 3/647, 161/1, Prime Minister's Tour in the Transkeian Territories, General Botha 1916, "Meeting at Lusikisiki, Pondo Chiefs Deputation," 13 September 1916; *TTGC 1916*, 57, 59; *PGC 1917*, 15; *TTGC 1918*, 36–37; *TTGC 1920*, 61–63; *TTGC 1921*, xxxiii–xxxiv; *TTGC 1924*, 152; *TTGC 1925*, 79, 83,158.

32. *PGC 1916*, 22.

33. Ibid., 22, Councillor Nongauza.

34. Kepe, *Environmental Entitlements in Mkambati*, 57–62; William Beinart, "Transkeian Smallholders and Agrarian Reform," *Journal of Contemporary African Studies* 11, no. 2 (1992): 184, 195n18; Union of South Africa, U.G.22–1932, *Report of the Native Economic Commission, 1930–32* (NEC), 3556, comments by CMT William T. Welsh; Mears, "Study in Native Administration," 215–16.

35. C. C. Henkel, *Tree Planting for Ornamental and Economic Purposes in the Transkeian Territories, South Africa* (Cape Town: Juta, 1894); *BBNA 1904*, RM Mt. Ayliff; *TTGC 1929*, 58; interview with Adolphus Qupa.

36. NEC, 2914–15, Rev. Williams; 3556, CMT William T. Welsh; 3729–30, Tennyson Mtywaku Makiwane; Mears, "Study in Native Administration," 215–16.

37. Interviews with Olga Tyekela; R. T. S. Mdaka.

38. *Annual Report 1901*, 154; CTA FCT 2/1/1/1, FCT to USA, 28 August 1894, describing the average size of wattle bundles.

39. These numbers are meant to be suggestive and are based on two estimates: (1) that each bundle comprised from 25 to 30 trees throughout the period, and (2) that wattle purchases were, on average, 4.5d per bundle (the median cost of the 3d-to-6d bundles available at plantations). However, the challenges of the data make this exercise rather speculative: over the years, official reports did not consistently break down plantation sales in such detail, they switched from recording numbers of bundles sold to actual cubic feet sold, and the size (length and circumference) and quantity of wattles in each bundle could vary, as could the price per bundle.

40. CTA NA 463, Papers of the Transkei General Council (TGC), "Transkei Wattle Plantations Annual Report, 1899," 28 February 1900, original emphasis.

41. Bundy, *Rise and Fall*, 124–25; M. H. Mason, "Dearth in the Transkei," *Nineteenth Century and After* 73 (January–June 1913): 677–78.

42. This partly reflected a gradual inflation of plantation wood prices by the mid-1920s. *TTGC 1926*, 114–16.

43. *Annual Report 1905*, 128–29; *TTGC 1909*, Report on Council Forests and Plantations (for 1908), 4.

44. CTA FCT 1/1/3/1, Superintendent of TGC Wattle Plantations to FCT, 21 April 1894; CTA FCT 2/1/1/2, FCT to USA, 30 October 1895; CTA FDU 1/4, 67,

FCT to DFO Umtata, 8 April 1904; *Annual Report 1904*, 154; CTA NA 648, B2356, A. G. Potter, by order of DFO Butterworth, to FCT, 13 September 1906.

45. Thus a recent description of colonial foresters in the Transkei as "prescient" in establishing plantations glosses over a much more complex and contested history. William Beinart and Peter Coates, *Environment and History: The Taming of Nature in the USA and South Africa* (London: Routledge, 1995), 48.

46. Southall, *South Africa's Transkei*, 72; Bundy, *Rise and Fall*, 121.

47. Beinart, *Political Economy of Pondoland*, 172.

48. CTA FCT 3/1/50, T601, RM Engcobo to CMT, 30 November 22. On the high prices of plantations, see also ibid., DFO Butterworth to FCT, 6 December 1922; FCT to CMT, 9 December 1922; FCT to DFO Butterworth, 5 April 1923.

49. *TTGC 1927*, discussion of Select Committee on Council Plantations, Item 10, Mr. R. D. Barry, 110.

50. *TTGC 1926*, 114–16, "Reduction of Charges of Plantation Produce," Councillor Sakwe. For other examples, see *TTGC 1930*, 225; *Annual Report 1932*, 48; *TTGC 1933*, 228.

51. Interview with Samuel Qina.

52. Interviews with Anderson Joyi; Dabulamanzi Gcanga; Samuel Qina.

53. Interview with William Jumba, Tsikitsiki Nodwayi, Festo Sonyoka, and Zwelivumile Quvile, Tabase, Umtata District, 4 February 1998.

54. Interviews with Samuel Qina; William Jumba et al.

55. Interview with Dabulamanzi Gcanga.

56. Interviews with Wele Boyana, Baziya Mission, Umtata District, 12 January 1998; William Jumba et al.

57. See Nancy Jacobs' critical interpretation of "populist" narratives in her informants' histories, in *Environment, Power, and Injustice*.

58. Tropp, "Displaced People, Replaced Narratives"; reports 2001–2002 on forestry restitution claims in the Baziya and Mbolompo area by the Transkei Land Service Organisation, available at http://www.tralso.co.za/reports.asp; "NGO Aims to Return 1000 Displaced Families," *Daily Dispatch Online*, 15 November 2000, http://www.dispatch.co.za/2000/11/15/easterncape/KDISPLAC.HTM.

59. *PGC 1916*, 22, Councillor Nongauza.

60. Interview with Samuel Qina.

61. See, for example, *Annual Report 1902*, Transkei, 148.

62. CTA CMT 3/648, 91/8/1, RM Kentani Frank Brownlee to CMT, 25 November 1912.

63. CTA AGR 750, F2839, USA to USNA, 22 September 1897; CTA FDU 1/5, 109, RM Qumbu to CMT, 12 January 1904.

64. Carlson, "Forestry in the Transkei," 303.

65. CTA CMT 3/520, "Minutes of Public Meeting held at Engcobo on the 14th May 1892"; CTA NA 321, CMT to USNA, letters of 4 August and 16 November 1893; CTA FCT 2/1/1/1, FCT to SLMA, "Forest Resources of the Transkeian Territories," 15 September 1893; CTA AGR 223, F1930, vol. 4, CM Elliot to USNA, 5

March 1895, enclosing "Extract from Minutes of a meeting with the Tembu Chief Dalindyebo and Major H. G. Elliot C.M.G. Chief Magistrate, at Umtata on the 4th March, 1895"; CTA 1/UTA C1/1/1, No. 124, Petition to Chairman Umtata DC, signed by Mdukiswa (Tabase), Ndwakuse (Esiqubadwini), Jeremiah Yengwa (Baziya Mission), and Nosali (Baziya), 4 June 1904; Sim, *Forests and Forest Flora*, 166–68; *PGC 1916*, 22, Councillor Nongauza; CTA FCT 3/1/50, T601, DFO Butterworth H. J. Ryan to FCT, 4 August 1922.

66. *Annual Report 1895*, 171; CTA FCT 2/1/1/2, No. 996/709, FCT to USA, date illegible 1896; FCT to USA, 19 February 1897; *Annual Report 1898*, 131; CTA FDU 1/2, 43, Nqadu Heights Plantation, FDU to plantation foreman, 14 August 1903; *Annual Report 1904*, 125; CTA FDU 1/1, 27A, FDU circular to Foresters, 29 July 1908; O. B. Miller, "Notes on the Distribution of Species in Natural Forests of the Transkeian Conservancy," *South African Journal of Natural History* 3, no. 2 (1921–22): 21; CTA FCT 3/1/8, H2000/5, FCT to CMT, 12 December 1936.

67. *Annual Report 1884*, 53–55, FCE D. E. Hutchins; Cape of Good Hope, A.6–1888, *Report of the Select Committee on Forests Bill* (1888), 68; *Annual Report 1889*, 95; CTA FCT 2/1/1/1, FCT to SLMA, 15 September 1893; Sim, *Forests and Forest Flora*, 166–68.

68. Interview with Samuel Qina; Godfrey, "Xhosa-English Dictionary," "H," 26, "K," 77, and "T," 26–27 and 92. See also Hunter, *Reaction to Conquest*, 74.

69. CTA FDU 1/2, 43, Nqadu Heights Plantation, FDU to plantation foreman, 14 August 1903.

70. CTA FKS 2/3/1, FKS to Forester T. James, Insikeni, 2 January 1896; CTA 1/TSO 1/2/1, Criminal Record Book, case 46 of 1899; CTA 1/TSO 1/1/16, case 105, R. v. Hlabizulu, 4 May 1900; CTA 1/TSO 1/1/18, case 70, R. v. Singala and 3 others, 14 May 1901; CTA NA 510, A262, RM Mqanduli to CMT, 28 December 1901; CTA NA 518, A387, RM Bizana to CMT, 15 August 1901; *BBNA 1902*, 73, RM Umzimkulu; CTA NA 559, Part I, A963, SNA to PM, 20 November 1902; CTA CMT 3/161, RM Tsolo to CMT, 29 January 1903; CTA CMT 3/40, RM Engcobo to CMT, 24 June 1903; Proclamation 135 of 1903, Part VII. See Tropp, "Roots and Rights," 331–40, for a discussion of kraalwood restrictions in KwaMatiwane from the early 1880s onward.

71. Foresters at the time claimed that colonial timber extraction in the region had precipitated this process. Throughout the late nineteenth century, in response to settler demand in the Cape, sawyers had focused their cutting on such trees as yellowwoods and sneezewood, enabling lemonwood and other "less desirable" species left behind to dominate. Sim, *Forests and Forest Flora*, 288. See Cawe and McKenzie, "Afromontane Forests of Transkei," 38, for a recent reassessment of such explanations.

72. *Annual Report 1903*, 124; Sim, *Forests and Forest Flora*, 288.

73. Henry E. Lowood, "The Calculating Forester: Quantification, Cameral Science, and the Emergence of Scientific Forestry Management in Germany," in *The Quantifying Spirit in the Eighteenth Century*, ed. Tore Frängsmyr, John L. Heilbron, and Robin E. Rider (Berkeley: University of California Press, 1990),

315–42; K. Sivaramakrishnan, "State Sciences and Development Histories: Encoding Local Forestry Knowledge in Bengal," *Development and Change* 31, no. 1 (January 2000): 61–89; James C. Scott, *Seeing Like a State: How Certain Schemes to Improve the Human Condition Have Failed* (New Haven, Conn.: Yale University Press, 1998), 11–30.

74. *Annual Report 1901*, 166–67; *Annual Report 1903*, 124–25; Sim, *Forests and Forest Flora*, 288.

75. *Annual Report 1886*, 18; *Annual Report 1888*, 41; Carlson, "Forestry in the Transkei"; CTA FDU 1/4, 67, FCT to FDU, 17 July 1906; Sim, *Forests and Forest Flora*, 10.

76. *Annual Report 1903*, 124–25.

77. Ibid.

78. CTA 1/UTA C1/1/1, No. 124, Petition to Chairman Umtata DC, signed by Mdukiswa (Tabase), Ndwakuse (Esiqubadwini), Jeremiah Yengwa (Baziya Mission), and Nosali (Baziya), 4 June 1904; FCT to RM Umtata, 25 June 1904.

79. *Annual Report 1904*, 125; *Annual Report 1906*, 8; CTA FDU 1/4, 67, FCT to FDU, 14 July 1906; Sim, *Forests and Forest Flora*, 288; CTA FCT 3/1/7, F110/10, Demarcations and Forests, Lusikisiki District, Asst. FCT to CCF, 11 December 1908; *Annual Report 1908*, 10; *Annual Report 1911*, 12.

80. *Annual Report 1908*, 10; CTA FDU 1/1, 29A, Qunu Plantation, FDU to FCT, 27 May 1915.

81. CTA FCT 3/1/52, T702/6, Working plans, Tsolo Central Reserve, 1924–30, FDU to FCT, 2 February 1926, forwarding reports from Apprentice J. T. de Lange, 15 December 1925, and Apprentice J. D. Smuts, 30 January 1926; CTA FCT 3/1/51, T700, FDU to FCT, 21 November 1927; FDU to FCT, 15 April 1937; Norman L. King, "Report on the Indigenous Forests in the Transkei and Suggestions for Their Management," 17–19, 21 August 1936.

82. *Annual Report 1905*, 105, Eastern Conservancy; CTA NA 463, Papers of Transkei General Council, H. L. Caplen, Butterworth, *Transkei Wattle Plantations Annual Report for 1899*, 28 August 1900.

83. CTA AGR 683, 42, DFO KWT to FCE, 6 April 1905; J. B. Hartley, Inspector, Location A, to RM Peddie, 25 April 1905; R. J. Dick, Special Magistrate KWT, to Civil Commissioner KWT, 12 May 1905; FCE to USA, 27 June 1905.

84. For example, *TTGC 1911*, ix; *TTGC 1925*, Report on Plantations, xviii; Carlson, "Weaning the Natives," 75–76.

85. Interview with Fumanekile and Ntombizanele Sithelo, Kaplan, Umtata District, 10 February 1998.

86. *TTGC 1925*, *Report on Plantations*, xviii, describing how limited local demand for eucalyptus was across the Transkei because "this is not favoured generally as firewood"; Lieberman, "Ethnobotanical Assessment of the Dwesa and Cwebe Nature Reserves," 63; Helen Meintjies, "Trends in Natural Resource Management: Policy and Practice in Southern Africa," Working Paper 22 (Johannesburg: Land and Agriculture Policy Centre, August 1995), 37.

87. Interview with Nozolile Kholwane, Fairfield, Umtata District, 23 February 1998.

88. Interviews with Dabulamanzi Gcanga; Emmie Fiko; Eunice Matomela; William Jumba et al.

89. Interview with Headman Sithelo.

90. Interview with Nozolile Kholwane.

91. Shone, *Forestry in Transkei*, 48. By contrast, see Colin T. Johnson, "A Preliminary Checklist of Xhosa Names for Trees Growing in Transkei," *Bothalia* 20, no. 2 (1990): 147–152.

92. For example, interviews with Dabulamanzi Gcanga; Emmie Fiko; Eunice Matomela.

93. *TTGC 1925*, Report on Plantations, xviii; *Annual Report 1934*, 13.

94. Thus colonial forest management was "made to fit" the realities of Africans' experiences with wood restrictions and their ideas about valuable species, much as Africans' perspectives on tsetse control in Northern Rhodesia impinged on officials' biomedical thinking and policies. Luise White, "Tsetse Visions: Narratives of Blood and Bugs in Colonial Northern Rhodesia, 1931–9," *Journal of African History* 36 (1995): 219–45.

95. *PGC 1916*, Councillor Jiyajiya, 22.

96. CTA FCT 3/1/52, T702/6, FCT to CCF, 28 September 1926.

97. CTA FDU 1/1, 27A, DFO Umtata to Foresters Miller, Muller, Kriel, Human, Roux, and Dreyer, 29 July 1908; *TTGC 1911*, 37; *TTGC 1913*, 71; interview with Olga Tyekela.

98. *Annual Report 1910*, 19–20; *Annual Report 1911*, 19; Mason, "Dearth in the Transkei," 677–78.

99. See, for example, *Annual Report 1911*, 22. Another option for at least some women at the turn of the century was to take on jobs at colonial tree farms in exchange for plantation firewood. CTA FCT 1/1/3/1, Forester H. L. Caplen, Umtinthloni, to FCT, 19 April 1894; *Annual Report 1899*, 101; CTA FCT 3/1/18, T8, "Demarcations, Engcobo District," Forester A. Donald to FDU, n.d. 1900; CTA 1/ECO 1/1/1/36, case 305, R. v. Teto, 23 September 1901; *Annual Report 1905*, 129.

100. Interviews with Samuel Qina; Rev. Matthew Gqweta.

101. The figures cited here derive from plantations run directly by the Forest Department, for which statistics are most consistently available during this period. Fuelwood sales from General Council plantations, much less precisely documented over time, also generally increased in these years. Officials first noted the "great demand" for firewood at these plantations in the early 1910s: *TTGC 1912*, iii–iv, "Report on Council Plantations," 13 February 1912.

102. African purchases were actually lower than these figures suggest, as this data also reflected wood bought by European settlers in the Transkei.

103. *Annual Report 1926*, 337.

104. Fortmann and Bruce, *Whose Trees?*; Meintjies, "Trends in Natural Resource Management," 37.

105. *Annual Report 1898*, 124; *Annual Report 1901*, 154; *Annual Report 1903*, 128, 138; *Annual Report 1904*, 142; *Annual Report 1905*, 105, Eastern Conservancy; *Annual Report 1906*, 14–15; *Annual Report 1907*, 10.

106. As Joanne Yawitch has noted, such problems persisted in many areas of rural South Africa in more recent decades. The creation of community woodlots, outgrowths of "betterment" and rural "development" schemes, has not always translated into easy wood access for many women. *Betterment: The Myth of Homeland Agriculture* (Johannesburg: South African Institute of Race Relations, 1981), 51.

107. *TTGC 1924*, 152, "Opening of Demarcated Forests."

108. *TTGC 1925*, 158, "Buying and Cutting of Wood from Demarcated Forests"; *TTGC 1925*, page e, "Summary of Replies," Minute no. 75; *TTGC 1926*, vi, "Summary of Replies," Minute no. 78.

109. *Annual Report 1931*, 20; population estimates are based on census data cited in Southall, *South Africa's Transkei*, 74.

110. CTA FCT 3/1/8, T246, FCT to Inspector of Forestry, Pietermaritzburg, "Native Forest Policy in the Transkei," n.d. 1932.

111. CTA CMT 3/1322, 24/C, vol. 3, Chairman DC Butterworth to CMT, 12 December 1934; RM Engcobo to CMT, letters of 5 January and 15 January 1935. See also CMT 3/1318, 24, F1, vol. 2, Director of Forestry A. O'Connor, "Forest Policy Transkei," 8 November 1943; C. S. Hubbard, "Afforestation and Fuel Supply in Relation to Development in the Transkeian Territories," *South African Medical Journal*, November 10, 1945, 407.

112. Moll, *No Blade of Grass*, 8–9; Southall, *South Africa's Transkei*, 80–82.

113. *Proceedings and Reports of Select Committees at the Session of the United Transkeian Territories General Council (UTTGC) 1934*, 76–77, CMT's statement, and lxiv, report from Supervisor of Plantations. In some areas, a popular response to such wood shortages and financial pressures was an intensified use of veld plants such as nkanga bush, which some officials cited as a major contributor to women's reduced plantation fuelwood purchases. *NEC*, vol. 6, 3552, CMT William T. Welsh; *UTTGC 1932*, F. R. B. Thompson, Director of Agriculture, "Report on Agriculture," lxx, and C. D. Nevill, Supervisor of Plantns, lxxvi–lxxvii; Hubbard, "Afforestation and Fuel Supply," 407–8.

114. Interviews with Liziwe and Monwabisi Ndzungu, Payne, Umtata District, 24 February 1998; Olga Tyekela; Nozolile Kholwane; CTA FCT 3/1/52, T702/6, FCT to CCF, 28 September 1926; CTA FCT 3/1/14, L3010, FCT to CCF, 26 June 1931; CTA FCT 3/1/51, T700, M.624, "Management of Indigenous Forests in the Transkei," J. Keet, for Director of Forestry, to Inspector of Forestry, Natal and Transkei, DFO Kokstad, DFO Umtata, and DFO Butterworth, 28 December 1933; CTA CMT 3/1322, 24/C, vol. 3, Secretary of Agriculture and Forestry to SNA, 20 January 1934; RM Tsomo to CMT, 14 December 1934; RM Libode to CMT, 11 December 1934; RM Mt. Frere to CMT, 15 December 1934; RM Umzimkulu to CMT, 11 December 1934; *Annual Report 1934*, 13; CTA CMT 3/1322, 24/C, vol. 3, W. M. du Plessis to SNA, 8 April 1935.

115. CTA CMT 3/1322, 24/C, vol. 3, Secretary of Agriculture and Forestry to SNA, schedule for Qumbu district, Etwa reserve, 20 January 1934.

116. *Annual Report 1934*, 13.

117. Beinart, "Soil Erosion, Conservationism"; Moll, *No Blade of Grass*; Hendricks, "Loose Planning and Rapid Resettlement."

118. CTA FCT 3/1/52, T702/6, FCT to CCF, 28 September 1926; *Annual Report 1926*, 337; CTA Archives of the DFO Kokstad (FKS) 4/1/1, Foresters' Reports, Bizana 1928–29; *NEC*, vol. 6, 3294, 3320, 3341–42, Sydney Gordon Butler, Principal of Tsolo Agricultural School, and Fred Roland Blythe Thompson, Principal of Teko Agricultural School; 3382, Alfred Owen Ballene Payn, M. P., 17 November 1930; 3787–88, John Guma and Nantiso Kula, representing Young Men's Agricultural Association; 3880–3885, Shadrack Sopola, Nqamakwe district; *TTGC 1931*, iii–iv, W. H. P. Freemantle, "Report of Recess Committee on Soil Erosion and Reclamation, 1930"; CTA FCT 3/1/14, L3010, FCT to CCF, 26 June 1931.

119. Beinart, "Soil Erosion, Conservationism and Ideas about Development"; Moll, "No Blade of Grass"; Hendricks, "Loose Planning and Rapid Resettlement"; Ashforth, *Politics of Official Discourse*; Evans, *Bureaucracy and Race*.

120. *Annual Report 1932*, 48, original emphasis.

121. *UTTGC 1933*, 63, Report of the Select Committee on Land Matters, Item no. 32, and minute no. 87, Report of Select Committee, section 11; CTA CMT 3/1319, 24/4, E. Clark, for SNA, to CMT, 15 December 1933; CTA CMT 3/1322, 24/C, vol. 3, "Free Removal of Forest Produce from Demarcated Crown Forests (Teza Servitude) and Closing of certain Demarcated Forests thereto," CMT to SNA, 25 July 1933.

122. CTA FCT 3/1/51, T700, M.624, "Management of Indigenous Forests in the Transkei," 11, J. Keet, for Director of Forestry, to Inspector of Forestry, Natal and Transkei, DFO Kokstad, DFO Umtata, and DFO Butterworth, 28 December 1933; CTA CMT 3/1319, 24/4, E. Clark, for SNA, to CMT, 15 December 1933.

123. CTA CMT 3/1322, 24/C, vol. 3, Secretary of Agriculture and Forestry W. M. du Plessis to SNA, 20 January 1934; GN 987 of 1935, amending GN 1605 of 1920.

124. CTA CMT 3/1322, 24/C, vol. 3, Secretary of Agriculture and Forestry W. M. du Plessis to SNA, 20 January 1934.

125. Ibid., Chairman DC Butterworth to CMT, 12 December 1934; RM Engcobo to CMT, letters of 5 January and 15 January 1935.

126. See, for example, ibid., RM Tsolo to CMT, 10 December 1934; RM Umzimkulu to CMT, 11 December 1934; RM Ngqeleni to CMT, 12 December 1934; RM Tsomo to CMT, 14 December 1934; RM Mt. Frere to CMT, 15 December 1934; RM Lusikisiki to CMT, 15 December 1934; RM St. Mark's to CMT, 21 December 1934; *Annual Report 1934*, 13.

127. GN 987 of 1935 and GN 248 of 1936.

128. *Annual Report 1938*, 18, 23; CTA CMT 3/1322, 24/C, vol. 3, FCT to CMT, 2 May 1939; RM Bizana to CMT, 20 June 1939.

129. Circular minute no. 24/4/2, CMT to various RMs across the Transkei, 4 May 1939, and responses from different magistrates over the next couple of months in same file; GN 1862 of 1939. For examples of routine closings in subsequent years, see GN 1876 of 1943; CTA CMT 3/1322, 24/C, vol. 3, FCT to F. J. Swan, Kentani, 20 February 1945; GN 1539 of 1950.

130. Hunt, *Colonial Lexicon*, 23–24.

131. Isaacman, *Cotton Is the Mother of Poverty*, 207–9.

132. Interviews with Cyprian Mvambo; R. T. S. Mdaka; Olga Tyekela; Jemimah Dubo; Dabulamanzi Gcanga; CTA FCT 3/1/50, T601, DFO Butterworth to FCT, 6 December 1922.

133. Interview with R. T. S. Mdaka.

134. Interview with Cyprian Mvambo.

135. For some representative examples, see CTA 1/TSO 1/2/5, case 96, R. v. Nolentyi and 3 others, 26 April 1907; CTA 1/TSO, 1/2/6, case 30, R. v. Sinnah Kumla and 11 others, 9 February 1910; case 117, R. v. Noyamile, 20 June 1911; CTA 1/TSO 1/2/8, Criminal Record Book, case 94, R. v. Mazoto and 5 others, 28 May 1913; case 268, R. v. Mavancolo and 8 others, 3 December 1913; R. v. Nora Nocuza and 5 others, 3–15 April 1914; CTA 1/TSO, 1/2/7, case 395, R. v. M. Mpayipeli and 6 others, 20 July–1 August 1916; CTA 1/TSO, 1/2/9, case 518, R. v. Saryan, 28 November 1916; CTA 1/TSO, 1/2/9, criminal cases 116–123, all held on 19 April 1917; CTA 1/TSO 5/1/8, file 31, vol. 5, "Enquiry held into charges made against Forest guard Paulus Komani during the case of R. vs. Witchell Jangi," case 164, 3 July 1920; CTA 1/TSO, 1/2/10, criminal cases 175–76 of 1920.

136. GN 1605 of 1920, section 2; interviews with Mambanjwa Kholwane; Lindiwe Gcanga et al.; Martha Ngadlela; Olga Tyekela; Liziwe and Monwabisi Ndzungu; Noheke Rangana; Albertina Ntwalana; Emmie Fiko; Eunice Matomela; CTA CMT 3/1322, 24/C, vol. 3, 8 April 1935, W. du Plessis to SNA.

137. CTA 1/ECO 4/1/4, Arthur J. Weasherle, Emjanyana, to RM Emjanyana, 15 August 1884; CTA 1/TSO 1/1/15, case 54, R. v. Mapama, 12 April 1899; *BBNA* 1902, 67, RM Qumbu E. Russell, 30 January 1902; CTA FCT 3/1/51, T700, M.624, J. Keet, to Inspector of Forestry, Natal and Transkei, "Management of Indigenous Forests in the Transkei," 28 December 1933; *Annual Report 1934*, 13; CTA CMT 3/1322, 24/C, vol. 3, RM Tsomo to CMT, 14 December 1934; RM Umzimkulu to CMT, 11 December 1934; 8 April 1935, W. du Plessis to SNA; FCT to CMT, 2 May 1939; FCT to F. J. Swan, Kentani, 20 February 1945.

138. *Annual Report 1892*, 90; CTA 1/UTA 1/1/1/22, case 524, R. v. Benjamin Sinukela and 8 others, 11 November 1898; *Annual Report 1900*, 156; CTA 1/TSO 1/1/19, case 226, R. v. Gqira and 16 others, 27 December 1901; Miller, "Notes on the Distribution of Species," 22; *TTGC* 1924, 87; CTA FCT 3/1/52, T702/6, Working plans, Tsolo Central Reserve, FCT to CCF, 29 August 1930; CTA CMT 3/1322, 24/C, vol. 3, Secretary of Agriculture and Forestry to SNA, 20 January 1934.

139. Belinda Bozzoli, with the assistance of Mmantho Nkotsoe, *Women of Phokeng: Consciousness, Life Strategy, and Migrancy in South Africa, 1900–1983*

(Portsmouth, N.H.: Heinemann, 1991), 45–47; Schmidt, *Peasants, Traders, and Wives*, 79, 84–85, 159; N. Jacobs, *Environment, Power, and Injustice*, 144.

140. Interview with Olga Tyekela.

141. See W. D. Hammond-Tooke, "Kinship, Locality, and Association: Hospitality Groups among the Cape Nguni," *Ethnology* 2, no. 3 (1963): 302–19.

142. Interviews with Alice N. Gcanga, Manzana, Engcobo District, 11 February 1998; Noheke Rangana; Cyprian Mvambo; Lindiwe Gcanga et al.; Nozolile Kholwane; Godfrey, "Xhosa-English Dictionary," "L," 20, and "B," 56; Joan A. Broster, *The Tembu: Their Beadwork, Songs and Dances* (Cape Town: Purnell, 1976), 86–87. For more recent examples of similar activities in other parts of the Transkei, see Hunter, *Reaction to Conquest*, 73–75, 87–92, 98, 103; W. D. Hammond-Tooke, *Bhaca Society: A People of the Transkeian Uplands South Africa* (Cape Town: Oxford University Press, 1962), 143–45; Heinz Kuckertz, *Creating Order: The Image of the Homestead in Mpondo Social Life* (Johannesburg: Witwatersrand University Press, 1990), 199, 208; G. Heron, "Household Production and the Organisation of Co-operative Labour in Shixini, Transkei" (M.A. thesis, Rhodes University, Grahamstown, 1989); Laura Cloete, "Domestic Strategies of Rural Transkeian Women," Development Studies Working Paper 54 (Institute of Social and Economic Research, Rhodes University, 1992), 10, 14–15, 18–19; Patrick A. McAllister, *Building the Homestead: Agriculture, Labour and Beer in South Africa's Transkei* (Leiden: African Studies Centre and Ashgate, 2001); Kepe, *Environmental Entitlements in Mkambati*, 61, 65–66, 78.

143. Interviews with Noheke Rangana; Lindiwe Gcanga et al.; Alice Gcanga; Nozolile Kholwane; Godfrey, "Xhosa-English Dictionary," "L," 20, and "B," 56, respectively; Kuckertz, *Creating Order*, 199, 208; Cloete, "Domestic Strategies of Rural Transkeian Women," 14–15. As William Beinart notes for Pondoland, women of wealthier families were in better positions to take advantage of such labor arrangements. *Political Economy of Pondoland*, 148. See also Kepe, *Environmental Entitlements in Mkambati*, 61, 65–66, 78.

144. Interview with Nozolile Kholwane. *Amarhewu* is a drink made from fermented grain.

145. Interview with Alice Gcanga.

146. *TTGC* 1920, 62–63, RM Tsolo.

CHAPTER 4

1. Mandala, *Work and Control in a Peasant Economy.*

2. Although I use the term *isuthu* here, *ibuma* and *ithonto* were also commonly used in the colonial Transkei.

3. Robert Godfrey defined *ungquphantsi* in this way in the 1940s: "the primitive Native hut, hemispherical in shape, formed of saplings, thatched over with grass reaching to the ground and plastered on the inside; the abakhwetha hut of

the present time retains this pattern." "Xhosa-English Dictionary," "N," 121.

4. Interview with Samuel Qina.

5. Interview with Fumanekile and Ntombizanele Sithelo; N. F. Dilika, R. V. Nikolova, and T. V. Jacobs, "Plants Used in the Circumcision Rites of the Xhosa Tribe in South Africa," in *Proceedings of the International Symposium on Medicinal and Aromatic Plants, August 27–30, 1995, Amherst MA USA*, ed. L. E. Craker, L. Nolan, and K. Shetty (Leuven, Belgium: International Society for Horticultural Science, 1995), 165–68; Godfrey I. Mzamane, "Some Medicinal, Magical and Edible Plants Used Among Some Bantu Tribes in South Africa," *Fort Hare Papers* 1, no. 1 (1945): 29–35.

6. CTA FCT 1/1/3/1, A. L. Raymond to FCT, 31 March 1894.

7. CTA CMT 3/40, RM Engcobo C. A. King to CMT, 24 June 1903. For a similar case, see CTA 1/TSO 1/1/12, case of 11 June 1895.

8. CTA FCT 3/1/50, T601/1, copy of case before RM Tsolo, R. v. Ntukaye, 22 May 1922; copy of case before RM Tsolo, R. v. Charlie Links (Lengisi or Lynx), 12 June 1922; DFO Mt. Frere to FCT, 19 September 1922. See also CTA 1/TSO 1/1/17, case 295, R. v. Zaza and 4 others, 27 September 1900; R. v. Zaza and 2 others, 17 October 1900.

9. Interview with Rev. G. Vika.

10. Anne Mager, "Youth Organisations and the Construction of Masculine Identities in the Ciskei and Transkei, 1945–1960," *Journal of Southern African Studies* 24, no. 4 (1998): 658. Mager's work builds from the prior ethnographic research of Philip and Iona Mayer: *Report of Research on the Self Organisation by Youth Among the Xhosa-Speaking Peoples of the Ciskei and Transkei, Part 1: The Red Xhosa* (Rhodes University, Institute of Social and Economic Research, November 1972). For a related discussion, see William Beinart, "The Origins of the *Indlavini*: Male Associations and Migrant Labour in the Transkei," in *Tradition and Transition in Southern Africa*, ed. Andrew Spiegel and Patrick McAllister (London: Transaction Publishers, 1991), 103–28.

11. Interview with R. T. S. Mdaka.

12. Interviews with Rev. G. Vika; Cyprian Mvambo.

13. Interviews with G. N. Mbabama, Umtata, 2 February 1998; Abel Somana, All Saint's Mission, Engcobo District, 5 February 1998; Tozama Gqweta, Baziya Mission, Umtata District, 12 January 1998; Dabulamanzi Gcanga; Cyprian Mvambo; Rev. Matthew Gqweta; Rev. G. Vika; Alice Gcanga; W. M. Ngombane; Mager, "Youth Organisations," 658–61.

14. Switzer, *Power and Resistance*, 10–12; Beinart and Bundy, *Hidden Struggles*, 106–9.

15. Interviews with R. T. S. Mdaka; Rev. Matthew Gqweta; Cyprian Mvambo.

16. Interview with Cyprian Mvambo.

17. Interview with R. T. S. Mdaka.

18. For examples of some bloody and fatal stick-fights in the region, see CTA 1/UTA 1/1/1/13, Circuit Court, case 133, R. v. Mtunzana and 34 others, 25 October

1893; CTA 1/TSO 3/1/7/5, 3/1/1907, RM Umtata to District Surgeon, Umtata, 1 March 1907, attaching deposition from subheadman Hluku in court of RM Tsolo, 27 February 1907; CTA 1/UTA 1/1/1/48, R. v. Maqiza, 29 April 1915.

19. Interview with Dabulamanzi Gcanga; CTA 1/TSO 2/1/3, Thomas Ngudle v. Matanzima Xayimpi, 10 September 1886; CTA 1/TSO 3/1/7/5, RM Umtata to District Surgeon, Umtata, 1 March 1907; RM Tsolo to RM Umtata, 8 March 1907; CTA 1/TSO 1/2/5, 117, R. v. Zenzile and 34 others, 16 May 1907; CTA 1/UTA 1/1/1/48, R. v. Maqiza, 29 April 1915.

20. Interview with R. T. S. Mdaka. For a similar example, see CTA 1/UTA 1/1/1/43, 2/1/1910, R. v. Silingo and others, 1 February 1910.

21. Interview with R. T. S. Mdaka.

22. UAR CMT 185, 3/22/3/12, Tsolo district, vol. 1, Ncolosi location, H. M. Tyali to RM Tsolo, 25 February 1895. See also CTA FDU 3/2, 29; Godfrey, "Xhosa-English Dictionary," "G," 63, "L," 42, "N," 178.

23. Interviews with Dabulamanzi Gcanga; Rev. Matthew Gqweta.

24. Interviews with Mvulayehlobo Jumba and Charlie Banti, Tabase, Umtata District, 27 January 1998; Headman Sithelo; Rev. Matthew Gqweta; Samuel Qina; Polisile Maka; Dabulamanzi Gcanga; Abel Somana; R. T. S. Mdaka; Cyprian Mvambo; Godfrey, "Xhosa-English Dictionary," "Z," 8, "N," 153–54 and 178, and "G," 63.

25. Interview with Headman Sithelo.

26. Interview with R. T. S. Mdaka.

27. Interviews with Tozama Gqweta; Rev. Matthew Gqweta; Samuel Qina; G. N. Mbabama; Rev. G. Vika; Dabulamanzi Gcanga; Headman Sithelo.

28. Interview with Dabulamanzi Gcanga.

29. Ibid.; interview with Rev. G. Vika.

30. See Christian M. Simon, "Dealing with Distress: A Medical Anthropological Analysis of the Search for Health in a Rural Transkeian Village" (M.A. thesis, Department of Anthropology, Rhodes University, December 1989).

31. Interview with William Jumba et. al; Godfrey, "Xhosa-English Dictionary," "G," 107.

32. Interview with Headman Sithelo.

33. Ibid.; Godfrey, "Xhosa-English Dictionary," "N," 178; Johnson, "Preliminary Checklist," 152.

34. Interviews with Headman Sithelo; Olga Tyekela.

35. Interview with Sampson and Notozamile Dyayiya.

36. For example, many indigenous remedies in the Transkei in general were included in C. C. Henkel, *Tree Planting for Ornamental and Economic Purposes in the Transkeian Territories, South Africa* (Cape Town: Juta, 1894), and Sim, *Forests and Forest Flora.*

37. *Annual Report 1888,* 62.

38. Interviews with Mlungisi Ngombane; Olga Tyekela; Eunice Matomela; Nozolile Kholwane; Liziwe and Monwabisi Ndzungu; Dabulamanzi Gcanga;

Emmie Fiko; Sampson and Notozamile Dyayiya; Headman Sithelo; Mvu-layehlobo Jumba and Charlie Banti; Joan A. Broster, *Amagqirha: Religion, Magic and Medicine in Transkei* (Cape Town: Via Afrika Limited, 1981), 104, 120; C. M. Lamla, "Present-Day Diviners (Ama-Gqira) in the Transkei" (M.A. thesis, University of Fort Hare, 1975), 144–45; Godfrey, "Xhosa-English Dictionary," "H," 65, "K,'" 104, and "Q," 90; Chris Simon and Masilo Lamla, "Merging Pharmacopoeia: Understanding the Historical Origins of Incorporative Pharmacopoeial Processes among Xhosa Healers in Southern Africa," *Journal of Ethnopharmacology* 33, no. 3 (July 1991): 240–41; R. B. Bhat and T. V. Jacobs, "Traditional Herbal Medicine in Transkei," *Journal of Ethnopharmacology* 48, no. 1 (August 1995): 8.

39. CTA 1/TSO 1/1/15, case 110, R. v. Umnyakama and Qoko, 23 May 1899; case 184, R. v. Mangati, 5 July 1899; case 283, R. v. Thomas C. P. Adams, Forester, 24 October 1899.

40. Interview with Sampson and Notozamile Dyayiya.

41. Interview with Eunice Matomela. *Isigcimamlilo (Pentanisia prunelloides)* has also often been used for stomach disorders, in childbirth, to reduce swelling, and for a variety of other illnesses: interviews with Nozolile Kholwane; Samuel Qina; Godfrey, "Xhosa-English Dictionary," "G," 17; Broster, *Amagqirha*, 121; Simon and Lamla, "Merging Pharmacopoeia," 240; John M. Watt and Maria Breyer-Brandwijk, *The Medicinal and Poisonous Plants of Southern Africa: Being an Account of Their Medicinal Uses, Chemical Composition, Pharmacological Effects and Toxicology in Man and Animal* (Edinburgh, E. & S. Livingstone, 1932), 902–3.

42. Interview with Olga Tyekela.

43. Interviews with Mlungisi Ngombane; Rev. G. Vika; Godfrey Callaway, *South Africa from Within: Made Known in the Letters of a Magistrate* (London: Society for Promoting Christian Knowledge, 1930), 104, on amaQwathi protecting their kraals from natural forces.

44. Interview with Eunice Matomela.

45. Interview with Liziwe and Monwabisi Ndzungu.

46. Broster, *Amagqirha*, 24–25. See also Godfrey Callaway, *Mxamli, the Feaster: A Pondomisi Tale of the Diocese of St. John's, South Africa* (London: Society for Promoting Christian Knowledge; New York: Macmillan, 1919), 65; and for similar practices in Mpondo communities, Hunter, *Reaction to Conquest*, 242–45.

47. Interview with Tozama Gqweta.

48. Godfrey, "Xhosa-English Dictionary," "B," 42.

49. Interview with Emmie Fiko. Broster also talks about this privileged knowledge in *Amagqirha*, 24–25.

50. Interview with Headman Sithelo.

51. *Ooqqirha*, the term used for medical professionals today, is distinct from *amagqirha*, the word for traditional diviners/healers.

52. Interview with Olga Tyekela.

53. Interview with William Jumba and Zwelivumile Quvile, Tabase, Umtata District, 4 February 1998; Godfrey, "Xhosa-English Dictionary," "T," 26–27.

54. W. Blohm, "Das Opfer und dessen Sinn bei den Xosa in Südafrika," *Archiv für Anthropologie* 23 (1933): 152, original in German, my translation.

55. Interviews with Fumanekile and Ntombizanele Sithelo; Lindiwe Gcanga et al.; R. N. Bolofo and C. T. Johnson, "The Identification of 'Isicakathi' and Its Medicinal Use in Transkei," *Bothalia* 18, no. 1 (1988): 125–30; Broster, *Tembu*, 5–6; Broster, *Amagqirha*, 105–6, 120; Godfrey, "Xhosa-English Dictionary," "T," 53; W. Blohm, "Schöpferische Kräfte in der Gesellschaft der Xosa-Gruppe," *Archiv für Anthropologie* 23 (1934): 190–94. See also John Henderson Soga's related chapter on "Mother and Infant" in *Ama-Xosa*, especially 290–98; P. A. W. Cook, *Social Organization and Ceremonial Institutions of the Bomvana* (Cape Town: Juta and Co., 1931), 48–50; Hammond-Tooke, *Bhaca Society*, 73–76.

56. Interview with Alice Gcanga; Soga, *Ama-Xosa*, 207.

57. Interview with Fumanekile and Ntombizanele Sithelo.

58. Many people also used different types of smoke to ritually protect other individuals, animals, and property. Lamla, "Present-Day Diviners," 55–58; interview with Tozama Gqweta.

59. Interview with Lindiwe Gcanga et al.; see also Bhat and Jacobs, "Traditional Herbal Medicine," 8.

60. Blohm, "Schöpferische Kräfte," 190–91, original in Xhosa, my English translation, partly based on Blohm's German translation.

61. For the now classic examination of songs as sources of historical interpretation, see Leroy Vail and Landeg White, *Power and the Praise Poem: Southern African Voices in History* (Charlottesville: University Press of Virginia, 1991).

62. Interview with William Jumba and Zwelivumile Quvile; Hammond-Tooke, *Tribes of Umtata District*, 55; Joan A. Broster, *Red Blanket Valley* (Johannesburg: Hugh Keartland, 1967), 122; Godfrey, "Xhosa-English Dictionary," "F," 68, "H," 65, and "N," 234. For similar practices among other groups in the Transkei, see Soga, *Ama-Xosa*, 293; Hunter, *Reaction to Conquest*, 152–54; Cook, *Social Organization*, 48–49; Hammond-Tooke, *Bhaca Society*, 75.

63. Interview with Eunice Matomela. Also, interviews with Dabulamanzi Gcanga; Mlungisi Ngombane. This species, *Hyperacanthus amoenus*, is also commonly called *uthongothi* or *umthongothi*.

64. Blohm, "Schöpferische Kräfte," 164, original in German, my translation.

65. For example, *umnquma* (wild olive, *Olea africana*) and *ulwathile* (false horsewood, *Hippobromus pauciflorus*) have historically been used for similar initiation ceremonies and ancestral rituals by different groups in the region in the early 1900s and in more recent decades. Interviews with Dabulamanzi Gcanga; Sampson Dyayiya and Notozamile Dyayiya; Broster, *Amagqirha*, 36, 91–92, 96, based on many years of researching Qwathi cultural practices. See Hunter, *Reaction to Conquest*, 243 and 250, for similar uses of *umthathi* and other species in ritual appeals to ancestors in Pondoland in the 1920s and early 1930s; Godfrey, "Xhosa-English Dictionary," "D," 36; Soga, *Ama-Xosa*, 254.

66. See, for example, Aneesa Kassam and Gemetchu Megerssa, "Sticks, Self, and Society in Booran Oromo: A Symbolic Interpretation," in *African Material*

Culture, ed. Mary Jo Arnoldi, Christraud M. Geary, and Kris L. Hardin (Bloomington: Indiana University Press, 1996), 145–66.

67. Beattie, *Ride through the Transkei*, 72–73; Manton Hirst, "The Healer's Art: Cape Nguni Diviners in the Townships of Grahamstown" (Ph.D. thesis, Department of Anthropology, Rhodes University, December 1990), 192; Broster, *Amagqirha*.

68. Interview with Dabulamanzi Gcanga.

69. Interview with Mvulayehlobo Jumba and Charlie Banti.

70. Tandi Somana, a research assistant.

71. Interview with Mlungisi Ngombane, Egerton, Umtata District, 25 February 1998. See also Godfrey, "Xhosa-English Dictionary," "N," 31; Soga, *Ama-Xosa*, 321. Robert Samuelson recorded very similar uses of *umphafa* among Zulu groups at the turn of the century, in *Long, Long Ago* (Durban: Knox, 1929), 390–92.

72. John M. Janzen, *The Quest for Therapy in Lower Zaire* (Berkeley: University of California Press, 1978).

73. CTA 1/UTA 1/1/1/5, case 23, R. v. Nomsimbi, 27 May 1879; CTA 1/TSO 1/1/3, case 7, R. v. Timani, 20 March 1885; CTA 1/TSO 2/1/3 Civil Cases, case 30, Hlanhlata v. Bikwapi and Siyol, 27 May 1886; CTA 1/ECO 1/3/1, Declarations, 1885–1901, Statement of Lina before RM Engcobo, n.d. 1889; CTA 1/ECO 1/1/2/1, Preparatory examinations in Circuit Court Cases, R. v. Jack, 11 October 1887; CTA 1/TSO 1/1/4, case 87, R. v. Matanzima Xayimpi, 30 November 1887; CTA 1/TSO 1/1/5, case 10, R. v. Nqambi, 22 February 1888; case 70, R. v. Solani, Jonah Bada and Mehlo Sokombela, 23 November 1888; CTA 1/TSO 1/1/7, case 32, R. v. Vinjwa, 8 May 1890; CTA 1/ECO 1/1/1/11, case 41, R. v. Charles, 2 February 1890; CTA 1/TSO 1/1/9, case 138, R. v. Thlakanyana, 29 August 1890; CTA 1/TSO 1/2/4, Criminal Record Book, case 108, R. v. Voboyi, 22 October 1891; CTA 1/TSO 1/1/9, case 109, R. v. Nonyenyeka, 12 July 1892; CTA 1/UTA 1/1/1/12, Preliminary examination in Circuit Court, R. v. Mgxabakazi, 7 September 1892; CTA 1/TSO 2/1/10, Criminal Cases, case 117, Sitshoto v. Dumayo, 11 November 1892; 1/TSO 1/1/11, case 134, R. v. Mgudlwa and Nkasana, 5 October 1894; CTA 1/UTA 1/1/1/16, case 360, R. v. Mdimsa and 3 others, 7 October 1894; CTA 1/UTA 1/1/1/18, case 78, R. v. Nyangwa, 6 March 1896; CTA 1/TSO 1/1/12, R. v. Nomagazi, 16 October 1896; CTA 1/UTA 1/1/1/24, case 540, R. v. Ngqondiyiya, 15 September 1899; CTA 1/UTA 1/1/1/24, case 645, R. v. Ty, 4 December 1899; CTA 1/UTA 1/1/1/25, case 152, R. v. Ntantiso, 23 March 1900; CTA 1/UTA 1/1/1/26, case 377, R. v. Tshoko and Manyana, 19 June 1900; CTA 1/ECO 1/1/1/36, R. v. Nomapelana, 19 June 1901; CTA 1/UTA 1/1/1/27, case 429, R. v. Batyi Mancayi, 12 July 1900; CTA 1/TSO 1/2/3, Criminal Record Book, case 139, R. v. Bara and 3 others, 11 June 1906; CTA 1/UTA 1/1/1/46, R. v. Magopeni, 22 October 1913; CTA 1/ECO 1/1/1/39, R. v. Ndabezimbi Pikani, 10 November 1924; CTA 1/ECO 1/1/1/39, R. v. Diniso Vivi, 8 December 1932.

74. CTA 1/ECO 1/1/1/11, case 139, R. v. Zindlani, 25 June 1890.

75. The species is also called "Mhlebe" by a witness: CTA 1/UTA 1/1/1/20, case 62, R. v. Nosesi Matikolo, 4 March 1897.

76. Also see David Graeber, "Love Magic and Political Morality in Central Madagascar, 1875–1990," in *Gendered Colonialisms in African History*, ed. Nancy Rose Hunt, Tessie P. Liu, and Jean Quataert (Malden, Mass.: Blackwell, 1997), 96–117.

77. CTA 1/TSO 2/1/3, Civil Cases, case 30, Hlanhlata v. Bikwapi and Siyo, 27 May 1886.

78. CTA 1/UTA 1/1/1/22, case 510, R. v. Xego, Xakaza and Gosa, 3 November 1896.

79. Godfrey, "Xhosa-English Dictionary," "V," 13, and see related plant use in "T," 22, and "H," 15. A few decades later, Joan Broster noted the continued use of such "love potions" in the Engcobo district. *Tembu*, 28–29.

80. Soga, *Ama-Xosa*, 173–74; see also Godfrey, "Xhosa-English Dictionary," "K," 22, and "M," 10.

81. Lamla, "Present-Day Diviners," 55–58.

82. Quotation is from Frank Brownlee, "Some Experiences of Native Superstition and Witchcraft," *Journal of the Royal African Society* 39, no. 154 (1940): 59. See also Anonymous, "Mhlakwapalwa at a Circuit Court: The Narrative of an Uncivilised Native (Translated)," *Cape Law Journal* 7 (1890), 231; *BBNA 1881*, 135–36, J. Oxley Oxland, British Resident, Eastern Pondoland, to SNA, 26 February 1881; Charles Brownlee, *Reminiscences of Kaffir Life and History* (Lovedale: Lovedale Press, 1896), 248; *BBNA 1909*, 58, RM Mount Fletcher; Frank Brownlee, *The Cattle Thief: The Story of Ntsukumbini* (London: Jonathan Cape, 1932), 37–38; F. Brownlee, "Nopongo and His Ninepence: The Psychology of a Native Litigant," *South African Law Journal* 50 (1933): 12–17; Hunter, *Reaction to Conquest*, 416; Godfrey, "Xhosa-English Dictionary," "H," 21.

83. Godfrey, "Xhosa-English Dictionary," "N," 51, and similar practices in "B," 66, "C," 22. See also Simon and Lamla, "Merging Pharmacopoeia," 242.

84. Interview with Headman Sithelo; Mlungisi Ngombane.

CHAPTER 5

1. Interview with Fumanekile and Ntombizanele Sithelo.

2. For example, *Annual Report 1888*, 61.

3. H. C. Schunke, "Kaffraria und die Östlichen Grenzdistrikte der Kapkolonie," *Doktor A. Petermanns Mitteilungen aus Justus Perthes' Geographischer Anstalt* (Gotha: Justus Perthes) 31 (1885): 169, original in German, my translation.

4. *Annual Report 1889*, 97.

5. *Annual Report 1893*, 139–40; CTA 3/40, FCT to CMT, 2 August 1893; CTA AGR 223, F1930, vol. 2, 13 September 1893, RM Tsolo and FCT to CMK; CTA AGR 766, F1930, 16 December 1893, FCT to USA; FCT to USA, 19 June 1894, CMK to USNA.

6. Pietermaritzburg Archives Repository (PAR), Henkel Papers, Accession A1238, 1832–1953, A2/1, n.d., "The [E]nchanted Vlei by the Oude Boschranger."

7. Mary Louise Pratt describes this "monarch-of-all-I-survey" genre and the "fantasy of dominance" embedded in Europeans' panoramic representations of African landscapes in the nineteenth century in "Scratches on the Face of the Country; Or, What Mr. Barrow Saw in the Land of the Bushmen," in *"Race," Writing and Difference*, ed. Henry Louis Gates, Jr. (Chicago: University of Chicago Press, 1986), 138–62, and *Imperial Eyes: Travel Writing and Transculturation* (London: Routledge, 1992).

8. As Sandra Greene notes in her work on the Anlo of southeastern Ghana, Europeans often "ridiculed, ignored, and/or selectively appropriated" local Africans' spiritual understandings of their landscapes in order to impose their own environmental and moral perspectives. Greene, *Sacred Sites and the Colonial Encounter*, 4.

9. On such transformations in the location and their impact on both the lives and memories of locally displaced residents, see Tropp, "Displaced People, Replaced Narratives."

10. Interview with Mlungisi Ngombane, Egerton, Umtata District, 25 February 1998. For some intriguingly similar stories surrounding sacred pools and the swallowing up of people, see B. B. Mukamuri, "Local Environmental Conservation Strategies: Karanga Religion, Politics and Environmental Control," *Environment and History* 1, no. 3 (1995): 297–311, especially 304–5.

11. Interviews with Fumanekile Sithelo; Headman Sithelo.

12. PAR, Henkel Papers, Accession A1238, A2/1, n.d., "The [E]nchanted Vlei." It is tempting to think that Henkel and his horse fell prey to the powers of the renowned *umdlebe* tree.

13. Anne Hutchings, *Zulu Medicinal Plants: An Inventory* (Pietermaritzburg: University of Natal Press, 1996), 176; Eve Palmer and Norah Pitman, *Trees of Southern Africa* (Cape Town: A. A. Balkema, 1972), 1:195, 1:217, and 3:1181–82; John M. Watt and Maria Breyer-Brandwijk, *The Medicinal and Poisonous Plants of Southern Africa: Being an Account of Their Medicinal Uses, Chemical Composition, Pharmacological Effects and Toxicology in Man and Animal* (Edinburgh, E. & S. Livingstone, 1932), 437. A related species, *Monadenium lugurdae*, is also highly toxic: Hutchings, 176–77; Watt and Breyer-Brandwijk, 424–26.

14. Sifiso Mxolisi Ndlovu, "African Public Intellectuals and the Production of Historical Knowledge in 'Pre-colonial' Times: A Case Study of the Making of the Archive of King Dingane kaSenzangakhona" (paper presented at the Triennial Conference of the European Association for Commonwealth Language and Literature Studies, University of Copenhagen, 21–27 March 2002). Ndlovu is citing the English translation of an account attributed to Dingane's premier court poet, Mshongweni, in Mazisi Kunene, *Emperor Shaka the Great: A Zulu Epic* (London: Heinemann, 1979).

15. Henry Callaway, *The Religious System of the Amazulu* (Natal: J. A. Blair, 1870), 422–23. The narrator of this account then goes on to explain how his own brother and others, while on a hunting expedition among the "Amanthlwenga" (a

Zulu term for amaTonga), made the fatal mistake of roasting their meat using *umdlebe* branches. The tree attacked the hunters with a horrible disease before killing some of the party. Ibid., 424–26. For analysis of the biographies and perspectives of Callaway and his informants, see Norman Etherington, "Missionary Doctors and African Healers in Mid-Victorian South Africa," *South African Historical Journal* 19 (1987): 77–91. On the Zulu term "amaNtlengwa," see A. T. Bryant, *Olden Times in Zululand and Natal* (London: Longmans, Green and Co., 1929), 84, 109, 280. Bryant also described the potency of a species called "umdhlebe" (which he identifies as *Synadenium arborescens.*) in Zulu medicine and magic in his *A Zulu-English Dictionary* (Pietermaritzburg: P. Davis and Sons, 1905), 55 (under "um-Bulelo"), 100 ("um-Dhlebe"), and 765 ("um-Dhlebe"), and in *Zulu Medicine and Medicine-Men* (Cape Town: C. Struik, 1966 repr. ed.), 20–21.

16. Palmer and Pitman, *Trees of Southern Africa*, 1:195, 1:217, and 3:1181–82.

17. Interview with Fumanekile and Ntombizanele Sithelo, describing the bark "clots" of *umdlebe* trees, a common feature of the pock ironwoods listed above; F. von Breitenbach and Jutta von Breitenbach, "Notes on the Natural Forests of Transkei," *Journal of Dendrology* 3, nos. 1 and 2 (1983): 35; K. H. Cooper and W. Swart, *Transkei Forest Survey* (Durban, Wildlife Society of Southern Africa, 1991), Appendix I, 53; Elsa Pooley, *The Complete Field Guide to the Trees of Natal, Zululand and Transkei* (Natal Flora Trust, 1993), 238, 410. From the nineteenth century to the present, none of the primary sources on floral distribution in these afromontane forests or the Transkei as a whole even mention *Synadenium cupulare.*

18. Godfrey, "Xhosa-English Dictionary," "D," 49; CTA 1/UTA 1/1/1/20, 4 March 1897, case no. 62, R. v. Nosesi Matikolo, involving a particular witchcraft accusation, as described by one witness: "she said I have come to fetch poison from you, I asked her what kind she said 'Mhlebe' and a piece of wild beast's skin with which she said I had bewitched her and that she had been ill for three days."

19. Interview with Headman Sithelo. For similar narratives among other Nguni groups, see W. D. Hammond-Tooke, "The Symbolic Structure of Cape Nguni Cosmology," in *Religion and Social Change in Southern Africa*, ed. Michael Whisson and Martin West (Cape Town: David Philip, 1975), 15–33; Callaway, *Religious System of the Amazulu*, 422–23.

20. Interview with Fumanekile and Ntombizanele Sithelo.

21. Robert Godfrey recorded various ways that people gave offerings to certain snake spirits, in order to protect themselves from harm and to secure important plants for medicinal uses: "Xhosa-English Dictionary," "G," 69, "L," 7, and "M," 3–4. See also Hunter, *Reaction to Conquest*, 312.

22. Interview with Fumanekile and Ntombizanele Sithelo. Some Zulu medical practitioners in the early 1900s associated *umdlebe* and other specific trees with blood because these species release "juice like blood" when their bark is cut. See J. B. McCord, "The Zulu Witch Doctor and Medicine-Man," *South African Journal of Science* 15, no. 5 (1918): 306–18, translating the account of Isaiah Wosiana of

the Mapumulo Mission Station. On similar water spirits that can change into multiple forms, see Hunter, *Reaction to Conquest*, 286–87; Godfrey, "Xhosa-English Dictionary," "C," 18; Soga, *Ama-Xosa*, 193–96.

23. Interview with Mlungisi Ngombane. Monica Hunter recounted a very similar narrative told during her research in Pondoland: "Informants told of a tree in Zululand which is valuable as medicine, but which causes death to any who approach it untreated. One wishing to approach should be treated, and throw a dog to the tree. The tree 'cries like a sheep.'" Hunter, *Reaction to Conquest*, 312. A. T. Bryant also recorded a similar Zulu practice in defining the term "umDhlebe" in his *Zulu-English Dictionary*, 100: "the doctor when cutting it, must first smear his hands with the bile of a goat, then approaching from the windward side, let fly his axe at the trunk of the tree and so chip out small pieces."

24. Broster, *Amagqirha*, 40; Broster, *Red Blanket Valley*, 60–61.

25. Broster, *Amagqirha*, 40; Hirst, "Healer's Art," 212.

26. The following story is excerpted from Callaway, *Religious System of the Amazulu*, 421–23.

27. Interviews with Fumanekile Sithelo; Mlungisi Ngombane.

28. Interviews with R. T. S. Mdaka, commenting in English; Mvulayehlobo Jumba and Charlie Banti; Wele Boyana; Headman Sithelo.

29. Tandi Somana, a research assistant.

30. Interview with Liziwe and Monwabisi Ndzungu.

31. Broster, *Red Blanket Valley*, 83, 93–94, 99; see also Broster, *Amagqirha*, chap. 4, which traces the "graduation" of an initiate *igqirha* at the Gulandoda forest. Another description of such practices in the Gulandoda forest is provided in M. W. Waters, *Cameos from the Kraal* (Cape Town: Juta, n.d.), 47–49.

32. See also Frans Prins, "Praise to the Bushman Ancestors of the Water: The Integration of San-Related Concepts in the Beliefs and Ritual of a Diviners' Training School in Tsolo, Eastern Cape," in *Miscast: Negotiating the Presence of the Bushmen*, ed. Pippa Skotnes (Cape Town: University of Cape Town Press, 1996), 211–23. Soga similarly described a sacred pool near the village of Komgha, "held in reverence as the scene of the chief Gcaleka's initiation into the office of diviner (witch-doctor)," in *Ama-Xosa*, 323. See also Cook, *Social Organization*, 128–29.

33. Hirst, "Healer's Art," 174–76.

34. W. D. Hammond-Tooke, "The Initiation of a Baca Isangoma Diviner," *African Studies* 14, no. 1 (1955): 16–22; Broster, *Amagqirha*, chap. 4; Broster, *Red Blanket Valley*; Lamla, "Present-Day Diviners," 96–98.

35. For example, Henkel wrote an unpublished and incomplete manuscript in his retirement, "Our Game and the Art of Shooting," cataloguing specific mammals and birds in the Eastern Cape and Transkei. PAR, Henkel Papers, Accession A1238, A2/1.

36. It is unlikely that buffaloes still inhabited the area. One German explorer in the Transkei noted in 1893 that a few buffaloes "are still supposed to exist in one

of the Zuurberg forests," but this was only secondhand information at best, quite possibly based on Henkel's earlier forestry surveys. H. C. Schunke, "The Transkeian Territories: Their Physical Geography and Ethnology," *Transactions of the South African Philosophical Society* 8, part 1 (1893): 9.

37. Hirst, "The Healer's Art," 176–85.

38. Ibid., 39; Broster, *Red Blanket Valley*, 95–96, 99–100; interview with Liziwe and Monwabisi Ndzungu. Godfrey Callaway recorded the presence of leopards in the Qelana area at the turn of the century. E. D. Sedding, ed., *Godfrey Callaway: Missionary in Kaffraria, 1892–1942* (London: Society for Promoting Christian Knowledge, 1945), 132–33. Evidence of the leopard being a common isilo in many forest areas can be found in Robert Godfrey's translation of *ibehlathi*: "it's a forest creature (seen by an initiate in his dreams), i.e. it's a leopard"; Godfrey, "Xhosa-English Dictionary," "H," 43. On the symbolic significance of the leopard and the lion, see Hirst, "Healer's Art," 203–6.

39. For discussion of various ancestral spirits appearing in the form of animals in the region, see Hammond-Tooke, "Symbolic Structure of Cape Nguni Cosmology"; Broster, *Amagqirha*, 24; Basil Holt, *The Tshezi of the Transkei: An Ethnographic Study* (Johannesburg, 1969), 284–85; Lamla, "Present-Day Diviners," 83; Hirst, "The Healer's Art," 186–97, which also describes elephants, several other smaller mammals, birds, and reptiles as forest *izilo*, and distinguishes between the different spatial categories (forest, river, grassland, sea, and homestead) into which animals are classified in "Cape Nguni" cosmologies; Prins, "Praise to the Bushman Ancestors," 217–23. Godfrey offered this illustrative definition of the "'smelling-out' doctor," or *isanuse*: "supreme in his profession, supposed to have supernatural powers derived from lions, leopards, elephants, pythons or Hottentot women in the other world, enabling him to supply charms to protect people from evil influences, and to 'smell out,' i.e. to find out, those who bewitch and their charms"; Godfrey, "Xhosa-English Dictionary," "N," 236, and see also "N," 261–62, and "T," 152.

40. Interviews with Liziwe and Monwabisi Ndzungu; Emmie Fiko; Adolphus Qupa. Different beliefs in various snakes in the Tsolo District and beyond are well-documented in Godfrey, "Xhosa-English Dictionary," "K," 12, 146, "M," 3–4, "N," 279–80, and "P," 12; Basil Holt, *Place-Names in the Transkeian Territories* (Johannesburg: Africana Museum, 1959), 3, describing "a poisonous snake regarded with superstitious dread by the Pondomise, because it is supposed to be used by witchdoctors"; Broster, *Amagqirha*, 58–60.

41. Broster, *Amagqirha*, 58–60, on the snake *umamlambo* and other "familiars"; Hirst, "The Healer's Art," 246–48, on distinctions between "familiars" and *izilo*.

42. Frank Brownlee, *The Transkeian Native Territories: Historical Records* (Lovedale: Lovedale Institution Press, 1923), 112; Godfrey, "Xhosa-English Dictionary," "K," 46, and "M," 14, describing such beliefs in different snake *izilo* among certain amaMpondomise clans and other Transkeian groups; Hammond-Tooke, *Command or Consensus*, 45, 58; Hirst, "The Healer's Art," 187–88, 194–95.

43. Hirst, "Healer's Art," 194–95; Godfrey, "Xhosa-English Dictionary," "G," 63; Jeffrey Peires, *The Dead Will Arise: Nongqawuse and the Great Xhosa Cattle-Killing Movement of 1856–7* (Johannesburg: Ravan Press, 1989), 81–87, describing the development of Sarhili's association with the python in the early 1850s, and his fame in praise poems as "the great python which encircles Hohita."

44. Prins, "Praise to the Bushman Ancestors," 218–19.

45. Godfrey, "Xhosa-English Dictionary," "C," 18, in which *ichanti* is defined as a water-dwelling creature "that assumes protean shapes and colours, and that cannot be handled with impunity by any save witch-doctors: *unechanti*, he has the chanti, i.e. he has been so fascinated by it as to become a likely candidate for initiation as a witch-doctor (*utwasa walemile ngechanti*)." See also CGH, *Report and Proceedings, with Appendices, of the Government Commission on Native Laws and Customs*, G.4–1883, Minutes of Evidence, Rev. J. A. Chalmers, 21 September, 1881, 141, describing *amagqirha* and citing an instance in which a young Xhosa man suddenly "saw the 'icanti' come out of a pool of water, and he was mesmerised by it and became speechless. This boy ultimately became a doctor." Henry Callaway described some similar practices among Zulu diviners: "On his initiation, he goes like one mad to a pool, and dives into it, seeking for snakes; having found them, he seizes them and comes out of the water with them, and entwines them still living about his body, that the people may see that he is indeed a diviner." Callaway, *Religious System of the Amazulu*, 299.

46. Hirst, "Healer's Art," 220; Manton Hirst, "A River of Metaphors: Interpreting the Xhosa Diviner's Myth," in *Culture and the Commonplace: Anthropological Essays in Honour of David Hammond-Tooke*, ed. Patrick McAllister (Johannesburg: Witwatersrand University Press, 1997). The burial of chiefs and clan leaders in many Xhosa-speaking groups in the Eastern Cape and Transkei were historically tied to particular natural places, often forests, banks of streams or rivers, or pools. Hirst, "Healer's Art," 171–72; Hammond-Tooke, *Command or Consensus*, 45, 58; Frank Brownlee, "Burial Places of Chiefs," *African Affairs* 43, no. 170 (January 1944): 23–24. Because of its potential to unleash harmful powers of the ancestors if not properly approached, *ichanti* has also historically been connected with bad luck and sorcery in many areas. See Peires, *Dead Will Arise*, 83–84, noting a case of an *iqgirha* being accused of using *ichanti* for nefarious purposes; Soga, *Ama-Xosa*, 193–96; Hunter, *Reaction to Conquest*, 286–87. For a fascinating discussion of the contemporary significance of *ichanti* and other snake familiars and spirits, see Sean Morrow and Nwabisa Vokwana, "'Oh! Hurry to the River': the Meaning of *uMamlambo* Models in the Tyumie Valley, Eastern Cape," *Kronos: The Journal of Cape History* 30 (2004): 184–99.

47. Hirst, "Healer's Art," 194–95; Godfrey, "Xhosa-English Dictionary," "K," 12. For a comparative case, see also the role of snakes as "guardians" of sacred pools in some Karanga religious practices, in Mukamuri, "Local Environmental Conservation Strategies," 304–5.

48. A. C. Jordan, *The Wrath of the Ancestors* (Lovedale: Lovedale Press, 1980), an English translation of the 1940 Xhosa novel *Ingqumbo yeminyanya*. Jordan was

born at Mbokotwana in 1906 and lived in the Tsolo District before attending college at Umtata and then leaving the region for a teaching post in the mid-1930s; Jeff Opland, "The Publication of A. C. Jordan's Xhosa novel, *Ingqumbo yeminyanya* (1940)," *Research in African Literatures* 21, no. 4 (1990): 135–57. Chinua Achebe's famous novel *Arrow of God* deals with similar symbols and issues in a different colonial context.

49. Luise White explores in detail how particular colonial realities were expressed and magnified in popular rumors in *Speaking with Vampires: Rumor and History in Colonial Africa* (Berkeley: University of California Press, 2000).

50. *Annual Report 1890*, 141.

51. "Baziya and Tabase," *Periodical Accounts Relating to the Missions of the Church of the United Brethren, Established among the Heathen* 34 (June 1887): 85.

52. Cory Library, Rhodes University, Grahamstown, South Africa, MS 14, 559, RM Tsolo, "Rough Diary of Every-day events," entry for 13 August 1887.

53. Godfrey, "Xhosa-English Dictionary," "G," 63, original emphasis.

54. Henkel's first report on these forests and an accompanying map of the region appeared in *Annual Report 1888*.

55. Thus, as Jennifer Cole and Karen Middleton have recently argued for Madagascar, rituals connecting people to ancestors were here also at the center of important struggles over colonial power. "Rethinking Ancestors and Colonial Power in Madagascar," *Africa* 71, no. 1 (2001): 1–37.

56. Sandra Greene explores the "significance of place as symbol" regarding a "sacred grove" in both the colonial and postcolonial periods, in "Sacred Terrain: Religion, Politics and Place in the History of Anloga (Ghana), *International Journal of African Historical Studies* 30, no. 1 (1997): 1–22, and in *Sacred Sites and the Colonial Encounter*, 109–31. See also Elizabeth Colson, "Places of Power and Shrines of the Land," in "The Making of African Landscapes," ed. Ute Luig and Achim von Oppen, special issue, *Paideuma: Mitteilungen zur Kulturkunde* 43 (1997): 47–57, and Michele Wagner's insights into the role of nature spirits and "earth-priests" in Tanzanian social and ecological history, in "Environment, Community and History."

CONCLUSION

1. Bruce Braun, *The Intemperate Rainforest: Nature, Culture, and Power on Canada's West Coast* (Minneapolis: University of Minnesota Press, 2002), 25–26.

2. One important aspect of the Native Trust regulations was that all undemarcated forests were officially transferred to the Trust, with state authorities empowered to assess and reserve any such lands for ecological "improvement" schemes, such as afforestation and soil and water conservation efforts. Native Trust and Land Act, No. 18 of 1936; General Trust Regulations, GN 494 of 1937; CMT 3/1333, 24/24/1, circular No. H.2200/I, "Forests on Land Transferred to the South African Native Trust," Director of Forestry, Pretoria, to all officers of the Division

of Forestry, 16 January 1937; FCT to Director of Forestry, Pretoria, 6 August 1937; *Annual Report 1937*, 10–11; Mears, "Study in Native Administration," 193–94, 212.

3. *Annual Report 1937*, 11, 57–58.

4. As in many other spheres in the historical development of "native policy" in South Africa, the Transkei was a "testing ground" and example for administrators to follow in managing other African reserves. It was the first area to experience the introduction of "betterment" and "rehabilitation" programs after 1939, when the state began to implement intrusive "overstocking" and soil erosion prevention schemes in African communities. See Evans, *Bureaucracy and Race*; Mamdani, *Citizen and Subject*; Saul Dubow, "Holding 'A Just Balance between White and Black': The Native Affairs Department in South Africa c. 1920–33," *Journal of Southern African Studies* 12, no. 2 (1986): 217–39; Moll, *No Blade of Grass*, 23–30; Hendricks, "Pillars of Apartheid."

5. This very same series of photographs, for instance, was reused in King, "Exploitation of the Indigenous Forests." See also King, "Historical Sketch"; Mears, "Study in Native Administration," 132; Darrow, "Forestry in the Eastern Cape Border Region"; Keet, *Historical Review*; Shone, *Forestry in Transkei*.

6. Melissa Leach and James Fairhead, "Fashioned Forest Pasts, Occluded Histories? International Environmental Analysis in West African Locales," in "Forests: Nature, People, Power," special issue, *Development and Change* 31 (2000): 35–59.

7. Georgina Thompson, "The Dynamics of Ecological Change in an Era of Political Transformations: An Environmental History of the Eastern Shores of Lake St. Lucia," in Dovers, Edgecombe, and Guest, *South Africa's Environmental History*, 191–212; Palmer, Timmermans, and Fay, *From Conflict to Negotiation*; Thembela Kepe and Ben Cousins, "Resource Tenure and Power Relations in Community Wildlife: The Case of Mkambati Area, South Africa," *Society and Natural Resources* 14, no. 10 (1 December 2001): 911–25; Kepe, *Environmental Entitlements in Mkambati*; Reports 2001–2002 on forestry restitution claims in the Baziya and Mbolompo area by the Transkei Land Service Organisation, available at http://www.tralso.co.za/reports.asp; "NGO Aims to Return 1000 Displaced Families," *Daily Dispatch Online*, 15 November 2000, http://www.dispatch.co.za/2000/11/15/easterncape/KDISPLAC.HTM.

8. "Farmers Kill 86 Dogs in Anti-Poaching Operation," *Electronic Mail and Guardian*, 19 August 1998; "ANC 'Inciting Racial Hatred Over Hunt,'" *Electronic Mail and Guardian*, 19 August 1998; "Dog Kill Starts Political Storm," *Daily Dispatch*, 20 August 1998; Ruth Edgecombe, "The Role of Environmental History in Applied Field Studies in the Centre of Environment and Development at the University of Natal in Pietermaritzburg: The Case of Traditional Hunting with Dogs" (paper presented at "African Environments: Past and Present," joint conference of the *Journal of Southern African Studies* and St. Antony's College, Oxford University, Oxford, England, 5–8 July 1999).

9. Ntshona, *Valuing the Commons*; International Institute for Environment and Development (London) and CSIR-Environmentek (Pretoria), *Instruments for Sustainable Private Sector Forestry*.

10. Palmer, Timmermans, and Fay, *From Conflict to Negotiation*; Thembela Kepe, *Waking Up from the Dream: The Pitfalls of "Fast-Track" Development on the Wild Coast* (Cape Town: Programme for Land and Agrarian Studies, University of the Western Cape, 2001).

11. Sustainable Livelihoods in Southern Africa Team, *Decentralisations in Practice in Southern Africa*; Sustainable Livelihoods in Southern Africa Team, *Rights Talk and Rights Practice*; Palmer, Timmermans, and Fay, *From Conflict to Negotiation*; Roger Southall and Zosa de Sas, "Containing the Chiefs: The ANC and Traditional Leaders in the Eastern Cape," Working Paper 39 (Fort Hare Institute of Social and Economic Research, August 2003).

12. Ntshona, *Valuing the Commons*, 23; Ntsebeza, "Decentralisation and Natural Resource Management in Rural South Africa"; J. F. Obiri and M. J. Lawes, "Challenges Facing New Forest Policies in South Africa: Attitudes of Forest Users toward Management of the Coastal Forests of the Eastern Cape Province" (paper presented at DWAF Natural Forests and Savanna Woodlands Symposium, 2002); G. P. von Maltitz and G. Fleming, "Status of Conservation of Indigenous Forests in South Africa," in *Towards Sustainable Management Based on Scientific Understanding of Forests and Woodlands: Proceedings of the Natural Forests and Savanna Woodlands Symposium II, September 5–9, 1999, Knysna, South Africa*, ed. A. H. W. Seydack, W. J. Vermeulen, and C. Vermeulen (Knysna, South Africa: Department of Water Affairs and Forestry, 2000); J. Evans, S. Shackleton, and G. von Maltitz, "Managing Woodlands under Communal Tenure: Institutional Issues," in Seydack, Vermeulen, and Vermeulen, *Towards Sustainable Management*.

13. Republic of South Africa, Department of Water Affairs and Forestry, *Towards a Policy for Sustainable Forest Management in South Africa—A Discussion Paper* (July 1995).

14. Republic of South Africa, Department of Water Affairs and Forestry, *Sustainable Forest Development in South Africa*.

15. See, for example, Obiri and Lawes, "Challenges Facing New Forest Policies"; Maltitz and Fleming, "Status of Conservation of Indigenous Forests."

16. The Gcaleka Xhosa chief Sarhili (chap. 1), for instance, has continued to serve as a "traditional" model of state resource control over the past century, as successive officials in the Cape, South African, and Transkeian homeland governments as well as some development consultants have promoted plans for resource management in what is today's Wild Coast National Park. Sim, *Forests and Forest Flora*, 7–8; *Annual Report 1931*, 19; CTA FCT 3/1/51, T700, M. 624, J. Keet, Director of Forestry, Transkei, to Inspector of Forestry, Natal and Transkei, "Management of Indigenous Forests in the Transkei," 28 December 1933; N. L. King, "Report on the Indigenous Forests in the Transkei and Suggestions for Their Management," 21 August 1936, 6–8; King, "Exploitation of the Indigenous Forests," 36–37; Kaiser Matanzima, "Chief Matanzima Talks to Us about Nature Conservation in the Transkei," *African Wildlife* 29, no. 2 (1975): 14–15; Keet, *Historical Review*, 236–37; Shone, *Forestry in Transkei*, 4; Sizwe Cawe, "Chief

Protects Forest 100 Years Ago," in "Coastal Campaign," supplement to the *Sunday Times*, 13 October 1991, 3; Land and Agriculture Policy Centre, *Overview of the Transkei Sub-Region*, 93, Appendix 2.16, "Restitution and Land Claims," regarding contested land claims along the Wild Coast: "While these communities have laid claim to Dwesa Reserve, there remains some doubt as to the historical validity of this claim as the forest is understood to have been protected as a royal preserve by King Sarili in the nineteenth century." As recently as 1996 and 1997, Minister of Water Affairs and Forestry Kader Asmal and other state officials even went so far as to propose the erection of a monument in the Dwesa-Cwebe area to honor Sarhili's "precedent"-setting conservationism. Such ideas were sharply criticized by local communities: "People's reactions were very emotional and angry. They ridiculed the notion of such protection [by Sarhili], other than against white woodcutters. They challenged 'experts' to repeat their 'fabricated' information about Sarili to the actual descendants of his subjects. They challenged outsiders . . . to come and learn from them about Sarili." Constitution Development Workshop, Dwebe Project, held at Rhodes University, Institute for Social and Economic Research, Grahamstown, minutes for 4 February 1997; quotation is from faxed letter, Andre Terblanche, *The Village Planner*, to Harold Winkler, Department of Land Affairs, 27 May 1996. Thanks to Derick Fay for these references.

17. Sustainable Livelihoods in Southern Africa Team, *Rights Talk and Rights Practice*; Palmer, Timmermans, and Fay, *From Conflict to Negotiation*; Ntshona, *Valuing the Commons*; Kepe, *Environmental Entitlements in Mkambati*; Michael Lipton, Frank Ellis, and Merle Lipton, eds., *Land, Labour and Livelihoods in Rural South Africa*, vol. 2, *KwaZulu-Natal and Northern Province* (Durban: Indicator Press, University of Natal, 1996); Andrew Ainslie, *Progress Report: Rural Livelihoods and Local Level Management of Natural Resources in the Peddie District* (Johannesburg: Land and Agriculture Policy Centre, 1996).

18. Ntshona, *Valuing the Commons*, 41–49; Palmer, Timmermans, and Fay, *From Conflict to Negotiation*; Institute of Social and Economic Research, Rhodes University, *Indigenous Knowledge, Conservation Reform*; Kepe, *Environmental Entitlements in Mkambati*, 45–51.

19. For a contemporary perspective on similar issues in the Western Cape, see Farieda Khan's description of the 1996 protests within the Muslim community of Cape Town against proposed development on the lower slopes of Table Mountain, the site of numerous Muslim graves and shrines. "The Roots of Environmental Racism and the Rise of Environmental Justice in the 1990s," in *Environmental Justice in South Africa*, ed. David A. McDonald (Athens: Ohio University Press; Cape Town: University of Cape Town Press, 2002), 15–48.

20. See, for example, the collected essays in McDonald, *Environmental Justice in South Africa*, and N. Jacobs, *Environment, Power, and Injustice*.

21. Greg Ruiters, "Race, Place, and Environmental Rights: A Radical Critique of Environmental Justice Discourse," in McDonald, *Environmental Justice in South Africa*, 112–26; Braun, *Intemperate Rainforest*, 5, 25, drawing from Michael

Mason, *Environmental Democracy* (London: Earthscan Publications, 1999); Charles Zerner, "Introduction: Toward a Broader Vision of Justice and Nature Conservation," in *People, Plants, and Justice: The Politics of Nature Conservation*, ed. Charles Zerner (New York: Columbia University Press, 2000), 3–20; Kepe, *Environmental Entitlements in Mkambati*, 84–85.

Bibliography

PUBLISHED GOVERNMENT SOURCES

Cape of Good Hope

Series

Department of Crown Lands and Public Works, *Reports of the Superintendent of Woods and Forests* and *Reports of the Conservators of Forests* (1881–1892)

Department of Lands, Mines, and Agriculture, *Reports of the Conservators of Forests* (1892–1893)

Department of Agriculture, *Reports of the Conservators of Forests* (1893–1906)

Forest Department, *Reports of the Chief Conservator of Forests* (1906–1909)

Department of Native Affairs, *Blue Books on Native Affairs* (BBNA) (1874–1909)

Other Papers

Ministerial Department of Native Affairs, *Report of S. A. Probart, Mission to Tambookieland in January 1876* (G.39–1876)

Reports from Chief Magistrates and Resident Magistrates in Basutoland, Transkei, etc. (A.25–1881)

Report of the Select Committee on Settlement of Tembuland (A.15–1882)

Report and Proceedings, with Appendices, of the Government Commission on Native Laws and Customs (G.4–1883)

Report and Proceedings of the Tembuland Commission (G.66–1883)

Ministerial Department of Crown Lands and Public Works, *Report of the Griqualand East Land Commission* (G.2–1884)

Despatches and Other Papers Relating to the Transkeian Territories (G.95–1884)

Report of the Select Committee on Forests Bill (A.6–1888)

Report of the Select Committee on the Forest Act (A.1–1899)

Report on the Extent, Value, and Administration of the Forests of the Transkei and Griqualand East (G.62–1893)

Report of the Select Committee on Crown Forests (A.12–1906)

Report of the Select Committee on the Umtata Water Supply Bill (A.4–1907)

Union of South Africa

Series

Forest Department, *Reports of the Chief Conservator of Forests*
Department of Agriculture and Forestry, *Reports of the Director of Forestry* (after 1932)

Other Papers

Report of the South African Native Affairs Commission, 1903–1905 (SANAC) (Cape Town, 1905)
Report of the Native Affairs Department for the Year Ended 31st December, 1912
Report of the Native Economic Commission, 1930–32 (NEC) (U.G.22–1932)
Report of the Native Affairs Commission Years 1939–1940 (U.G.42–1941)

Transkeian Territories

Series

Proceedings and Reports of Select Committees at the Session of the Pondoland General Council (PGC)
Proceedings and Reports of Select Committees at the Session of the Transkeian Territories General Council (TTGC)
Proceedings and Reports of Select Committees at the Session of the United Transkeian Territories General Council (UTTGC)

Republic of South Africa

Department of Water Affairs and Forestry. *Sustainable Forest Development in South Africa: The Policy of the Government of National Unity: White Paper.* Pretoria, 1996.
Department of Water Affairs and Forestry. *Towards A Policy for Sustainable Forest Management in South Africa—A Discussion Paper.* July 1995.

ARCHIVAL SOURCES

South Africa

Cape Town Archives Repository (CTA), Cape Town

Archives of the Secretary for Agriculture (AGR)
Archives of the Chief Magistrate of East Griqualand (CMK)
Archives of the Chief Magistrate of the Transkeian Territories (CMT)

Archives of the Conservator of Forests, Eastern Conservancy (FCE)
Archives of the Conservator of Forests, Transkeian Conservancy (FCT)
Archives of the District Forest Officer Kokstad (FKS)
Archives of the District Forest Officer Umtata (FDU)
Archives of the Secretary for Native Affairs (NA)
Archives of the Prime Minister's Office (PMO)
Papers of the Resident Magistrate of the Engcobo District (1/ECO)
Papers of the Resident Magistrate of the Kentani District (1/KNT)
Papers of the Resident Magistrate of the Mount Fletcher District (1/MTF)
Papers of the Resident Magistrate of the Mqanduli District (1/MQL)
Papers of the Resident Magistrate of the Tsolo District (1/TSO)
Papers of the Resident Magistrate of the Umtata District (1/UTA)

National Archives Repository (NAR), Pretoria

Archives of the Native Affairs Department (NTS)
Archives of the Forest Department (FOR)

Pietermaritzburg Archives Repository (PAR), Pietermaritzburg

Accession A1238, Henkel Papers

Umtata Archives Repository (UAR), Umtata

Archives of the Chief Magistrate of the Transkeian Territories (CMT)

Cory Library, Grahamstown

MS 14, 559, Resident Magistrate of Tsolo, "Rough Diary of Every-day Events" (1886–87)
MS 1157, Walter Stanford, "Statement Made by Tembu Chief to W. E. Stanford" (1908)

South African Library, Cape Town, Manuscripts Collection (SAL)

MSB 783, Robert Godfrey, unpublished manuscript, 3rd edition of "A Xhosa-English Dictionary"

University of Cape Town, Manuscripts and Archives Library (UCT)

BC293, Sir W. E. M. Stanford Papers (Stanford Papers)
BC543, Bain and Lister Family Papers

Great Britain

Public Record Office (PRO), Kew Gardens

Colonial Office files: CO 48, correspondence files, series 478, 503, and 525

PERIODICALS AND NEWSPAPERS

Christian Express
Daily Dispatch
East London Daily Dispatch and Frontier Advertiser
Electronic Mail and Guardian
Imvo Zabantsundu
Izwi la Bantu
Kokstad Advertiser
Periodical Accounts Relating to the Missions of the Church of the United Brethren, Established among the Heathen (Periodical Accounts)
South African Outlook
Territorial News
Transkeian Gazette
Umtata Herald

BOOKS AND ARTICLES

Adas, Michael. "From Footdragging to Flight: The Evasive History of Peasant Avoidance Strategies in South and South East Asia." *Journal of Peasant Studies* 13, no. 2 (1986): 64–86.

Adenaike, Carol Keyes, and Jan Vansina, eds. *In Pursuit of History: Fieldwork in Africa.* Portsmouth, N.H.: Heinemann, 1996.

Ainslie, Andrew. *Progress Report: Rural Livelihoods and Local Level Management of Natural Resources in the Peddie District.* Johannesburg: Land and Agriculture Policy Centre, 1996.

Amanor, Kojo Sebastian. "Managing the Fallow: Weeding Technology and Environmental Knowledge in the Krobo District of Ghana." In "Indigenous Agricultural Systems and Development," special issue, *Agriculture and Human Values* 8, no. 1 (Winter-Spring 1991): 5–13.

———. *The New Frontier: Farmer Responses to Land Degradation: A West African Study.* Atlantic Highlands, N.J.: Zed Books, 1994.

Anderson, David M. and Richard Grove, eds. *Conservation in Africa: People, Policies and Practice.* Cambridge: Cambridge University Press, 1987.

Anonymous. "Mhlakwapalwa at a Circuit Court: The Narrative of an Uncivilised Native (Translated)." *Cape Law Journal* 7 (1890): 225–35.

Ashforth, Adam. *The Politics of Official Discourse in Twentieth-Century South Africa.* Oxford: Clarendon Press, 1990.

Ayliff, John, and Joseph Whiteside. *History of the Abambo, Generally Known as Fingos.* Butterworth, Transkei: 1912.

Beattie, T. R. *A Ride through the Transkei.* Kingwilliamstown: S. E. Rowles, 1891.

Beinart, William. "African History and Environmental History." *African Affairs* 99, no. 395 (April 2000): 269–302.

————. "Chieftaincy and the Concept of Articulation: South Africa ca. 1900–1950." *Canadian Journal of African Studies* 19, no. 1 (1985): 91–98.

————. "Conflict in Qumbu: Rural Consciousness, Ethnicity and Violence in the Colonial Transkei." In *Hidden Struggles in Rural South Africa: Politics and Popular Movements in the Transkei and Eastern Cape, 1890–1930*, edited by William Beinart and Colin Bundy, 106–37. London: James Currey, 1987.

————. "Environmental Origins of the Pondoland Revolt." In *South Africa's Environmental History*, edited by Stephen Dovers, Ruth Edgecombe, and Bill Guest, 76–89. Cape Town: David Philip, 2002; Athens: Ohio University Press, 2003.

————. "Introduction: The Politics of Colonial Conservation." *Journal of Southern African Studies* 11, no. 1 (1984): 143–62.

————. "The Night of the Jackal: Sheep, Pastures and Predators in the Cape." *Past and Present* 158 (February 1998): 172–206.

————. "The Origins of the *Indlavini*: Male Associations and Migrant Labour in the Transkei." In *Tradition and Transition in Southern Africa*, edited by Andrew Spiegel and Patrick McAllister, 103–28. London: Transaction Publishers, 1991.

————. *The Political Economy of Pondoland 1860–1930*. Cambridge: Cambridge University Press, 1982.

————. *The Rise of Conservation in South Africa: Settlers, Livestock, and the Environment 1770–1950*. Oxford: Oxford University Press, 2003.

————. "Settler Accumulation in East Griqualand." In *Putting a Plough to the Ground: Accumulation and Dispossession in Rural South Africa, 1850–1930*, edited by William Beinart, Peter Delius, and Stanley Trapido, 259–310. Johannesburg: Ravan Press, 1986.

————. "Soil Erosion, Animals and Pasture over the Longer Term: Environmental Destruction in Southern Africa." In *The Lie of the Land*, edited by Melissa Leach and Robin Mearns, 54–72. London: International African Institute, 1996.

————. "Soil Erosion, Conservationism and Ideas about Development: A Southern African Exploration, 1900–1960." *Journal of Southern African Studies* 11, no. 1 (October 1984): 52–83.

————. "Transkeian Smallholders and Agrarian Reform." *Journal of Contemporary African Studies* 11, no. 2 (1992): 178–99.

————. "Vets, Viruses and Environmentalism at the Cape." In *Ecology and Empire*, edited by Tom Griffiths and Libby Robin, 87–101. Edinburgh: Keele University Press, 1997.

Beinart, William, and Colin Bundy. *Hidden Struggles in Rural South Africa: Politics and Popular Movements in the Transkei and Eastern Cape, 1890–1930*. London: James Currey, 1987.

Beinart, William, and Peter Coates. *Environment and History: The Taming of Nature in the USA and South Africa*. London: Routledge, 1995.

Beinart, William, Peter Delius, and Stanley Trapido, eds. *Putting a Plough to the Ground: Accumulation and Dispossession in Rural South Africa, 1850–1930*. Johannesburg: Ravan Press, 1986.

Beinart, William, and JoAnn McGregor, eds. *Social History and African Environments*. Athens: Ohio University Press; Oxford: James Currey, 2003.

Berman, Bruce. *Control and Crisis in Colonial Kenya: The Dialectic of Domination* (London: James Currey, 1990).

———. "Ethnicity, Patronage and the African State: The Politics of Uncivil Nationalism." *African Affairs* 97, no. 388 (July 1998): 324–25.

Berry, Sara. *Chiefs Know Their Boundaries: Essays on Property, Power, and the Past in Asante, 1896–1996*. Portsmouth, N.H.: Heinemann, 2001.

———. *No Condition Is Permanent: The Social Dynamics of Agrarian Change in Sub-Saharan Africa*. Madison: University of Wisconsin Press, 1993.

———. "Resource Access and Management as Historical Processes: Conceptual and Methodological Issues." In *Access, Control and Management of Natural Resources in Sub-Saharan Africa—Methodological Considerations*, edited by Christian Lund and Henrik Secher Marcussen. Occasional Paper 13. International Development Studies, Roskilde University, Denmark, 1994.

———. "Social Institutions and Access to Resources." *Africa* 59, no. 1 (1989): 41–55.

Bhat, R. B., and T. V. Jacobs. "Traditional Herbal Medicine in Transkei." *Journal of Ethnopharmacology* 48, no. 1 (August 1995): 7–12.

Blohm, W. "Das Opfer und dessen Sinn bei den Xosa in Südafrika." *Archiv für Anthropologie* 23 (1933): 150–53.

———. "Schöpferische Kräfte in der Gesellschaft der Xosa-Gruppe." *Archiv für Anthropologie* 23 (1934): 159–95.

Bolofo, R. N., and C. T. Johnson. "The Identification of 'Isicakathi' and Its Medicinal Use in Transkei." *Bothalia* 18, no. 1 (1988): 125–30.

Bozzoli, Belinda. "Marxism, Feminism and South African Studies." *Journal of Southern African Studies* 9, no. 2 (April 1983): 139–71.

Bozzoli, Belinda, and Peter Delius. "Radical History and South African Society." In *History from South Africa: Alternative Visions and Practices*, edited by Joshua Brown, Patrick Manning, Karin Shapiro, and Jon Wiener, 3–25. Philadelphia: Temple University Press, 1991.

Bozzoli, Belinda, with the assistance of Mmantho Nkotsoe. *Women of Phokeng: Consciousness, Life Strategy, and Migrancy in South Africa, 1900–1983*. Portsmouth, N.H.: Heinemann, 1991.

Braun, Bruce. *The Intemperate Rainforest: Nature, Culture, and Power on Canada's West Coast*. Minneapolis: University of Minnesota Press, 2002.

Brosius, J. Peter. "Analyses and Interventions: Anthropological Engagements with Environmentalism." *Current Anthropology* 40, no. 3 (1999): 277–309.

———. "Endangered Forest, Endangered People: Environmentalist Representations of Indigenous Knowledge." *Human Ecology* 25, no. 1 (1997): 47–69.

Broster, Joan A. *Amagqirha: Religion, Magic and Medicine in Transkei*. Cape Town: Via Afrika Limited, 1981.

———. *Red Blanket Valley*. Johannesburg: Hugh Keartland, 1967.

————. *The Tembu: Their Beadwork, Songs and Dances.* Cape Town: Purnell, 1976.

Brown, Joshua, Patrick Manning, Karin Shapiro, and Jon Wiener, eds. *History from South Africa: Alternative Visions and Practices.* Philadelphia: Temple University Press, 1991.

Brown, Karen. "The Conservation and Utilisation of the Natural World: Silviculture in the Cape Colony, c. 1902–1910." *Environment and History* 7, no. 4 (2001): 427–47.

Brownlee, Charles. *Reminiscences of Kaffir Life and History.* Lovedale: Lovedale Press, 1896.

Brownlee, Frank. "Burial Places of Chiefs." *African Affairs* 43, no. 170 (January 1944): 23–24.

————. *The Cattle Thief: The Story of Ntsukumbini.* London: Jonathan Cape, 1932.

————. "Nopongo and His Ninepence: The Psychology of a Native Litigant." *South African Law Journal* 50 (1933): 12–17.

————. "Some Experiences of Native Superstition and Witchcraft." *Journal of the Royal African Society* 39, no. 154 (1940): 54–60.

————. *The Transkeian Native Territories: Historical Records.* Lovedale: Lovedale Institution Press, 1923.

Bryant, A. T. *Olden Times in Zululand and Natal.* London: Longmans, Green and Co., 1929.

————. *A Zulu-English Dictionary.* Pietermaritzburg: P. Davis and Sons, 1905.

————. *Zulu Medicine and Medicine-Men.* Cape Town: C. Struik, 1966 repr. ed.

Bundy, Colin. "Mr. Rhodes and the Poisoned Goods: Popular Opposition to the Glen Grey Council System, 1894–1906." In *Hidden Struggles in Rural South Africa: Politics and Popular Movements in the Transkei and Eastern Cape, 1890–1930,* edited by William Beinart and Colin Bundy, 138–65. London: James Currey, 1987.

————. *The Rise and Fall of the South African Peasantry.* London: Heinemann, 1979.

Callaway, Godfrey. *Mxamli, the Feaster: A Pondomisi Tale of the Diocese of St. John's, South Africa.* London: Society for Promoting Christian Knowledge; New York: Macmillan, 1919.

————. *Pioneers in Pondoland.* Lovedale: Lovedale Press, 1939.

————. *South Africa from Within: Made Known in the Letters of a Magistrate.* London: Society for Promoting Christian Knowledge, 1930.

Callaway, Henry. *The Religious System of the Amazulu.* Natal: J. A. Blair, 1870.

Carlson, Knut A. "Forestry in the Transkei." *Cape Agricultural Journal,* June 11, 1896, 303–6.

————. *Transplanted: Being the Adventures of a Pioneer Forester in South Africa.* Pretoria: Minerva Drukpers, 1947.

————. "Weaning the Natives from the Natural Forests in the Native Territories." *Journal of the South African Forestry Association* 2 (April 1939): 75–76, "Correspondence" section.

Carney, Judith, and Michael Watts. "Disciplining Women?: Rice, Mechanization, and the Evolution of Mandinka Gender Relations in Senegambia." *Signs* 16, no. 4 (1991): 651–81.

Carton, Benedict. *Blood from Your Children: The Colonial Origins of Generational Conflict in South Africa.* Charlottesville: University Press of Virginia, 2000.

Cawe, Sizwe. "Chief Protects Forest 100 Years Ago." In "Coastal Campaign," supplement to the *Sunday Times*, 13 October 1991, 3.

Cawe, Sizwe, and Bruce McKenzie. "The Afromontane Forests of Transkei, Southern Africa. II: A Floristic Classification." *South African Journal of Botany* 55, no. 1 (1989): 31–39.

Chanock, Martin. *Law, Custom and Social Order: The Colonial Experience in Malawi and Zambia.* Cambridge: Cambridge University Press, 1985.

———. *The Making of South African Legal Culture 1902–1936: Fear, Favour and Prejudice.* Cambridge: Cambridge University Press, 2001.

Cingo, W. D. *Ibali laba Tembu.* Palmerton, Pondoland: Mission Printing Press, 1927.

Cloete, Laura. "Domestic Strategies of Rural Transkeian Women." Development Studies Working Paper 54. Institute of Social and Economic Research, Rhodes University, 1992.

Cohen, David, and E. S. Atieno Odhiambo. *Siaya: The Historical Anthropology of an African Landscape.* Athens: Ohio University Press, 1989.

Cole, Jennifer, and Karen Middleton. "Rethinking Ancestors and Colonial Power in Madagascar." *Africa* 71, no. 1 (2001): 1–37.

Colson, Elizabeth. "Places of Power and Shrines of the Land." In "The Making of African Landscapes," edited by Ute Luig and Achim von Oppen, 47–57. Special issue, *Paideuma: Mitteilungen zur Kulturkunde* 43 (1997).

Cook, P. A. W. *Social Organization and Ceremonial Institutions of the Bomvana.* Cape Town: Juta and Co., 1931.

Cooper, Frederick. "Conflict and Connection: Rethinking Colonial African History." *American Historical Review* 99, no. 5 (December 1994): 1516–45.

Cooper, Frederick, and Ann Laura Stoler, eds. *Tensions of Empire: Colonial Cultures in a Bourgeois World.* Berkeley: University of California Press, 1997.

Cooper, K. H., and W. Swart. *Transkei Forest Survey.* Durban: Wildlife Society of Southern Africa, 1991.

Crais, Clifton. *The Politics of Evil: Magic, State Power, and the Political Imagination in South Africa.* Cambridge: Cambridge University Press, 2002.

Crehan, Kate. *The Fractured Community: Landscapes of Power and Gender in Rural Zambia.* Berkeley: University of California Press, 1997.

———. "'Tribes' and the People Who Read Books: Managing History in Colonial Zambia." *Journal of Southern African Studies* 23, no. 2 (1997): 203–18.

Cronon, William. "A Place for Stories: Nature, History and Narrative." *Journal of American History* 78 (March 1992): 1347–76.

Crummey, Donald, ed. *Banditry, Rebellion and Social Protest in Africa*. London: James Currey, 1986.

Darrow, W. K. "Forestry in the Eastern Cape Border Region." Bulletin 51, South African Department of Forestry, 1973.

Delius, Peter. *A Lion Amongst the Cattle: Reconstruction and Resistance in the Northern Transvaal*. Portsmouth, N.H.: Heinemann, 1996.

Delius, Peter, and Stefan Schirmer. "Soil Conservation in a Racially Ordered Society: South Africa 1930–1970." *Journal of Southern African Studies* 26, no. 4 (December 2000): 655–74.

de Wet, Chris. *Moving Together, Drifting Apart: Betterment Planning and Villagisation in a South African Homeland*. Johannesburg: Witwatersrand University Press, 1995.

Dilika, N. F., R. V. Nikolova, and T. V. Jacobs. "Plants Used in the Circumcision Rites of the Xhosa Tribe in South Africa." In *Proceedings of the International Symposium on Medicinal and Aromatic Plants, August 27–30, 1995, Amherst MA USA*, edited by L. E. Craker, L. Nolan, and K. Shetty, 165–68. Leuven, Belgium: International Society for Horticultural Science, 1995.

Dovers, Stephen, Ruth Edgecombe, and Bill Guest, eds. *South Africa's Environmental History: Cases and Comparisons*. Cape Town: David Philip, 2002; Athens: Ohio University Press, 2003.

Dubow, Saul. "Holding 'A Just Balance between White and Black': The Native Affairs Department in South Africa c. 1920–33." *Journal of Southern African Studies* 12, no. 2 (April 1986): 217–39.

———. *Racial Segregation and the Origins of Apartheid in South Africa, 1919–1936*. Oxford: Oxford University Press, 1989.

Eberhard, Anton, and Clive van Horen. *Poverty and Power: Energy and the South African State*. London: Pluto Press, 1995.

Ellen, Roy, Peter Parkes, and Alan Bicker, eds. *Indigenous Environmental Knowledge and its Transformations: Critical Anthropological Perspectives*. Amsterdam: Harwood Academic, 2000.

Etherington, Norman. "Missionary Doctors and African Healers in Mid-Victorian South Africa." *South African Historical Journal* 19 (1987): 77–91.

Evans, Ivan. *Bureaucracy and Race: Native Administration in South Africa*. Berkeley: University of California Press, 1997.

Evans, J., S. Shackleton, and G. von Maltitz. "Managing Woodlands under Communal Tenure: Institutional Issues." In *Towards Sustainable Management Based on Scientific Understanding of Forests and Woodlands: Proceedings of the Natural Forests and Savanna Woodlands Symposium II, September 5–9, 1999, Knysna, South Africa*, edited by A. H. W. Seydack, W. J. Vermeulen, and C. Vermeulen. Knysna, South Africa: Department of Water Affairs and Forestry, 2000.

Fairhead, James, and Melissa Leach. *Misreading the African Landscape: Society and Ecology in a Forest-Savanna Mosaic*. Cambridge: Cambridge University Press, 1996.

Feeley, J. M. "The Early Farmers of Transkei, Southern Africa before A.D. 1870." *Cambridge Monographs in African Archaeology* 24, Bar International Series 378 (1987).

Fortmann, Louise. "The Tree Tenure Factor in Agroforestry with Particular Reference to Africa." In *Whose Trees?: Proprietary Dimensions of Forestry*, edited by Louise Fortmann and John Bruce, 16–33. Boulder: Westview Press, 1988.

Fortmann, Louise, and John Bruce, eds. *Whose Trees?: Proprietary Dimensions of Forestry.* Boulder: Westview Press, 1988.

Gengenbach, Heidi. "'I'll Bury You in the Border!': Women's Land Struggles in Post-War Facazisse (Magude District), Mozambique." *Journal of Southern African Studies* 24, no. 1 (March 1998): 7–36.

Giblin, James. *The Politics of Environmental Control in Northeastern Tanzania, 1840–1940.* Philadelphia: University of Pennsylvania, 1992.

Giles-Vernick, Tamara. *Cutting the Vines of the Past: Environmental Histories of the Central African Rain Forest.* Charlottesville: University Press of Virginia, 2002.

———. "Na lege ti guiriri (On the Road of History): Mapping Out the Past and Present in M'Bres Region, Central African Republic." *Ethnohistory* 43, no. 2 (Spring 1996): 245–75.

———. "We Wander Like Birds: Migration, Indigeneity, and the Fabrication of Frontiers in the Sangha River Basin of Equatorial Africa." *Environmental History* 4, no. 2 (April 1999): 168–97.

Gilfoyle, Dan. "The Heartwater Mystery: Veterinary and Popular Ideas about Tick-Borne Animal Diseases at the Cape, c. 1877–1910." *Kronos: The Journal of Cape History* 29 (2003): 139–60.

Goheen, Mitzi. "Chiefs, Sub-Chiefs and Local Control: Negotiations over Land, Struggles over Meaning." *Africa* 62, no. 3 (1992): 389–411.

Graeber, David. "Love Magic and Political Morality in Central Madagascar, 1875–1990." In *Gendered Colonialisms in African History*, edited by Nancy Rose Hunt, Tessie P. Liu, and Jean Quataert, 96–117. Malden, Mass.: Blackwell, 1997.

Greene, Sandra. *Sacred Sites and the Colonial Encounter: A History of Meaning and Memory in Ghana.* Bloomington: Indiana University Press, 2002.

———. "Sacred Terrain: Religion, Politics and Place in the History of Anloga (Ghana)." *International Journal of African Historical Studies* 30, no. 1 (1997): 1–22.

Griffiths, Tom, and Libby Robin, eds. *Ecology and Empire: Environmental History of Settler Societies.* Edinburgh: Keele University Press, 1997.

Grove, Richard. "Colonial Conservation, Ecological Hegemony and Popular Resistance: Towards a Global Synthesis." In *Imperialism and the Natural World*, edited by John Mackenzie, 5–50. Manchester: Manchester University Press, 1990.

———. "Conserving Eden: The (European) East India Companies and their Environmental Policies on St. Helena, Mauritius and in Western India, 1660 to 1854." *Comparative Studies in Society and History* 35, no. 2 (April 1993): 318–51.

———. "Early Themes in African Conservation: The Cape in the Nineteenth Century." In *Conservation in Africa*, edited by David M. Anderson and Richard Grove, 21–39. Cambridge: Cambridge University Press, 1987.

———. *Green Imperialism: Colonial Expansion, Tropical Island Edens and the Origins of Environmentalism, 1600–1860.* Cambridge: Cambridge University Press, 1995.

———. "Scotland in South Africa: John Croumbie Brown and the Roots of Settler Environmentalism." In *Ecology and Empire*, edited by Tom Griffiths and Libby Robin, 139–53. Edinburgh: Keele University Press, 1997.

———. "Scottish Missionaries, Evangelical Discourses and the Origins of Conservation Thinking in Southern Africa 1820–1900." *Journal of Southern African Studies* 15, no. 2 (January 1989): 163–87.

Gupta, Akhil. *Postcolonial Developments: Agriculture in the Making of Modern India.* Durham, N.C.: Duke University Press, 1998.

Ham, C., and J. M. Theron. "Community Forestry and Woodlot Development in South Africa: The Past, Present and Future." *Southern African Forestry Journal* 184 (March 1999): 71–79.

Hamilton, Carolyn. *Terrific Majesty: The Powers of Shaka Zulu and the Limits of Historical Invention.* Cambridge, Mass.: Harvard University Press, 1998.

Hammond-Tooke, W. D. *Bhaca Society: A People of the Transkeian Uplands South Africa.* Cape Town: Oxford University Press, 1962.

———. *Command or Consensus: The Development of Transkeian Local Government.* Cape Town: David Philip, 1975.

———. "The Initiation of a Baca Isangoma Diviner." *African Studies* 14, no. 1 (1955): 16–22.

———. "Kinship, Locality, and Association: Hospitality Groups among the Cape Nguni." *Ethnology* 2, no. 3 (1963): 302–19.

———. "The Symbolic Structure of Cape Nguni Cosmology." In *Religion and Social Change in Southern Africa*, edited by Michael Whisson and Martin West, 15–33. Cape Town: David Philip, 1975.

———. "The Transkeian Council System 1895–1955: An Appraisal." *Journal of African History* 9, no. 3 (1968): 455–77.

———. *The Tribes of Umtata District.* Ethnological Publications 35. Pretoria: Union of South Africa, Department of Native Affairs, 1956.

Harms, Robert. *Games Against Nature: An Eco-Cultural History of the Nunu of Equatorial Africa.* New Haven, Conn.: Yale University Press, 1987.

Hendricks, Fred. "Loose Planning and Rapid Resettlement: The Politics of Conservation and Control in Transkei, South Africa, 1950–1970." *Journal of Southern African Studies* 15, no. 2 (January 1989): 306–25.

Henkel, C. C. *Tree Planting for Ornamental and Economic Purposes in the Transkeian Territories, South Africa.* Cape Town: Juta, 1894.

Hey, Douglas. "The History and Status of Nature Conservation in South Africa." In *A History of Scientific Endeavour in South Africa*, edited by A. C. Brown, 132–63. Cape Town: Royal Society of South Africa, 1977.

Hirst, Manton. "A River of Metaphors: Interpreting the Xhosa Diviner's Myth." In *Culture and the Commonplace: Anthropological Essays in Honour of David Hammond-Tooke*, edited by Patrick McAllister. Johannesburg: Witwatersrand University Press, 1997.

Hofmeyr, Isabel. *"We Spend Our Years as a Tale That Is Told": Oral Historical Narrative in a South African Chiefdom*. Portsmouth, N.H.: Heinemann, 1993.

Holt, Basil. *Place-Names in the Transkeian Territories*. Johannesburg: Africana Museum, 1959.

———. *The Tshezi of the Transkei: An Ethnographic Study*. Johannesburg, 1969.

Hubbard, C. S. "Afforestation and Fuel Supply in Relation to Development in the Transkeian Territories." *South African Medical Journal*, November 10, 1945, 407–8.

Hunt, Nancy Rose. *A Colonial Lexicon of Birth Ritual, Medicalization, and Mobility in the Congo*. Durham, N.C.: Duke University Press, 1999.

Hunter, Monica. *Reaction to Conquest: Effects of Contacts with Europeans on the Pondo of South Africa*. London: Oxford University Press, 1936.

Hutchings, Anne. *Zulu Medicinal Plants: An Inventory*. Pietermaritzburg: University of Natal Press, 1996.

Institute of Social and Economic Research, Rhodes University. *Indigenous Knowledge, Conservation Reform, Natural Resource Management and Rural Development in the Dwesa and Cwebe Nature Reserves and Neighboring Village Settlements*. Interim Report for the Human Sciences Research Council. Grahamstown, South Africa: 1997.

International Institute for Environment and Development (London) and CSIR-Environmentek (Pretoria). *Instruments for Sustainable Private Sector Forestry*. CD-ROM. London and Pretoria, 2003.

Isaacman, Allen. "Chiefs, Rural Differentiation and Peasant Protest: The Mozambican Forced Cotton Regime 1938–61." *African Economic History* 14 (1985): 15–56.

———. *Cotton Is the Mother of Poverty: Peasants, Work, and Rural Struggle in Colonial Mozambique, 1938–1961*. Portsmouth, N.H.: Heinemann, 1996.

———. "Peasants and Rural Social Protest in Africa." *African Studies Review* 33, no. 2 (September 1990): 1–120.

Jacobs, Jane. "Earth Honoring: Western Desires and Indigenous Knowledges." In *Writing Women and Space: Colonial and Postcolonial Geographies*, edited by Alison Blunt and Gillian Rose, 169–96. New York: Guilford Press, 1994.

Jacobs, Nancy. *Environment, Power, and Injustice: A South African History*. Cambridge: Cambridge University Press, 2003.

Janzen, John M. *The Quest for Therapy in Lower Zaire*. Berkeley: University of California Press, 1978.

Johnson, Colin T. "A Preliminary Checklist of Xhosa Names for Trees Growing in Transkei." *Bothalia* 20, no. 2 (1990): 147–152.

Johnson, Colin T., and Sizwe Cawe. "Analysis of the Tree Taxa in Transkei." *South African Journal of Botany* 53, no. 5 (1987): 388–94.

Jordan, A. C. *The Wrath of the Ancestors.* Lovedale: Lovedale Press, 1980.

Kassam, Aneesa, and Gemetchu Megerssa. "Sticks, Self, and Society in Booran Oromo: A Symbolic Interpretation." In *African Material Culture,* edited by Mary Jo Arnoldi, Christraud M. Geary, and Kris L. Hardin, 145–66. Bloomington: Indiana University Press, 1996.

Keet, J. D. M. *Historical Review of the Development of Forestry in South Africa.* Pretoria: Department of Environmental Affairs, 1984.

Kepe, Thembela. *Environmental Entitlements in Mkambati: Livelihoods, Social Institutions and Environmental Change on the Wild Coast of the Eastern Cape.* Research Report 1. Cape Town: Programme for Land and Agrarian Studies, University of the Western Cape, 1997.

———. *Waking Up from the Dream: The Pitfalls of "Fast-Track" Development on the Wild Coast.* Cape Town: Programme for Land and Agrarian Studies, University of the Western Cape, 2001.

Kepe, Thembela, and Ben Cousins. "Resource Tenure and Power Relations in Community Wildlife: The Case of Mkambati Area, South Africa." *Society and Natural Resources* 14, no. 10 (1 December 2001): 911–25.

Kepe, Thembela, and Ian Scoones. "Creating Grasslands: Social Institutions and Environmental Change in Mkambati Area, South Africa." *Human Ecology* 27, no. 1 (1999): 29–53.

Khan, Farieda. "Rewriting South Africa's Conservation History—The Role of the Native Farmer's Association." *Journal of Southern African Studies,* 20, no. 4 (December 1994): 499–516.

———. "The Roots of Environmental Racism and the Rise of Environmental Justice in the 1990s." In *Environmental Justice in South Africa,* edited by David A. McDonald, 15–48. Athens: Ohio University Press; Cape Town: University of Cape Town Press, 2002.

King, N. L. "The Exploitation of the Indigenous Forests of South Africa." *Journal of the South African Forestry Association* 6 (April 1941): 26–48.

———. "Historical Sketch of the Development of Forestry in South Africa." *Journal of the South African Forestry Association* 1 (October 1938): 4–16.

Kingon, Rev. John Robert Lewis. "Some Place-Names of Tsolo." *Reports of the South African Association for the Advancement of Science* 4 (1916): 603–19.

Kratz, Corinne. "We've Always Done It Like This ... Except for a Few Details: 'Tradition' and 'Innovation' in Okiek Ceremonies." *Comparative Studies in Society and History* 35, no. 1 (1993): 30–65.

Kreike, Emmanuel. *Re-Creating Eden: Land Use, Environment, and Society in Southern Angola and Northern Namibia.* Portsmouth, N.H.: Heinemann, 2004.

Kropf, Albert. A *Kaffir-English Dictionary*. Stutterheim: Lovedale Mission Press, 1899.

Kuckertz, Heinz. *Creating Order: The Image of the Homestead in Mpondo Social Life*. Johannesburg: Witwatersrand University Press, 1990.

Kunene, Mazisi. *Emperor Shaka the Great: A Zulu Epic*. London: Heinemann, 1979.

la Hausse, Paul. "Oral History and South African Historians." In *History from South Africa: Alternative Visions and Practices*, edited by Joshua Brown, Patrick Manning, Karin Shapiro, and Jon Wiener, 342–50. Philadelphia: Temple University Press, 1991.

Land and Agriculture Policy Centre. *Overview of the Transkei Sub-Region of the Eastern Cape Province*. Johannesburg: Land and Agriculture Policy Centre, February 1995.

Leach, Gerald, and Robin Mearns. *Beyond the Woodfuel Crisis: People, Land and Trees in Africa*. London: Earthscan Publications, 1988.

Leach, Melissa. *Rainforest Relations: Gender and Resource Use Among the Mende of Gola, Sierra Leone*. Washington, D.C.: Smithsonian Institution Press, 1994.

Leach, Melissa, and James Fairhead. "Fashioned Forest Pasts, Occluded Histories? International Environmental Analysis in West African Locales." In "Forests: Nature, People, Power," special issue, *Development and Change* 31 (2000): 35–59.

Leach, Melissa, and Cathy Green. "Gender and Environmental History: From Representation of Women and Nature to Gender Analysis of Ecology and Politics." *Environment and History* 3, no. 3 (1997): 343–70.

Leach, Melissa, and Robin Mearns, eds. *The Lie of the Land: Challenging Received Wisdom on the African Environment*. London: International African Institute, 1996.

Leach, Melissa, Robin Mearns, and Ian Scoones. "Environmental Entitlements: Dynamics and Institutions in Community-Based Natural Resource Management." *World Development* 27, no. 2 (1999): 225–47.

Lieberman, Daniel. "Ethnobotanical Assessment of the Dwesa and Cwebe Nature Reserves." In *Indigenous Knowledge, Conservation Reform...*, Institute of Social and Economic Research, 40–84. Grahamstown, South Africa: 1997.

Lipton, Michael, Frank Ellis, and Merle Lipton, eds. *Land, Labour and Livelihoods in Rural South Africa*, vol. 2, *KwaZulu-Natal and Northern Province*. Durban: Indicator Press, University of Natal, 1996.

Lonsdale, John. "The Moral Economy of Mau Mau: Wealth, Poverty and Civic Virtue in Kikuyu Political Thought." In *Unhappy Valley: Conflict in Kenya and Africa*, edited by Bruce Berman and John Lonsdale, chap. 12. Athens: Ohio University Press, 1992.

Lowood, Henry E. "The Calculating Forester: Quantification, Cameral Science, and the Emergence of Scientific Forestry Management in Germany." In *The Quantifying Spirit in the Eighteenth Century*, edited by Tore Frängsmyr, John

L. Heilbron, and Robin E. Rider, 315–42. Berkeley: University of California Press, 1990.

Luckhoff, H. A. "The Story of Forestry and its People." In *Our Green Heritage*, edited by W. F. E. Immelman, C. L. Wicht, and D. P. Ackerman, chap. 4. Cape Town: Tafelberg, 1973.

Luig, Ute, and Achim von Oppen, eds. "The Making of African Landscapes." Special issue, *Paideuma: Mitteilungen zur Kulturkunde* 43 (1997).

Maack, Pamela. "'We Don't Want Terraces!': Protest and Identity under Uluguru Land Usage Scheme." In *Custodians of the Land: Ecology and Culture in the History of Tanzania*, edited by Gregory Maddox, James Giblin, and Isaria Kimambo, 152–74. Athens: Ohio University Press, 1996.

Mackenzie, Fiona. *Land, Ecology and Resistance in Kenya, 1880–1952*. Edinburgh: Edinburgh University Press, 1998.

Mackenzie, John M., ed. *Imperialism and the Natural World*. Manchester: Manchester University Press, 1990.

Maddox, Gregory, James Giblin, and Isaria Kimambo, eds. *Custodians of the Land: Ecology and Culture in the History of Tanzania*. Athens: Ohio University Press, 1996.

Mafeje, Archie. "Religion, Class and Ideology in South Africa." In *Religion and Social Change in Southern Africa*, edited by Michael Whisson and Martin West, 164–84. Cape Town: David Philip, 1975.

Mager, Anne. "'The People Get Fenced': Gender, Rehabilitation and African Nationalism in the Ciskei and Border Region, 1945–1955." *Journal of Southern African Studies* 18, no. 4 (December 1992): 761–82.

———. "Youth Organisations and the Construction of Masculine Identities in the Ciskei and Transkei, 1945–1960." *Journal of Southern African Studies* 24, no. 4 (December 1998): 653–67.

Mamdani, Mahmood. *Citizen and Subject: Contemporary Africa and the Legacy of Late Colonialism*. Princeton, N.J.: Princeton University Press, 1996.

Mandala, Elias. *Work and Control in a Peasant Economy: A History of the Lower Tchiri Valley in Malawi, 1859–1960*. Madison: University of Wisconsin, 1990.

Mann, Kristin, and Richard Roberts, eds. *Law in Colonial Africa*. Portsmouth, N.H.: Heinemann, 1991.

Marks, Shula. *The Ambiguities of Dependence in South Africa: Class, Nationalism, and the State in Twentieth-Century Natal*. Johannesburg: Ravan Press, 1986.

———. *Reluctant Rebellion: The 1906–8 Disturbances in Natal*. Oxford: Clarendon Press, 1970.

Mason, M. H. "Dearth in the Transkei." *Nineteenth Century and After* 73 (January–June 1913): 667–81.

Mason, Michael. *Environmental Democracy*. London: Earthscan Publications, 1999.

Matanzima, Kaiser. "Chief Matanzima Talks to Us about Nature Conservation in the Transkei." *African Wildlife* 29, no. 2 (1975): 14–15

Mayer, Philip, and Iona Mayer. *Report of Research on the Self Organisation by Youth Among the Xhosa-Speaking Peoples of the Ciskei and Transkei, Part 1: The Red Xhosa*. Rhodes University, Institute of Social and Economic Research, November 1972.

Mbeki, Govan. *South Africa: The Peasants' Revolt*. Harmondsworth: Penguin Press, 1964.

Mbilinyi, Marjorie. "'I'd Have Been a Man': Politics and the Labor Process in Producing Personal Narratives." In *Interpreting Women's Lives: Feminist Theory and Personal Narratives*, edited by Personal Narratives Group, 204–27. Bloomington: Indiana University Press, 1989.

McAllister, Patrick A. *Building the Homestead: Agriculture, Labour and Beer in South Africa's Transkei*. Leiden: African Studies Centre and Ashgate, 2001.

———, ed. *Culture and the Commonplace: Anthropological Essays in Honour of David Hammond-Tooke*. Johannesburg: Witwatersrand University Press, 1997.

———. "Resistance to 'Betterment' in the Transkei: A Case Study from Willowvale District." *Journal of Southern African Studies* 15, no. 2 (January 1989): 346–68.

McCann, James. *People of the Plow: An Agricultural History of Ethiopia, 1800–1990*. Madison: University of Wisconsin Press, 1995.

McClendon, Thomas. *Genders and Generations Apart in South Africa: Labor Tenants and Customary Law in Segregation-Era Natal, 1920s to 1940s*. Portsmouth, N.H.: Heinemann, 2002.

McCord, J. B. "The Zulu Witch Doctor and Medicine-Man." *South African Journal of Science* 15, no. 5 (1918): 306–18.

McDonald, David A., ed. *Environmental Justice in South Africa*. Athens: Ohio University Press; Cape Town: University of Cape Town Press, 2002.

McGregor, JoAnn. "Conservation, Control and Ecological Change: The Politics and Ecology of Colonial Conservation in Shurugwi, Zimbabwe." *Environment and History* 1, no. 3 (1995): 257–79.

McLaren, James. *Concise Kaffir-English Dictionary*. London: Longmans, Green, and Co., 1915.

Mearns, Robin. "Institutions and Natural Resource Management: Access to and Control over Woodfuel in East Africa." In *People and Environment in Africa*, edited by Tony Binns, 103–14. Chichester: John Wiley and Sons, 1995.

Meintjies, Helen. "Trends in Natural Resource Management: Policy and Practice in Southern Africa." Working Paper 22. Johannesburg: Land and Agriculture Policy Centre, August 1995.

Miller, O. B. "Notes on the Distribution of Species in Natural Forests of the Transkeian Conservancy." *South African Journal of Natural History* 3, no. 2 (1921–22): 20–23.

Minkley, Gary, and Ciraj Rassool. "Orality, Memory, and Social History in South Africa." In *Negotiating the Past: The Making of Memory in South Africa*, edited by Sarah Nuttall and Carli Coetzee. Cape Town: Oxford University Press, 1998.

Moll, Terrence. *No Blade of Grass: Rural Production and State Intervention in Transkei, 1925–1960*. Cambridge African Occasional Papers 6. Cambridge: African Studies Centre, 1988.

Monson, Jamie. "Canoe-Building under Colonialism: Forestry and Food Policies in the Inner Kilombero Valley 1920–40." In *Custodians of the Land: Ecology and Culture in the History of Tanzania*, edited by Gregory Maddox, James Giblin, and Isaria Kimambo, 200–12. Athens: Ohio University Press, 1996.

Moore, Donald S. "Marxism, Culture, and Political Ecology: Environmental Struggles in Zimbabwe's Eastern Highlands." In *Liberation Ecologies: Environment, Development, Social Movements*, edited by Richard Peet and Michael Watts, 125–47. New York: Routledge, 1996.

———. "Subaltern Struggles and the Politics of Place: Remapping Resistance in Zimbabwe's Eastern Highlands." *Cultural Anthropology* 13, no. 3 (August 1998): 344–81.

Moore, Henrietta, and Megan Vaughan. *Cutting Down Trees: Gender, Nutrition, and Agricultural Change in the Northern Province of Zambia, 1890–1990*. Portsmouth, N.H.: Heinemann, 1994.

Moore, Sally Falk. *Social Facts and Fabrications: "Customary" Law on Kilimanjaro, 1880–1980*. Cambridge: Cambridge University Press, 1986.

Morrow, Sean, and Nwabisa Vokwana. "'Oh! Hurry to the River': The Meaning of uMamlambo Models in the Tyumie Valley, Eastern Cape." *Kronos: The Journal of Cape History* 30 (2004): 184–99.

Mukamuri, B. B. "Local Environmental Conservation Strategies: Karanga Religion, Politics and Environmental Control." *Environment and History* 1, no. 3 (1995): 297–311.

Munslow, Barry, with Yemi Katerere, Adriaan Ferf, and Phil O'Keefe. *The Fuelwood Trap: A Study of the SADCC Region*. London: Earthscan Publications, 1988.

Mzamane, Godfrey I. "Some Medicinal, Magical and Edible Plants Used Among Some Bantu Tribes in South Africa." *Fort Hare Papers* 1, no. 1 (June 1945): 29–35.

Neumann, Roderick. *Imposing Wilderness: Struggles over Livelihood and Nature Preservation in Africa*. Berkeley: University of California Press, 1998.

Newell, Diane. *Tangled Webs of History: Indians and the Law in Canada's Pacific Coast Fisheries*. Toronto: University of Toronto Press, 1993.

Ntsebeza, Lungisile. *Land Tenure Reform, Traditional Authorities and Rural Local Government in Post-Apartheid South Africa: Case Studies from the Eastern Cape*. Research Report 3. Cape Town: Programme for Land and Agrarian Studies, University of the Western Cape, 1999.

Ntshona, Zolile. *Valuing the Commons: Rural Livelihoods and Communal Rangeland Resource in the Maluti District, Eastern Cape*. Research Report 13. Cape Town: Programme for Land and Agrarian Studies, University of the Western Cape, November 2002.

Opland, Jeff. "The Publication of A. C. Jordan's Xhosa novel, *Ingqumbo yeminyanya* (1940)." *Research in African Literatures* 21, no. 4 (Winter 1990): 135–57.

Palmer, Eve, and Norah Pitman. *Trees of Southern Africa.* 3 vols. Cape Town: A. A. Balkema, 1972.

Palmer, Robin, Herman Timmermans, and Derick Fay, eds. *From Conflict to Negotiation: Nature-Based Development on South Africa's Wild Coast.* Pretoria: Human Sciences Research Council; Grahamstown: Institute of Social and Economic Research, Rhodes University, 2002.

Parry, Richard. "'In a Sense Citizens, But Not Altogether Citizens . . .': Rhodes, Race, and the Ideology of Segregation at the Cape in the Late Nineteenth Century." *Canadian Journal of African Studies* 17, no. 3 (1983): 377–91.

Peires, Jeffrey. *The Dead Will Arise: Nongqawuse and the Great Xhosa Cattle-Killing Movement of 1856–7.* Johannesburg: Ravan Press, 1989.

———. *The House of Phalo: A History of the Xhosa People in the Days of Their Independence.* Berkeley: University of California Press, 1981.

Pels, Peter. "The Anthropology of Colonialism: Culture, History and the Emergence of Western Governmentality." *Annual Review of Anthropology* 26 (1997): 163–83.

———. "The Pidginization of Luguru Politics: Administrative Ethnography and the Paradoxes of Indirect Rule." *American Ethnologist* 23, no. 4 (1996): 738–61.

Peluso, Nancy Lee, and Peter Vandergeest. "Genealogies of the Political Forest and Customary Rights in Indonesia, Malaysia, and Thailand." *Journal of Asian Studies* 60, no. 3 (August 2001): 761–812.

Peters, Pauline. *Dividing the Commons: Politics, Policy, and Culture in Botswana.* Charlottesville: University Press of Virginia, 1994.

———. "Struggles over Water, Struggles over Meaning: Cattle, Water and the State in Botswana." *Africa* 54, no. 3 (1984): 29–49.

Peterson, Derek. "Translating the Word: Dialogism and Debate in Two Gikuyu Dictionaries." *Journal of Religious History* 23, no. 1 (February 1999): 31–50.

Phoofolo, Pule. "Epidemics and Revolutions: The Rinderpest Epidemic in Late Nineteenth-Century Southern Africa." *Past and Present* 138 (February 1993): 112–43.

Pooley, Elsa. *The Complete Field Guide to the Trees of Natal, Zululand and Transkei.* Pietermaritzburg: Natal Flora Trust Publications, 1993.

Pratt, Mary Louise. *Imperial Eyes: Travel Writing and Transculturation.* London: Routledge, 1992.

———. "Scratches on the Face of the Country; Or, What Mr. Barrow Saw in the Land of the Bushmen." In *"Race," Writing and Difference*, edited by Henry Louis Gates, Jr., 138–62. Chicago: University of Chicago Press, 1986.

Prins, Frans. "Praise to the Bushman Ancestors of the Water: The Integration of San-Related Concepts in the Beliefs and Ritual of a Diviners' Training School in Tsolo, Eastern Cape." In *Miscast: Negotiating the Presence of the Bushmen*, edited by Pippa Skotnes, 211–23. Cape Town: University of Cape Town Press, 1996.

Ranger, Terence. "The Invention of Tradition in Colonial Africa." In *The Invention of Tradition*, edited by Eric Hobsbawm and Terence Ranger, 211–62. Cambridge: Cambridge University Press, 1983.

———. "The Invention of Tradition Revisited: The Case of Colonial Africa." In *Legitimacy and the State in Twentieth-Century Africa*, edited by Terence Ranger and Olufemi Vaughan, 62–111. London: Macmillan Press, 1993.

———. *Voices from the Rocks: Nature, Culture and History in the Matopos Hills of Zimbabwe.* Oxford: James Currey, 1999.

———. "Women and Environment in African Religion: The Case of Zimbabwe." In *Social History and African Environments*, edited by William Beinart and JoAnn McGregor, 72–86. Athens: Ohio University Press; Oxford: James Currey, 2003.

Redding, Sean. "Beer Brewing in Umtata: Women, Migrant Labor, and Social Control in a Rural Town." In *Liquor and Labor in Southern Africa*, edited by Jonathan Crush and Charles Ambler, 235–51. Athens: Ohio University Press, 1992.

———. "A Blood-Stained Tax: Poll Tax and the Bambatha Rebellion in South Africa." *African Studies Review* 43, no. 2 (September 2000): 29–54.

———. "Government Witchcraft: Taxation, the Supernatural, and the Mpondo Revolt in the Transkei, South Africa, 1955–1963." *African Affairs* 95, no. 381 (October 1996): 555–79.

———. "Legal Minors and Social Children: Rural African Women and Taxation in the Transkei, South Africa." *African Studies Review* 36, no. 3 (December 1993): 49–74.

———. "Sorcery and Sovereignty: Taxation, Witchcraft, and Political Symbols in the 1880 Transkeian Rebellion." *Journal of Southern African Studies* 22, no. 2 (June 1996): 249–69.

———. "South African Women and Migration in Umtata, Transkei, 1880–1935." In *Courtyards, Markets, City Streets: Urban Women in Africa*, edited by Kathleen Sheldon, 31–46. Boulder: Westview Press, 1996.

Ribot, Jesse C. "From Exclusion to Participation: Turning Senegal's Forestry Policy Around?" *World Development* 23, no. 9 (1995): 1587–99.

———. "Theorizing Access: Forest Profits along Senegal's Charcoal Commodity Chain." *Development and Change* 29, no. 2 (April 1998): 307–41.

Ribot, Jesse C., and Nancy Lee Peluso. "A Theory of Access." *Rural Sociology* 68, no. 2 (2003): 153–81.

Richards, Paul. "Ecological Change and the Politics of African Land Use." *African Studies Review* 26, no. 2 (June 1983): 1–72.

———. *Indigenous Agricultural Revolution: Ecology and Food Production in West Africa.* London: Unwin, 1985.

Roberts, Richard. "The Case of Faama Mademba Sy and the Ambiguities of Legal Jurisdiction in Early Colonial French Soudan." In *Law in Colonial Africa*, edited by Kristin Mann and Richard Roberts, 185–98. Portsmouth, N.H.: Heinemann, 1991.

Rocheleau, Dianne. "Women, Trees, and Tenure: Implications for Agroforestry." In *Whose Trees?: Proprietary Dimensions of Forestry*, edited by Louise Fortmann and John Bruce, 254–72. Boulder: Westview Press, 1988.

Ross, Robert. *Adam Kok's Griquas: A Study in the Development of Stratification in South Africa*. Cambridge: Cambridge University Press, 1976.

Roux, Edward. *Time Longer than Rope: A History of the Black Man's Struggle for Freedom in South Africa*. 2nd ed. Madison: University of Wisconsin Press, 1964.

Ruiters, Greg. "Race, Place, and Environmental Rights: A Radical Critique of Environmental Justice Discourse." In *Environmental Justice in South Africa*, edited by David A. McDonald, 112–26. Athens: Ohio University Press; Cape Town: University of Cape Town Press, 2002.

Samuelson, R. C. A. *Long, Long Ago*. Durban: Knox, 1929.

Saunders, Christopher. *The Annexation of the Transkeian Territories*. Archives Year Book for South African History 39. Pretoria: Government Printer, 1978.

———. *The Making of the South African Past: Major Historians on Race and Class*. Cape Town: David Philip, 1988.

———. "The Transkeian Rebellion of 1880–81: A Case-Study of Transkeian Resistance to White Control." *South African Historical Journal* 8 (1976): 32–39.

Schmidt, Elizabeth. *Peasants, Traders, and Wives: Shona Women in the History of Zimbabwe, 1870–1939*. Portsmouth, N.H.: Heinemann, 1992.

Schoenbrun, David Lee. *A Green Place, A Good Place: Agrarian Change, Gender, and Social Identity in the Great Lakes Region to the 15th Century*. Portsmouth, N.H.: Heinemann, 1998.

Schroeder, Richard A. *Shady Practices: Agroforestry and Gender Politics in The Gambia*. Berkeley: University of California Press, 1999.

Schunke, H. C. "Kaffraria und die Östlichen Grenzdistrikte der Kapkolonie." *Doktor A. Petermanns Mitteilungen aus Justus Perthes' Geographischer Anstalt* (Gotha: Justus Perthes) 31 (1885): 161–71, 201–09, tafel 9.

———. "The Transkeian Territories: Their Physical Geography and Ethnology." *Transactions of the South African Philosophical Society* 8, part I (1893): 1–15.

Scott, James C. *Seeing Like a State: How Certain Schemes to Improve the Human Condition Have Failed*. New Haven, Conn.: Yale University Press, 1998.

Sedding, E. D., ed. *Godfrey Callaway: Missionary in Kaffraria, 1892–1942*. London: Society for Promoting Christian Knowledge, 1945.

Shackleton, Charlie, Sheona Shackleton, and Ben Cousins. "The Role of Land-Based Strategies in Rural Livelihoods: The Contribution of Arable Production, Animal Husbandry and Natural Resource Harvesting in Communal Areas in South Africa." *Development Southern Africa* 18, no. 5 (December 2001): 581–604.

Shone, A. K. *Forestry in Transkei*. Umtata: Transkei Department of Agriculture and Forestry, 1985.

Sim, Thomas R. *The Forests and Forest Flora of the Cape Colony of Good Hope*. Aberdeen: Taylor and Henderson, 1907.

Simon, Chris, and Masilo Lamla. "Merging Pharmacopoeia: Understanding the Historical Origins of Incorporative Pharmacopoeial Processes among Xhosa Healers in Southern Africa." *Journal of Ethnopharmacology* 33, no. 3 (July 1991): 237–242.

Sivaramakrishnan, K. "Colonialism and Forestry in India: Imagining the Past in Present Politics." *Comparative Studies in Society and History* 37, no. 1 (January 1995): 3–40.

———. "State Sciences and Development Histories: Encoding Local Forestry Knowledge in Bengal." *Development and Change* 31, no. 1 (January 2000): 61–89.

Smith, Ken. *The Changing Past: Trends in South African Historical Writing.* Athens: Ohio University Press, 1988.

Soga, John Henderson. *The Ama-Xosa: Life and Customs.* Lovedale, South Africa: Lovedale Press, 1932.

Southall, Roger. *South Africa's Transkei: The Political Economy of an "Independent" Bantustan.* New York: Monthly Review Press, 1983.

Southall, Roger, and Zosa de Sas. "Containing the Chiefs: The ANC and Traditional Leaders in the Eastern Cape." Working Paper 39. Fort Hare Institute of Social and Economic Research, August 2003.

Spear, Thomas. *Mountain Farmers.* Berkeley: University of California Press, 1997.

Spiegel, Andrew D. "Struggling with Tradition in South Africa: The Multivocality of Images of the Past." In *Social Construction of the Past: Representation as Power,* edited by George Bond and Angela Gilliam, 185–202. London: Routledge, 1994.

Stanford, W. E. *The Reminiscences of Sir Walter Stanford.* Edited by J. W. Macquarrie. 2 vols. Cape Town: van Riebeeck Society, 1958–62.

Stapleton, Timothy. "The Expansion of a Pseudo-Ethnicity in the Eastern Cape: Reconsidering the Fingo 'Exodus' of 1865." *International Journal of African Historical Studies* 29, no. 2 (1996): 233–50.

———. "Oral Evidence in a Pseudo-Ethnicity: The Fingo Debate." *History in Africa* 22 (1995): 358–69.

Stoler, Ann Laura, and Frederick Cooper. "Between Metropole and Colony: Rethinking a Research Agenda." In *Tensions of Empire: Colonial Cultures in a Bourgeois World,* edited by Frederick Cooper and Ann Laura Stoler, 1–56. Berkeley: University of California Press, 1997.

Sullivan, Sian. "Gender, Ethnographic Myths and Community-Based Conservation in a Former Namibian 'Homeland.'" In *Rethinking Pastoralism in Africa: Gender, Culture and the Myth of the Patriarchal Pastoralist,* edited by Dorothy Hodgson, 142–64. Oxford: James Currey, 2000.

Sustainable Livelihoods in Southern Africa. *Natural Resources and Sustainable Livelihoods in South Africa: Policy and Institutional Framework.* Research Briefing 1. Cape Town: Programme for Land and Agrarian Studies, University of the Western Cape, September 2001.

Sustainable Livelihoods in Southern Africa Team. *Decentralisations in Practice in Southern Africa*. Cape Town: Programme for Land and Agrarian Studies, University of the Western Cape, 2003.

———. *Rights Talk and Rights Practice: Challenges for Southern Africa*. Cape Town: Programme for Land and Agrarian Studies, University of the Western Cape, 2003.

Switzer, Les. *Power and Resistance in an African Society: The Ciskei Xhosa and the Making of South Africa*. Madison: University of Wisconsin Press, 1993.

Tapscott, Chris. "Changing Discourses of Development in South Africa." In *Power of Development*, ed. Jonathan Crush, 176–91. New York: Routledge, 1995.

———. "The Rise of 'Development' Thinking in Transkei." *Journal of Contemporary African Studies* 11, no. 2 (1992): 55–75.

Thomas, Nicholas. "The Inversion of Tradition." *American Ethnologist* 19, no. 2 (May 1992): 213–32.

Thomas-Slayter, Barbara, and Dianne Rocheleau, eds. *Gender, Environment, and Development in Kenya: A Grassroots Perspective*. Boulder: Lynne Rienner, 1995.

Thompson, Georgina. "The Dynamics of Ecological Change in an Era of Political Transformations: An Environmental History of the Eastern Shores of Lake St. Lucia." In *South Africa's Environmental History*, edited by Stephen Dovers, Ruth Edgecombe, and Bill Guest, 191–212. Cape Town: David Philip, 2002; Athens: Ohio University Press, 2003.

Tonkin, Elizabeth. *Narrating Our Pasts: The Social Construction of Oral History*. Cambridge: Cambridge University Press, 1992.

Tropp, Jacob. "The Contested Nature of Colonial Landscapes: Historical Perspectives on Livestock and Environments in the Transkei." *Kronos: The Journal of Cape History* 30 (November 2004): 118–37.

———. "Displaced People, Replaced Narratives: Forest Conflicts and Historical Perspectives in the Tsolo District, Transkei." *Journal of Southern African Studies* 29, no. 1 (March 2003): 207–33.

———. "Dogs, Poison and the Meaning of Colonial Intervention in the Transkei, South Africa." *Journal of African History* 43, no. 3 (2002): 451–72.

———. "The Python and the Crying Tree: Interpreting Tales of Environmental and Colonial Power in the Transkei." *International Journal of African Historical Studies* 36, no. 3 (2003): 511–32.

Trouillot, Michel-Rolph. *Silencing the Past: Power and the Production of History*. Boston: Beacon Press, 1995.

Tsing, Anna. "Becoming a Tribal Elder and Other Green Development Fantasies." In *Transforming the Indonesian Uplands: Marginality, Power and Production*, edited by Tania Murray Li, 159–60. Amsterdam: Harwood Academic Publishers, 1999.

Vail, Leroy, ed. *The Creation of Tribalism in Southern Africa*. London: James Currey, 1989.

Vail, Leroy, and Landeg White. *Power and the Praise Poem: Southern African Voices in History*. Charlottesville: University Press of Virginia, 1991.

van Horen, Clive, and Anton Eberhard. "Energy, Environment and the Rural Poor in South Africa." *Development Southern Africa* 12, no. 2 (April 1995): 197–211.

van Onselen, Charles. "Reactions to Rinderpest in Southern Africa, 1896–97." *Journal of African History* 13, no. 3 (1972): 473–88.

Vansina, Jan. "Epilogue: Fieldwork in History." In *In Pursuit of History: Fieldwork in Africa*, edited by Carol Keyes Adenaike and Jan Vansina, 127–40. Portsmouth, N.H.: Heinemann, 1996.

van Sittert, Lance. "'Keeping the Enemy at Bay': The Extermination of Wild Carnivora in the Cape Colony, 1889–1910." *Environmental History* 3, no. 3 (July 1998): 333–56.

———. "The Nature of Power: Cape Environmental History, the History of Ideas and Neoliberal Historiography." *Journal of African History* 45, no. 2 (2004): 305–13.

———. "'Our Irrepressible Fellow Colonist': The Biological Invasion of Prickly Pear (*Opuntia ficus-indica*) in the Eastern Cape, c. 1890–c. 1910." In *South Africa's Environmental History*, edited by Stephen Dovers, Ruth Edgecombe, and Bill Guest, 139–59. Cape Town: David Philip, 2002; Athens: Ohio University Press, 2003.

———. "'The Seed Blows About in Every Breeze': Noxious Weed Eradication in the Cape Colony, 1860–1909." *Journal of Southern African Studies* 26, no. 4 (December 2000): 655–74.

von Breitenbach, F., and Jutta von Breitenbach. "Notes on the Natural Forests of Transkei." *Journal of Dendrology* 3, nos. 1 and 2 (1983): 17–53.

von Maltitz, G. P., and G. Fleming. "Status of Conservation of Indigenous Forests in South Africa." In *Towards Sustainable Management Based on Scientific Understanding of Forests and Woodlands: Proceedings of the Natural Forests and Savanna Woodlands Symposium II, September 5–9, 1999, Knysna, South Africa*, edited by A. H. W. Seydack, W. J. Vermeulen, and C. Vermeulen. Knysna, South Africa: Department of Water Affairs and Forestry, 2000.

Wagner, Michele. "Environment, Community and History: 'Nature in the Mind' in Nineteenth and Early Twentieth-Century Buha, Tanzania." in *Custodians of the Land: Ecology and Culture in the History of Tanzania*, edited by Gregory Maddox, James Giblin, and Isaria Kimambo, 175–99. Athens: Ohio University Press, 1996.

Walker, Cherryl. "Gender and the Development of the Migrant Labour System c. 1850–1930: An Overview." In *Women and Gender in Southern Africa to 1945*, edited by Cherryl Walker, 168-96. Cape Town: David Philip, 1990.

Waters, M. W. *Cameos from the Kraal*. Cape Town: Juta, n.d.

Watt, John M., and Maria Breyer-Brandwijk. *The Medicinal and Poisonous Plants of Southern Africa: Being an Account of Their Medicinal Uses, Chemical Composition, Pharmacological Effects and Toxicology in Man and Animal*. Edinburgh: E. & S. Livingstone, 1932.

Webster, Alan. "Unmasking the Fingo: The War of 1835 Revisited." In *The Mfecane Aftermath: Reconstructive Debates in Southern African History*, edited by Carolyn Hamilton, 241–76. Johannesburg: Witwatersrand University Press, 1995.

Westaway, Ashley, and Gary Minkley. "Rural Restitution: The Production of Historical Truth Outside the Academy." Working Paper 45. Fort Hare Institute of
Social and Economic Research, August 2003.

White, Luise. *Speaking with Vampires: Rumor and History in Colonial Africa.*
Berkeley: University of California Press, 2000.

———. "'They Could Make Their Victims Dull': Genders and Genres, Fantasies
and Cures in Colonial Southern Uganda." *American Historical Review* 100, no.
5 (December 1995): 1379–1402.

———. "True Stories: Narrative, Event, History, and Blood in the Lake Victoria
Basin." In *African Words, African Voices: Critical Practices in Oral History*, edited by Luise White, Stephan Miescher, and David Cohen, 281–99. Bloomington: Indiana University Press, 2001.

———. "Tsetse Visions: Narratives of Blood and Bugs in Colonial Northern
Rhodesia, 1931–9." *Journal of African History* 36, no. 2 (1995): 219–45.

Willems-Braun, Bruce. "Buried Epistemologies: The Politics of Nature in (Post)co-
lonial British Columbia." *Annals of the Association of American Geographers*
81, no. 1 (1997): 3–31.

Wilson, K. "Trees in Fields in Southern Zimbabwe." *Journal of Southern African
Studies* 15, no. 2 (January 1989): 369–83.

———. "'Water Used to be Scattered in the Landscape': Local Understandings of
Soil Erosion and Land Use Planning in Southern Zimbabwe." *Environment
and History* 1, no. 3 (1995): 281–96.

Witt, Harald. "The Emergence of Privately Grown Industrial Tree Plantations."
In *South Africa's Environmental History*, edited by Stephen Dovers, Ruth
Edgecombe, and Bill Guest, 93–94. Cape Town: David Philip, 2002; Athens:
Ohio University Press, 2003.

Yawitch, Joanne. *Betterment: The Myth of Homeland Agriculture.* Johannesburg:
South African Institute of Race Relations, 1981.

Yngstrom, Ingrid. "Representations of Custom, Social Identity and Environmental Relations in Central Tanzania." In *Social History and African Environments*, edited by William Beinart and JoAnn McGregor, 175–95. Athens: Ohio
University Press; Oxford: James Currey, 2003.

Zerner, Charles. "Introduction: Toward a Broader Vision of Justice and Nature
Conservation." In *People, Plants, and Justice: The Politics of Nature Conservation*, ed. Charles Zerner, 3–20. New York: Columbia University Press,
2000.

UNPUBLISHED THESES AND PAPERS

Cawe, Sizwe. "Coastal Forest Survey: A Classification of the Coastal Forests of
Transkei and an Assessment of Their Timber Potential." Unpublished paper,
Umtata, University of Transkei, 1990.

————. "A Quantitative and Qualitative Survey of the Inland Forests of Transkei." M.S. thesis, Department of Botany, University of Transkei, January 1986.

Charman, Andrew J. E. "Progressive Élite in Bunga Politics: African Farmers in the Transkeian Territories, 1903–1948." Ph.D. thesis, Cambridge University, 1999.

Crickmay, Venter and Associates. "Transkei Forestry Strategic Development Plan: The Future of Forestry in Transkei." Unpublished paper, prepared for the Transkei Department of Agriculture and Forestry, Pietermaritzburg, 1993.

Edgecombe, Ruth. "The Role of Environmental History in Applied Field Studies in the Centre of Environment and Development at the University of Natal in Pietermaritzburg: The Case of Traditional Hunting with Dogs." Paper presented at "African Environments: Past and Present," joint conference of the *Journal of Southern African Studies* and St. Antony's College, Oxford University, Oxford, England, 5–8 July 1999.

Fay, Derick. "Paternalist and Commercial Visions of Forest Conservation in South Africa: Intergovernmental Disputes and African Agency in the Coastal Transkei, ca. 1890–1936." Paper presented at "Producing and Consuming Narratives," American Society for Environmental History annual meeting, Denver, Colorado, 20–24 March 2002.

Hendricks, Fred. "The Pillars of Apartheid: Land Tenure, Rural Planning and the Chieftaincy." Ph.D. thesis, Uppsala University, 1990.

Heron, G. "Household Production and the Organisation of Co-operative Labour in Shixini, Transkei." M.A. thesis, Rhodes University, Grahamstown, 1989.

Hirst, Manton. "The Healer's Art: Cape Nguni Diviners in the Townships of Grahamstown." Ph.D. thesis, Department of Anthropology, Rhodes University, December 1990.

Lamla, C. M. "Present-Day Diviners (Ama-Gqira) in the Transkei." M.A. thesis, Department of African Studies, University of Fort Hare, 1975.

McKenzie, Bruce. "Ecological Considerations of Some Past and Present Land Use Practices in Transkei." Ph.D. thesis, Botany Department, University of Cape Town, 1984.

Mears, W. J. G. "A Study in Native Administration: The Transkeian Territories, 1894–1943." D.Litt., University of South Africa, 1947.

Ndima, Mlungisi. "A History of the Qwathi People from Earliest Times to 1910." M.A. thesis, Rhodes University, 1988.

Ndlovu, Sifiso Mxolisi. "African Public Intellectuals and the Production of Historical Knowledge in 'Pre-colonial' Times: A Case Study of the Making of the Archive of King Dingane kaSenzangakhona." Paper presented at the Triennial Conference of the European Association for Commonwealth Language and Literature Studies, University of Copenhagen, 21–27 March 2002.

Ntsebeza, Lungisile. "Decentralisation and Natural Resource Management in Rural South Africa: Problems and Prospects." Paper presented at the ISACP bi-annual conference, Victoria Falls, Zimbabwe, 17–21 June 2002.

Obiri, J. F., and M. J. Lawes. "Challenges Facing New Forest Policies in South Africa: Attitudes of Forest Users Toward Management of the Coastal Forests of the Eastern Cape Province." Paper presented at DWAF Natural Forests and Savanna Woodlands Symposium, 2002.

Simon, Christian M. "Dealing with Distress: A Medical Anthropological Analysis of the Search for Health in a Rural Transkeian Village." M.A. thesis, Department of Anthropology, Rhodes University, December 1989.

Tropp, Jacob. "Roots and Rights in the Transkei: Colonialism, Natural Resources, and Social Change, 1880–1940." Ph.D. thesis, Department of History, University of Minnesota, June 2002.

Wagenaar, Elsie J. C. "A History of the Thembu and Their Relationship with the Cape, 1850–1900." Ph.D. thesis, Rhodes University, 1988.

CITED INTERVIEWS

Boyana, Wele. Baziya Mission, Umtata District, 12 January 1998.

Cutshwa, Vandras. Tabase Mission, Umtata District, 4 February 1998.

Dubo, Jemimah. Payne, Umtata District, 24 February 1998.

Dyayiya, Sampson, and Notozamile Dyayiya. Silverton, Umtata District, 27 February 1998.

Fiko, Emmie. Egerton, Umtata District, 23 February 1998.

Gcanga, Alice. Manzana, Engcobo District, 11 February 1998.

Gcanga, Dabulamanzi. Manzana, Engcobo District, 5 February 1998.

Gcanga, Lindiwe, Nobantu Nomganga, and Mrs. Ncedani. Manzana, Engcobo District, 5 February 1998.

Gqweta, A. K. Baziya Mission, Umtata District, 8 January 1998.

Gqweta, Rev. Matthew. Baziya Mission, Umtata District, 12 January 1998.

Gqweta, Tozama. Baziya Mission, Umtata District, 12 January 1998.

Joyi, Anderson. Mputi, Umtata District, 23 December 1997.

Jumba, Mvulayehlobo, and Charlie Banti. Tabase, Umtata District, 27 January 1998.

Jumba, William, and Zwelivumile Quvile. Tabase, Umtata District, 4 February 1998.

Jumba, William, Tsikitsiki Nodwayi, Festo Sonyoka, and Zwelivumile Quvile. Tabase, Umtata District, 4 February 1998.

Kholwane, Mambanjwa. Preston Farms, Umtata District, 24 February 1998.

Kholwane, Nozolile. Fairfield, Umtata District, 23 February 1998.

Maka, Polisile. Manzana, Engcobo District, 28 January 1998.

Matomela, Eunice. Springvale, Umtata District, 23 February 1998.

Mbabama, G. N. Umtata, 2 February 1998.

Mdaka, R. T. S. Manzana, Engcobo District, 11 February 1998.

Mvambo, Cyprian. Umtata, 3 February 1998.

Ndzungu, Liziwe, and Monwabisi Ndzungu. Payne, Umtata District, 24 February 1998.

Ngadlela, Martha. Manzana, Engcobo District, 11 February 1998.

Ngombane, Mlungisi. Egerton, Umtata District, 25 February 1998.

Ngombane, W. M. Mputi, Umtata District, 8 January 1998.

Ntwalana, Albertina. Lindile, Umtata District, 24 February 1998.

Qina, Samuel. Tabase, Umtata District, 27 January 1998.

Qupa, Adolphus. Baziya Mission Station, Umtata District, 8 January 1998.

Rangana, Noheke. Manzana, Engcobo District, 28 January 1998.

Sithelo, Fumanekile, and Ntombizanele Sithelo. Kaplan, Umtata District, 10 February 1998.

Sithelo, Headman. Kaplan, Umtata District, 25 February 1998.

Somana, Abel. All Saint's Mission, Engcobo District, 5 February 1998.

Tyekela, Olga. Fairfield, Umtata District, 23 February 1998.

Vika, Rev. G. Etyeni, Tsolo District, 2 February 1998.

Index

abakhwetha, forest conservation and, 126–32, 139
access analysis, environmental relations and, 8
Adams, Thomas, 45–46, 86
African National Congress (ANC), 1, 164
Agriculture and Forestry, Department of, 108–9
amahlathi, 13
amayeza, 133
axes, fuelwood and, 85–86

Bambatha rebellion, 92
Banti, Charlie, 140
Bantustan system, 6
Barry, R. D., 101
Baur, R., 81
Beinart, William, 7, 37
Berry, Sara, 8, 13
black stinkwood, 147
black wattle (*Acacia mearnsii*), 107
Blohm, W., 136, 137–38, 139
Boyana, Wele, 101–2
Braun, Bruce, 160
Broster, Joan, 135, 153, 155
  Red Blanket Valley, 153–54
Brownlee, Frank, 143
buffer zones, 34–35
Bundy, Colin, 7

Callaway, Henry, 150
Cape Agricultural Journal, 68
Cape Colony, 2–3, 16–17, 34, 66
Cape python, 156

Caplen, H. L., 74
Carlson, Knut, 68
Carmichael, Walter, 120
charms, 141–44
colonial government
  appointed headmen and, 36–37
  legal institutions and, 41–43
  policies of, 34
  resource control and, 40–42, 45–52, 93–95
  in Transkei, 15–17
council system, 52–53
Crehan, Kate, 19

Dalasile (chief), 33
Dalindyebo (paramount chief), 40–42, 67
death rituals, 140–41
Dingane (chief), 150
domestic tree production, 98
Drakensberg Range, 15
du Plessis, W. M., 115–16
Dyayiya, Sampson amd Notozamile, 134

Eastern Cape Province, 1–2, 165–66
East Griqualand, 16, 33, 53
economic restructuring. *See under* gender
Elliot, Henry, 42, 75–76, 87
  on hut tax, 77
Emigrant Thembuland, 16
"Enchanted Vlei, The" (Henkel), 147–48
Engcobo District, 34, 36, 42–43, 58–59, 143, 147

environmental authority
  headmen and, 32
  official views of, 31–32
environmental disputes, recent, 163–65
environmental justice, concept of, 166
environmental relations, study of, 8
environmental rights, gender and,
    11–13
eucalyptus trees, 108

Fiko, Emmie, 135
Fingoland, 16, 38, 53, 105
Fletcher, Mount, 15
Fodo (headman), 37–38
forest access, 13–14, 162
  chiefs, headmen and, 54
  colonialism and, 3–4, 7–9
  cultural importance of, 162
  economic and ecological stresses
    and, 82–84, 161–62
  gender and, 81–82
  partitioning of, 34–35
  "subsistence" and, 72–73
Forest Act of 1888, 65–72, 95, 158
forest administration policies, 17–18, 21
Forest Department, 38, 54, 161–63
  free permit system and, 74–75
  versus Native Affairs, Department
    of, 65–72, 161
  strategies, 65, 89, 95–96
forest guards, African, 31–32, 48–50
  headmen and, 51–52
forest management, customary author-
    ity and, 56
forest officers, European, 31–32, 45–46,
    48–50
Forests and Forest Flora of the Colony
    of the Cape of Good Hope (Sim),
    93
free permit system, 91–92, 93
  dual control and, 72–73
  Forest Department and, 74–75
  headmen and, 73–74
  Henkel and, 73–74
  hut tax and, 76–77
  Native Affairs, Department of, and,
    74–75

  negotiations over, 79
  popular responses to, 74–76
  Proclamation 209 of 1890 and, 72
fuelwood, 80–88, 107–21, 161–62
  access and availability, 112–16
  axes and, 85–86
  colonial policies, 81, 116–17
  everyday negotiations, 117–20
  government sales of, 109–11
  Henkel and, 85
  supplies, 111–12
  women and, 109–10, 118–20, 162
  work parties and, 119

Gazini (umgzina, Curtisia dentata),
    142
Gcaleka Xhosa, 156
Gcanga, Alice, 136
Gcanga, Dabulamanzi, 58, 108, 130,
    131–32
Gcanga, Vetu, 57
gender
  economic restructuring and, 63–64,
    79–80, 100
  environmental rights and, 11–13
  forest use and, 81–82, 98–99
Giles-Vernick, Tamara, 14
Glen Grey Act, 52–53
Godfrey, Robert, 22–23, 135, 143–44,
    156, 157
Gqogqora, 25–26
Grove, Richard, 7, 17

Hammond-Tooke, William, 70–71
headmen, 31–32
  appointed, 36–37
  conflicts and, 38–40
  forest guards and, 51–52
  free permit system and, 73–74
headmen's forest system, 55–61, 161
  corruption in, 59, 62
  problems of, 62
  women and, 59–61
healing and protection
  tree species and, 132–41
  umthombothi (sandalwood, Aca-
    lypha glabrata) and, 136

Henkel, Caesar, 44, 45, 54, 65, 85
    African beliefs and, 146–48
    on African forest guards, 48
    on dual control, 68–69, 72–73
    "Enchanted Vlei, The," 147–48
    free permit system and, 73–74
    on hut tax, 78
    Nocu Forest and, 150, 155–56
    snake isilo and, 156–58
Heywood, A. W., 47, 71, 86, 91–92
    lemonwood tree and, 104–5, 106
Hirst, Manton, 154, 156
Hunt, Nancy Rose, 9
hut tax, 64, 76–77
hut wattle, 108
    purchase of, 98–100

Idutywa Reserve, 16, 101
igqirha, 135, 142, 152, 154–55
    Nome and, 151–53
Imvo Zabantsundu, 66
Innes, James Rose, 69
interviews, 3
iqwili (Alepidea amatymbica), 133
isindiyandiya (bastard sneezewood,
        Bersama tysoniana), 142–44
isuthu, 126–27
izidumo (water tree, Ilex mitis), 143–44

Jameson, Leander S., 54
Jiyajiya (councillor), 109
Jordan, A. C., 156
    Wrath of the Ancestors, 156–57
Joyi, Anderson, 101

Kambi, 38
Keet, J. D., 115
Kentani District, 96
Kholwane, Nozolile, 108, 119
kraalwood problem, 104–7
KwaMatiwane, defined, 19–21

labor reserve economics, 79–80, 84
Lamla, C. M., 143
landscapes, ritual and, 145, 159
lemonwood tree (umvete or uvete, Xy-
        malos monospora), 104–7

Lengisi, Charlie, 126–27
Lister, Joseph, 91, 93–94, 106
love potions, 142–43

Macingwane, Joseph, 48
Mager, Anne, 127–28
majola (mole snake, inkwakhwa), 156
Makaula (chief), 44, 67
male initiation, 126–32
Mandela, Nelson, 1
Mangala (chief), 97
Masiza, L. W., 56
Matomela, Eunice, 134, 135, 138–39
Mbeki, Thabo, 1
Mdaka, R. T. S., 60–61, 118, 128–31, 153
Mditshwa (chief), 33
medicines. See amayeza
Mfengu, 33–34, 37–38, 39–40
Mgudhlwa (chief), 75–76
Mhlontlo (chief), 33
missionaries, 33
Mkwenkwe (headman), 56
Moiketsi, Lebenya, 55–56
Moravian missionaries, 22
Mpondomise, 34, 39–40

Nationalist Party, 6
National Trust legislation, 162
Native Affairs, Department of, 54, 55
    versus Forest Department, 65–72,
        161
    free permit system and, 74–75
    priorities, 65–66
Native Economic Commission, 114
Natives Land Act, 95
Natives Trust and Land Act, 18
Ndayi, Sam, 51
Ndlebe, Bikwe, 51
Ndzungu, Liziwe, 153, 155
Ndzungu, Monwabisi, 135, 155
negotiation, 9–10
Ngangelizwe (paramount chief), 33, 41
Ngombane, Mlungisi, 140–41, 149,
        151–52
Ngudle, Thomas, 38–40, 44
Ninth Frontier War, 16
Nkala, Paul, 38

Nkonya (forest guard), 49, 51
Nocu Forest, 149, 155, 159
Nodwayi, Tsikitsiki, 101, 132–33
Nome, 151–59
    igqirha and, 151–53
    ritual significance of, 154–55
Nompintsho (forest guard), 50
Nongauza (councillor), 97, 102
Nota (councillor), 60
Nqweniso (headman), 44

Pamla, Charles, 60
Peluso, Nancy, 8, 13
periodization, 18
Pondoland, 16, 97
Prins, Frans, 156
Proclamation 43 of 1906, 92
Proclamation 209 of 1890, 65–66, 72
    popular resentment against, 66–67,
        70
Proclamation 299 of 1908, 55
Proclamation 308 of 1890, 65–66
    popular resentment against, 66–67,
        70
Proclamation 388 of 1896, 67–68
pythons, 155–56

Qina, Samuel, 101, 126–27
Qumbu District, 37
Qwathi, 35
Qwathi chiefdom, 33

Rangana, Noheke, 118
Raymond, A. L., 127
rebellion of 1880–81, 16, 21, 33–34, 149,
        158
    economic disparities and, 36
    postwar resettlement, 33–38
Red Blanket Valley (Broster), 153–54
Redding, Sean, 76–77, 78
research context, 4–7, 10–11, 13–14
research methods and sources, 19,
        21–26
    oral history, 23–26, 145
resource control
    African authority and custom, 40–42
    bribery, extortion and, 50–51
        colonial government and, 40–42,
            45–52
        socioeconomic and ecological pres-
            sures and, 46–48
Rhodes, Cecil, 67
Ribot, Jesse, 8, 13
rinderpest, 46, 71, 83

Sakwe (councillor), 101
Sarhili (chief), 65
Schreiner, William P., 71
Schunke, H. C., 146–47
Scott, John, 42–43
Shaka, 150
Silimela (chief), 42–43
Sim, Thomas, 93
    Forests and Forest Flora of the
        Colony of the Cape of Good
        Hope, 93
Simmons, B. R., 114
Sithelo, Fumanekile and
        Ntombizanele, 127, 137, 151
Soga, J. H., 142–43
Somana, Tandi, 24
source materials, 21–26
South African War, 47, 71
"Stages of Forest Destruction," 163
Stanford, A. H., 35
Stanford, Walter, 35–36, 43, 70–72, 81,
        92, 146
    fuelwood and, 84
stick fighting, 128–32
"subsistence" wood rights, 79, 86–88,
        161

tariff system, 92
teza. See fuelwood
Thembu, 34, 37–38
Thembuland, 33, 40–41
    chiefs' negotiation strategies in,
        41–43
Transkei
    African societies in, 15–16, 17
    colonialism in, 2–3, 15–17
    definition, 2
    environmental relations, study of, 8
    historical writing on, 7

as labor reserve, 3, 79–80
natural resource access, 6
physical geography, 14–15
resource use, society and culture
and, 14
Transkeian Conservancy, 38, 64
Transkeian Territories General Council (TTGC), 53, 54, 55–56, 101
trees, 13, 162
amayeza and, 133
black stinkwood, 147
black wattle (*Acacia mearnsii*), 107
charms and, 141–44
cultural importance of, 125–26
domestic production of, 98
eucalyptus, 107
Gazini (umgzina, *Curtisia dentata*),
142
healing and protection and, 132–41
iqwili (*Alepidea amatymbica*), 133
isindiyandiya (bastard sneezewood,
Bersama tysoniana), 142–44
izidumo (water tree, *Ilex mitis*),
143–44
lemonwood (umvete or uvete, *Xymalos monospora*), 104–7
love potions and, 142–43
male initiation and, 126–32
Sindiandiya (isindiyandiya, bastard
sneezewood, *Bersama tysoniana*),
142
stick fighting and, 128–32
ugqonci (*Trichocladus ellipticus*),
130
uluthongothi (gardenia, *Hyperacanthus amoenus*), 138–39
umabophe (*Plumbago capensis*),
142–43
umceya (real yellowwood, *Podocarpus latifolius*), 109
umdlebe (Dead Man's Tree,
Synadenium cupulare and other
species), 150–53
umhlonyane (wild wormwood,
Artemisia caffra), 133, 137, 142–43
umkhoba (common yellowwood,
Podocarpus falcatus), 109

umkhwenkwe (cheesewood, *Pittosporum viridiflorum*), 133, 137
umnonono (hard pear, *Strychnos
henningsii*), 130, 133
umnqayi (blackwood, *Gymnosporia
peduncularis*), 130, 140
umphafa (buffalo thorn, *Zizyphus
mucronata*), 140–41
umthathi (sneezewood, *Ptaeroxylon
obliquum*), 133, 135–36, 137
umthombothi (sandalwood, *Acalypha glabrata*), 137
umvete (lemonwood, *Xymalos
monospora*)
umzane (white ironwood, *Vepris
lanceolata*), 130, 137
uvelabahleke, 142
uvete (lemonwood, *Xymalos monospora*)
"weaning the native" and, 103–9
Tsolo District, 19, 34–35, 44, 48,
127–28, 156–57
Tyekela, Olga, 119, 134, 136

ugqonci (*Trichocladus ellipticus*), 108
ukubethelela, 135–36
ukufutha practices, 137–38
uluthongothi (gardenia, *Hyperacanthus amoenus*), 138–39
umabophe (*Plumbago capensis*), 142–43
umceya (real yellowwood, *Podocarpus
latifolius*), 109
umdlebe (Dead Man's Tree, *Synadenium cupulare* and other species),
150–53
umhlonyane (wild wormwood,
Artemisia caffra), 133, 137, 142–43
umkhoba (common yellowwood,
Podocarpus falcatus), 109
umkhwenkwe (cheesewood, *Pittosporum viridiflorum*), 133, 137
umnonono (hard pear, *Strychnos henningsii*), 133
umnqayi (blackwood, *Gymnosporia
peduncularis*), male initiation, 140
umphafa (buffalo thorn, *Zizyphus mucronata*), 140–41

Umtata District, 34–35, 37, 46, 48, 67
umthathi (sneezewood, *Ptaeroxylon
    obliquum*), 102, 103–4, 133, 135–36,
    137
umthombothi (sandalwood, *Acalypha
    glabrata*), 137
umvete (lemonwood, *Xymalos mono-
    spora*)
umzane (white ironwood, *Vepris
    lanceolata*), 137
Umzimkulu District, 78
Union of South Africa, establishment
    of, 95
United Transkeian Territories General
    Council, 115
uvete (lemonwood, *Xymalos monospora*)

van Sittert, Lance, 7

Water Affairs and Forestry, Depart-
    ment of (DWAF), 165
wattle, 98–100, 108

"weaning the native," 91, 120–21
    popular perspectives on, 95–102
    problems with, 96
    tree species and, 103–9
    wood scarcity and, 96–98
women
    forest access and, 59–61
    fuelwood and, 80–88, 109–10,
        118–20, 162
    "subsistence" wood rights and,
        86–88
    work parties and, 119
wood, uses of, 3
Wrath of the Ancestors (Jordan),
    156–57

Xhosa language, 22–23

yellowwood, 104, 108–9

ZiZi Mfengu, 148–49
Zuurberg, 15